Graduate Texts in Physics

Series Editors

Kurt H. Becker, NYU Polytechnic School of Engineering, Brooklyn, NY, USA

Jean-Marc Di Meglio, Matière et Systèmes Complexes, Bâtiment Condorcet, Université Paris Diderot, Paris, France

Morten Hjorth-Jensen, Department of Physics, Blindern, University of Oslo, Oslo, Norway

Bill Munro, NTT Basic Research Laboratories, Atsugi, Japan

William T. Rhodes, Department of Computer and Electrical Engineering and Computer Science, Florida Atlantic University, Boca Raton, FL, USA

Susan Scott, Australian National University, Acton, Australia

H. Eugene Stanley, Center for Polymer Studies, Physics Department, Boston University, Boston, MA, USA

Martin Stutzmann, Walter Schottky Institute, Technical University of Munich, Garching, Germany

Andreas Wipf, Institute of Theoretical Physics, Friedrich-Schiller-University Jena, Jena, Germany

Graduate Texts in Physics publishes core learning/teaching material for graduate- and advanced-level undergraduate courses on topics of current and emerging fields within physics, both pure and applied. These textbooks serve students at the MS- or PhD-level and their instructors as comprehensive sources of principles, definitions, derivations, experiments and applications (as relevant) for their mastery and teaching, respectively. International in scope and relevance, the textbooks correspond to course syllabi sufficiently to serve as required reading. Their didactic style, comprehensiveness and coverage of fundamental material also make them suitable as introductions or references for scientists entering, or requiring timely knowledge of, a research field.

More information about this series at http://www.springer.com/series/8431

Thomas P. Pearsall

Quantum Photonics

Second Edition

Thomas P. Pearsall
Paris, France

ISSN 1868-4513 ISSN 1868-4521 (electronic)
Graduate Texts in Physics
ISBN 978-3-030-47327-3 ISBN 978-3-030-47325-9 (eBook)
https://doi.org/10.1007/978-3-030-47325-9

1st edition: © Springer International Publishing AG 2017
2nd edition: © Springer Nature Switzerland AG 2020
This work is subject to copyright. All rights are reserved by the Publisher, whether the whole or part of the material is concerned, specifically the rights of translation, reprinting, reuse of illustrations, recitation, broadcasting, reproduction on microfilms or in any other physical way, and transmission or information storage and retrieval, electronic adaptation, computer software, or by similar or dissimilar methodology now known or hereafter developed.
The use of general descriptive names, registered names, trademarks, service marks, etc. in this publication does not imply, even in the absence of a specific statement, that such names are exempt from the relevant protective laws and regulations and therefore free for general use.
The publisher, the authors and the editors are safe to assume that the advice and information in this book are believed to be true and accurate at the date of publication. Neither the publisher nor the authors or the editors give a warranty, expressed or implied, with respect to the material contained herein or for any errors or omissions that may have been made. The publisher remains neutral with regard to jurisdictional claims in published maps and institutional affiliations.

This Springer imprint is published by the registered company Springer Nature Switzerland AG
The registered company address is: Gewerbestrasse 11, 6330 Cham, Switzerland

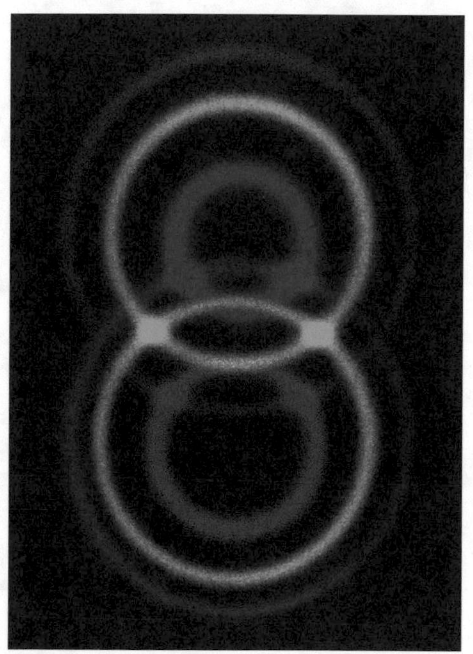

Simultaneous production of photon pairs.

Spontaneous parametric down conversion is a quantum photonic process that produces pairs of photons having energy and momentum that are precisely determined by the non-linear optical environment in which this production takes place. Each photon of a pair is emitted along the surface of a cone. In this photograph different photon pairs are color-coded. The circles correspond to the cross-section of the emission cone for each photon. For the pair in green, the energy of each photon is the same, but their polarizations are orthogonal. Where the cones intersect, the polarization of each photon is a superposition of states, and the photons are said to be entangled. Such photons are the workhorses of numerous quantum photonic experiments and applications.(Reproduced by the kind permission of the Physics Faculty of the University of Vienna, ©Faculty for Physics, University of Vienna. Picture credit: Paul Kwiat & Michael Reck)

Preface

Electrons and photons are fundamental particles that we encounter every day, in plain view, inside and outside the laboratory. Photonics treats the interaction of electrons and photons in the laboratory and in devices like lasers and photodetectors. *Quantum Photonics* is a textbook and reference for the behavior of electrons and photons in an environment that puts their quantum mechanical behavior in the foreground.

A century ago at the dawn of the quantum age, Louis de Broglie considered the experimental discovery of Max Planck that the energy of a photon is proportional to its frequency, and the theoretical proposal of Albert Einstein that energy is equal to mass times the speed of light squared. As a result he stated his belief that a particle with mass m should also have a frequency f given by $f = \frac{mc^2}{h}$. He was not awarded the Nobel Prize for this conjecture, but for something completely different: the proposal that an electron has a wavelength inversely proportional to its momentum. In 2019, while this second edition was in preparation, the International Bureau of Weights and Measures (BIPM) issued new definitions for the fundamental physical constants and achieved a major conceptual transformation in the way these constants are understood and used. When the BIPM considered a standard for mass, it adopted the conjecture of de Broglie as a postulate and defined the standard mass using the speed of light, Planck's constant and the frequency of the hyperfine optical transition in Cs-133.

Following the insight of Louis de Broglie, we now ascribe to all elementary particles both wave and particle behavior. Two important manifestations of quantum behavior are tunneling and interference. A single electron or single photon can be made to interfere with itself by causing the wavefunction to follow two different trajectories, followed by recombination of the wavefunction at an observation point. An example of photon tunneling can be observed by pressing your finger against the outside of a partially filled water glass, making it possible to see your finger through the water, an example of frustrated total internal reflection.

When a single electron encounters a tunneling structure, its quantum wavefunction is separated into two separate spatial regions, representing the probabilities for transmission through, or reflection from, the structure. When a single photon encounters a beam splitter, its quantum wavefunction is similarly separated into two spatial regions, representing probabilities for transmission and reflection. The effect in both cases is the delocalization of the particle-wave. The manipulation of delocalized single photons is the basis of the emerging technology of secure communications via quantum cryptography.

Understanding the non-local behavior of electron and photon quanta is the central theme of this textbook. Single photons are treated by quantizing the electromagnetic fields. This introduces photon creation and annihilation operators and the number states which permit simple and straightforward access to both measurable physical properties and fundamental understanding of quantum photon behavior. The number states can be combined in specific sums to produce the coherent modes of the macroscopic electromagnetic field. Coherent modes are formed from a Poisson distribution of number states.

The distribution of photons in a beam of light is measured as fluctuations or noise in the detected signal. Measurements of photon noise distinguish between emission of highly correlated photons from a thermal source and uncorrelated photons from a coherent laser source, based entirely on the distribution of quantum number states that constitute the photon beam. Fluctuations in quantum electron transport can be explored in an entirely analogous way, using creation and annihilation operators to represent single electron states. In the quantum photonics world, the noise is the signal of interest for measuring quantum interference and correlation. These two phenomena are fundamental expressions of quantum behavior.

Achieving control and deriving benefit from the non-local behavior of quanta is an important frontier of $photonics = optics + electronics$. The quantum photonics frontier is often surprising and still largely unexplored. This textbook thus strives to provide an introduction and foundation for those seeking to make the next contributions.

Paris, France Thomas P. Pearsall

Acknowledgements

Quantum Photonics is the result of the contributions of many people. On one hand are the scientific advances that form the content of the text, but there are also many contributions of colleagues to the project of turning this material into a textbook that will help to educate scientists of the future.

First of all, I would like to thank Professor Yoshikazu Takeda for the invitation to spend a short sabbatical at the University of Nagoya, where this book first took shape. And, I would like to thank Professor Erik Gornik for the opportunity to present and discuss some of this material with students and faculty at the Technische Universität, Wien.

Next, I owe a debt of gratitude to my colleagues who volunteered to read, criticize, and improve the manuscript: Gerald Bastard of the Ecole Normale Supérieure, Michiel de Dood of the University of Leiden, Ulf Gennser of LPN-CNRS, Moty Heiblum of the Weizmann Institute, Rio Howard, Jean-Pierre Huignard, of Thales, Thomas Udem of the Max-Planck Institut für Quantenoptik, Chris van de Walle of U.C. Santa Barbara, and Howard Wiseman of Griffith University.

I would also like to acknowledge useful discussions of quantum photonics experimental measurements with Mark Beck of Reed College and Jan-Peter Meyn of the University of Erlangen.

The MATLAB computational routines included in this text were developed by the author. In addition, I benefitted with helpful discussions and evaluation of simulation software written by Stephan Birner, Paul Harrison, and Richard Lesar.

Finally, I would like to thank Alain Aspect and Philippe Grangier for inviting me to audit their course on quantum optics at the Ecole Polytechnique. I would like to give my thanks to Paul Kwiat of the University of Illinois and Alexander Ling of the National University of Singapore for contributing stunning photographs of spontaneous parametric down conversion, and to Gwendal Fève of the Ecole Normal Supérieure, Jim Chelikowsky of the University of Texas, Jeff Zhe-Yu Ou of Purdue University, and Christoph Westbrook of the Institute of Optics for important technical discussions.

Contents

About the Author

Thomas P. Pearsall has a distinguished career in photonic science and technology, where he has made major contributions to fiber-optic telecommunications and silicon photonics.

As a graduate of Cornell University, he worked in research at Thomson/CSF and Bell Labs for nearly two decades. In 1990, Pearsall was named Boeing-Johnson Chair and Professor at the University of Washington. From 1998 to 2002, he directed research on planar photonic crystals at Corning in Fontainebleau, France.

During his career, he has invented and developed the semiconductor materials and the lasers, LEDs, and photodetectors that are ubiquitous in optical fiber telecommunication networks around the globe. He has also designed and demonstrated key elements of silicon-based photonics, using strained-silicon and silicon-based photonic crystal materials.

In 2003, Pearsall started EPIC, The European Photonics Industry Consortium. EPIC has been a leading contributor to the launch and the development of the European Technology Platform Photonics-21.

He is a fellow of the American Physical Society and a fellow of the IEEE. e-mail: pearsall@ieee.org

Chapter 1
Electrons

Abstract Electrons are wave-particles with mass, charge and magnetic moment. An electron carries the unit charge of the electro-magnetic field. An electron can have a wide range of kinetic energies, depending on its velocity. It can be accelerated, in which case it also radiates photons. Electrons can interact with matter through mechanical impact or electro-magnetic influence. Only twenty-seven years separate the discovery of the electron as a particle carrying electric charge and the revolutionary proposal that it also displays wave-like properties. The electron is thus a fundamental particle with both wave and particle aspects. The Pauli exclusion principle says that no two electrons can occupy the same state (same place, same time, same state with the same magnetic moment), and we derive the Fermi-Dirac statistical distribution, which gives the probability that a state can be occupied by an electron. The wavelength—momentum relationship that follows from the de Broglie's hypothesis leads naturally to the Schrödinger wave equation which carries important information concerning the energy and momentum of the electron.

1.1 Introduction

The first two chapters of this book introduce the important properties of electrons and photons. These are fundamental, indivisible quantum particles that are also common to our everyday experience in the macroscopic world. For example, electrons and photons work together to produce Lasers, LEDS, full color displays and cameras.

Electrons and photons display both wave-like and particle-like behavior. Electrons are wave-particles with mass. An electron carries the unit charge of the electro-magnetic field. An electron can have a wide range of kinetic energies, depending on its velocity. It can be accelerated, in which case it also radiates photons. Electrons can interact with matter through mechanical impact or electro-magnetic influence. A photon is a quantum counter of the excitation of the electro-magnetic field. A photon can thus be created (absorbing energy). Once created a photon has a fixed energy that is proportional to its frequency. A photon can be destroyed (releasing all of its energy). Under special circumstances, a pair of pho-

© Springer Nature Switzerland AG 2020

T. P. Pearsall, *Quantum Photonics*, Graduate Texts in Physics,
https://doi.org/10.1007/978-3-030-47325-9_1

tons can experience interference of their mutual wavefunctions. Otherwise, photons do not interact with each other.

The electron is an elementary particle characterised by its mass ($m_e = 9 \times 10^{-31}$ kg), charge ($q = -1.6 \times 10^{-19}$ C), and magnetic moment ($\mu_e = -0.928 \times 10^{-23}\,\mathrm{J\,T^{-1}}$).

Here is a brief history:

Shortly after the big bang, when the universe had cooled quite a bit, stable electron particles first appeared in our universe.

1897: J.J. Thomson deduces the presence of the electron from studies of light radiated by a gas undergoing an electric discharge. He uses this apparatus to measure the mass-to-charge ratio of the electron by measuring the deflection of a beam of electrons in a magnetic field. Thomson wins the Nobel Prize in 1906.

1909: R. Millikan measures the charge of an electron in a classic experiment. By balancing the gravitational force on an oil drop by the opposing attraction of an electric field, Millikan shows that the attraction is quantized in units of electronic charge. Millikan wins the Nobel Prize for this work in 1923.

1913: Niels Bohr proposes that electrons circulate in stable circular orbits having well-defined energies around a compact and massive nucleus. Bohr is awarded the Nobel Prize in 1922.

1914: James Franck and Gustav Hertz demonstrate that the energies of electrons in an atom are quantized. Einstein is reported to have remarked, "This work is so beautiful, it makes you cry." They receive the Nobel Prize in 1926.

1915: A. Sommerfeld develops the Bohr-Sommerfeld model of the atom, introducing elliptical orbits giving quantization of angular momentum within a single energy level. Sommerfeld does not win a the Nobel Prize. However, four of Sommerfeld's PhD students: Werner Heisenberg, Wolfgang Pauli, Peter Debye, and Hans Bethe, and three of Sommerfeld's post-docs Linus Pauling, Isidor Rabi and Max von Laue, went on to win Nobel Prizes.

1924: de Broglie proposes that electrons in motion have a wavelength, and that stable electron orbits have a circumference equal to an integral number of such wavelengths. He claims credit for explaining the physical basis of the quantization of angular momentum of electrons in atoms proposed by Bohr and Sommerfeld. This idea is quickly generalized to show that all particles have a wavelength equal to Planck's constant divided by the momentum of the particle. The de Broglie wavelength is the first crude estimate of the "size" of an electron, and it is seen that this "size" is variable, depending on its momentum (and thus, its energy) relative to the observer. De Broglie wins the Nobel Prize in 1929.

1925: Wolfgang Pauli proposes the exclusion principle in order to explain why the electrons of an atom were not all in the lowest possible energy level. This principle says that no two electrons can occupy the same quantum state. As the atomic number rises, additional electrons may populate only the unoccupied states at higher energies. Pauli had to wait until 1945 to receive the Nobel Prize for this contribution.

1925: Clinton Davisson and his former PhD student Lester Germer at Bell Labs in New York City directly observe angle-dependent diffraction of electrons from the surface of crystalline nickel. Electron microscopy is born. Only Davisson gets the Nobel Prize in 1937.

1926: Fermi-Dirac statistics and Fermi level. Fermi wins the Nobel Prize in (1938) for the discovery of nuclear fission, the basis of the atomic bomb and nuclear power. Dirac shared the 1933 Nobel Prize with Schrödinger for new quantum mechanical theory.

1973: Hans Dehmelt uses a molecular trap developed independently by Wolfgang Paul that permits a single electron to be isolated and studied. Dehmelt and Paul share the Nobel Prize in 1989. In his Nobel lecture, Dehmelt quoted a phrase from Einstein, "You know, it would be sufficient to really understand the electron."

1.2 De Broglie's Second Idea

Louis de Broglie was born in 1892 to a French noble family. His older brother Maurice, born in 1875, was an accomplished experimental physicist who was invited to attend the first Solvay Conference in 1911, often celebrated as the starting point for quantum physics (see Fig. 1.1). Louis was a theorist and worked on his doctoral thesis in the early 1920s at the University of Paris. Léon Brillouin was a classmate and colleague of Louis during these years. Brillouin's father, Marcel also participated in the 1911 Solvay Conference. Thus, de Broglie and Brillouin were steeped in and inspired by the quantum revolution which was taking off at the same time as their graduate studies.

De Broglie defended his thesis in late November of 1924. The cover page is shown in Fig. 1.2.

The thesis is short, about 100 pages in all. Almost all of the chapters are concerned with the effect of special relativity on the properties of various fundamental particles such as the energy and phase of a propagating light beam.

In Chap. 3 there is an abrupt change of subject, and de Broglie addresses the hypothesis developed by Bohr to explain the existence of discrete atomic energy levels. Seven years earlier, Bohr proposed that the electrons in atoms traveled in stable orbits, thus allowing atoms to have long lifetimes, an experimental truth we all recognize. The condition originally proposed by Bohr was:

$$m_0 \omega R^2 = n \frac{h}{2\pi}, \tag{1.1}$$

where m is the mass of the electron, ω the angular frequency of rotation around the atom, and R the radius of its orbit. For a circular orbit, $\omega = \frac{\mathrm{V}}{R}$, and Bohr's condition becomes:

Fig. 1.1 The first Solvay Conference, held in the Hôtel Métropôle in Brussels (see http://en.wikipedia.org/wiki/Solvay_Conference). Marcel Brillouin, the father of Léon Brillouin is seated 2nd from the *left*, next to Eugène Solvay, the wealthy Belgian industrialist who thought up the idea of a conference of the most brilliant scientists of his day. Maurice de Broglie, the older brother of Louis, is standing 6th from the *left*. Albert Einstein, standing 2nd from the *right* attended this as well as many subsequent editions of this conference. (Photograph by Benjamin Couprie, 1911, public domain, https://commons.wikimedia.org/wiki/File:1911_Solvay_conference.jpg)

$$m_0 \mathrm{v} R = n\frac{h}{2\pi}. \tag{1.2}$$

where v is the linear speed of the electron particle.

This has the simple interpretation that the angular momentum of the electron ($= m\mathrm{v}R$) is quantized in units of h-bar $= \hbar = \frac{h}{2\pi}$. However, in 1924 there was no idea about why this quantization occurred, or what properties of the electron assured this behavior.

On page 44 of his thesis, de Broglie offered an interpretation that was consistent with his everyday experience: the Bohr condition was similar to the behavior of waves of water in a closed circular tank. Stable states occur when there are standing waves. The condition for the existence of a standing wave is that the length of the circuit be an integral number of wavelengths of the standing wave. There are only certain fixed lengths of the canal that can support standing waves. The possible canal lengths are given by the relationship $L = n\lambda$. The argument of de Broglie contains no equations. It appears on page 44 of his thesis and is shown in Fig. 1.3.

If we substitute the resonance condition of de Broglie into (1.2) (remember that $R = \frac{L}{2\pi}$) we get:

Série A. N° 988
N° d'ordre :
1819

THÈSES

PRÉSENTÉES

A LA FACULTÉ DES SCIENCES
DE L'UNIVERSITÉ DE PARIS

POUR OBTENIR

LE GRADE DE DOCTEUR ÈS SCIENCES PHYSIQUES

PAR

Louis de BROGLIE

1re THÈSE. — Recherches sur la théorie des quanta.

2e THÈSE. — Propositions données par la Faculté.

Soutenues le 25 Novembre 1924 devant la Commission d'examen

MM. J. PERRIN....... *Président.*
CARTAN.........
MAUGUIN....... } *Examinateurs.*
Paul LANGEVIN.

BIBLIOTHÈQUE DE L'UNIVERSITÉ

PARIS
MASSON ET Cie, ÉDITEURS
LIBRAIRES DE L'ACADÉMIE DE MÉDECINE
120, BOULEVARD SAINT-GERMAIN

1924

D. 55435

Fig. 1.2 Cover page for the doctoral thesis of Louis de Broglie. Each doctoral candidate had to write on two subjects: one chosen by the candidate, and one assigned by the faculty. The title of his chosen subject is: "Research on the Theory of Quanta"

44 LOUIS DE BROGLIE

un des rayons de son onde de phase, celle-ci doit courir le long de la trajectoire avec une fréquence constante (puisque l'énergie totale est constante) et une vitesse variable dont nous avons appris à calculer la valeur. La propagation est donc analogue à celle d'une onde liquide dans un canal fermé sur lui-même et de profondeur variable. Il est physiquement évident que, pour avoir un régime stable, la longueur du canal doit être en résonance avec l'onde; autrement dit, les portions d'onde qui se suivent à une distance égale à un multiple entier de la longueur l du canal et qui se trouvent par suite au même point de celui-ci, doivent être en phase. La condition de résonance est $l = n\lambda$ si la longueur d'onde est constante et $\oint \frac{\nu}{V} dl = n$ (entier) dans le cas général.

Translation of the highlighted portion:

"The propagation (of the electron) is therefore analogous to that of a wave of liquid in a tank that forms a closed path. In order to have a stable condition for the wave, it is physically evident that the length of the tank must be in resonance with the wave. In other words, the portions of the wave that are located a full length l of the tank behind preceding portion of the wave must be in phase with the preceding portion. The condition for resonance is $l = n\lambda$."

Fig. 1.3 The proposition by de Broglie in his thesis: the stable orbits of electrons in atoms are like stationary waves of water in a closed circular tank

$$
\begin{aligned}
m_0 \mathrm{v}\left(\frac{L}{2\pi}\right) &= n\frac{h}{2\pi} \\
m_0 \mathrm{v}(n\lambda) &= nh \\
m_0 \mathrm{v} &= \frac{h}{\lambda}
\end{aligned}
\tag{1.3}
$$

Equation (1.3) says that the electron has momentum that is inversely proportional to its wavelength. This simple conclusion does not appear in de Broglie's thesis, nor does the extension of this result to free electrons or other particles like photons. However, de Broglie let the cat out the bag so to speak, for which he was awarded the Nobel Prize (1929). He claimed credit in his thesis for "the first plausible physical explanation for the condition of stable orbits as proposed by Bohr and Sommerfeld." The principal result of his idea was to open the way for the development of Schrödinger's wave equation, and the first quantitative description of the behavior of electrons and atoms.

I find that the most interesting part of de Broglie's reasoning to be the notion that because quantization exists, **there must be** an associated wave behavior.

In practice it is quite difficult to measure the momentum of an electron, but quite easy to determine its energy. The wavelength of an electron, (or any other particle-wave) is expressed in terms of its energy and mass as:

$$\lambda = \frac{h}{\sqrt{2mE}} \tag{1.4}$$

Example 1.1 The wavelength of an electron accelerated by a potential of 1 eV is:

$$1\ \text{eV} = 1.6 \times 10^{-19}\ \text{J}$$

$$\lambda = \frac{h}{\sqrt{2mE}} = \frac{6.6 \times 10^{-34}\ \text{J s}}{\sqrt{2 \cdot 9 \times 10^{-31}\ \text{kg} \cdot 1.6 \times 10^{-19}\ \text{J}}} = 12\ \text{Å}$$

The de Broglie wavelength associated with a particle is variable and depends on the relative difference in energy state between the observer and the particle in question. For example, if you were able to travel alongside the 1 eV electron at the same speed (about 6×10^3 km s^{-1}, a small fraction of the speed of light), The electron would appear to be nearly stationary, and its wavelength would be very long. Different observers will attribute different wavelengths to the same electron depending on their relative energy difference.

Equation (1.4) works both ways. If a particle is confined physically in space by potential barriers, then this puts an upper limit on the wavelength that the particle can have. In turn this means that the energy of the particle is increased:

$$E = \frac{h^2}{2m\lambda^2} = \frac{\hbar^2 k^2}{2m}, \text{ where } \hbar = \frac{h}{2\pi} \text{ and } k = \frac{2\pi}{\lambda} \tag{1.5}$$

From the macroscopic point of view, such a particle would appear to have a net momentum of zero. However, the particle "motion" consists of 2 counter-propagating de Broglie waves each having momentum $\frac{h}{\lambda}$.

The de Broglie wavelength applies to any particle including atoms and molecules, even to photons. If we apply the relationship to Planck's equation, for example,

$$E = h\nu = \frac{h}{\lambda}c = pc, \tag{1.6}$$

where p is the momentum of the photon, and $c =$ the speed of light.

Louis de Broglie and Léon Brillouin (who studied the motion of electron waves in a periodic environment like a crystal) both attended the fifth Solvay Conference was held in Brussels in 1927. Nearly 20 years after the first conference, Mr. Solvay had now built a beautiful library in the Leopold Park in Brussels. The participants' photo is taken outside this building (see Fig. 1.4). They represent in a large part a new generation of physicists whose work would revolutionize the understanding and analysis of quantum phenomena.

Fig. 1.4 The Solvay Conference of 1927. The conference topic was electrons and photons. This meeting could be considered to mark the beginning of Quantum Photonics. Two-thirds of the participants would eventually win the Nobel Prize. While there are still a few people who attended the 1st Solvay conference, notably Curie, Einstein and Bohr, the participants here for the first time represent a new generation of physicists, who developed modern quantum mechanics. Among them are Louis de Broglie, brother of Maurice and Léon Brillouin, son of Marcel. (Photograph by Benjamin Couprie, 1927, public domain, https://commons.wikimedia.org/wiki/File:1927_Solvay_conference.jpg) (*Back row* L–R: A. Piccard, E. Henriot, P. Ehrenfest, E. Herzen, T. de Donder, E. Schrödinger, E Verschaffelt, W. Pauli, W. Heisenberg, R.H. Fowler, L. Brillouin, *Middle row* L-R: P. Debye, M. Knudsen, W.L. Bragg, H.A. Kramers. P.A.M. Dirac, A.H. Compton, L.V. de Broglie, M. Born, N. Bohr, *Front row* L-R: I. Langmuir, M. Planck, M. Curie, H.A. Lorentz, A. Einstein, P. Langevin, C. E Guye, C.T.R. Wilson, O.W. Richardson)

1.3 The Pauli Exclusion Principle

The year is 1925, and the physics community has been stunned in 1924 by the proposition of de Broglie on one hand, and on the other the experimental verification of electron diffraction at Bell Labs by Davisson and Germer. Electron waves are real and measureable.

An excellent description of the development of the Pauli exclusion principle has been given by Pauli himself:

> The history of the discovery of the «exclusion principle», for which I have received the honor of the Nobel Prize award in the year 1945, goes back to my student days in Munich... It was at the University of Munich that I was introduced by Sommerfeld to

the structure of the atom - somewhat strange from the point of view of classical physics. The series of whole numbers 2, 8, 18, 32, giving the lengths of the periods in the natural system of chemical elements, was zealously discussed in Munich, including the remark of the Swedish physicist, Rydberg, that these numbers are of the simple form $2n^2$, if n takes on all integer values.

A new phase of my scientific life began when I met Niels Bohr personally for the first time. This was in 1922, when he gave a series of guest lectures at Göttingen, in which he reported on his theoretical investigations on the Periodic System of Elements. The question, as to why all electrons for an atom in its ground state were not bound in the innermost shell, had already been emphasized by Bohr as a fundamental problem in his earlier works. In his Göttingen lectures he treated particularly the closing of this innermost K-shell in the helium atom and its essential connection with the two non-combining spectra of helium, the ortho- and para-helium spectra. However, no convincing explanation for this phenomenon could be given on the basis of classical mechanics. It made a strong impression on me that Bohr at that time and in later discussions was looking for a general explanation which should hold for the closing of every electron shell and in which the number 2 was considered to be as essential as 8. Excerpts from the Nobel Lecture by Wolfgang Pauli, December 13, 1946.

Pauli based his exclusion principle on experimental studies of the splitting of atomic spectra in magnetic fields. These experiments show the number of states in each electron shell. However, he was baffled by the factor 2 in the Rydberg formula $2n^2$. To solve this he also proposed a new quantum number, which was later identified by Goudsmid and Uhlenberg as an intrinsic magnetic moment called spin. With the aid of this concept, Pauli could propose that each known state for electrons in an atom could be occupied by at most one electron. The principal result of this idea is that atomic states are filled successively by electrons, each having a unique combination of 4 quantum numbers.

Electron spin is only one example of how Pauli looked for the "missing link" in physics. By 1930, he was studying the particle tracks of beta-decay, trying to understand the apparent violations of energy and momentum conservation. In a letter to the attendees of a physics conference in Tübingen, Germany, he proposed a "desperate remedy": the existence of a new neutral particle, nearly impossible to detect, as a solution. As before, he did not give a name to the proposed quantum. Enrico Fermi thought up the name neutrino shortly thereafter. The neutrino group is today a keystone of the standard model, and the subject of the Nobel Prize awarded in 2015.

1.4 Fermi-Dirac Statistics

The Pauli exclusion principle means that no two electrons can be found in the same place at the same time with the same quantum numbers. This feature is a fundamental difference between electrons from photons. Two photons having identical properties of frequency and polarization are indistinguishable. They can be made to occupy the same space at the same time, leading to interference between them. Such interference is not possible for electrons.

The fundamental difference between the behavior of electrons and that of photons can be seen on a macroscopic level as well as on the quantum level. Current in a photonic device is carried by electrons, typically in such large numbers that their behavior is better described by statistics of the ensemble rather than by following the trajectory of individual quanta. Any given material, such a silicon crystal, contains electrons. Under equilibrium conditions, these electrons occupy stable states in the material. Using principles of statistical mechanics, we can determine the probability that a given state, having a corresponding energy, is occupied by an electron. The Fermi energy is the energy at which the probability is 50% that a state is occupied. In an ideal semiconductor, for example, the Fermi level lies in the forbidden gap that separates the valence band, where the majority of states are occupied by electrons, from the conduction band where the majority of states are unoccupied.

The Fermi-Dirac statistical distribution is based on the fundamental quantum properties of electrons. It is used to characterize the behavior of electrons in matter, including the current-voltage relationship in macroscopic electronic devices like transistors, and photonic devices like lasers and LEDs.

In a statistical ensemble, each state has a probability of occupation P given by:

$$P = e^{\dfrac{\Omega + \mu N - E}{kT}} \tag{1.7}$$

N number of particles in the state in question
μ chemical potential
E energy of the state
k Boltzmann's constant
Ω is the grand potential:

$$\Omega = -kT \ln\left(\sum_{\text{states}} e^{\dfrac{\mu N - E}{kT}}\right) \tag{1.8}$$

For the case of an energy level that can filled by an electron, using the Pauli exclusion principle, there are two possibilities: either the state is unoccupied ($N = 0$, $E = 0$), or it is occupied by 1 electron ($N = 1$). As a result,

$$\Omega = -kT \ln\left(1 + e^{\dfrac{\mu - E}{kT}}\right) \tag{1.9}$$

The occupation probability density as a function of energy is given by

$$n(E) = \frac{\partial \Omega}{\partial \mu} = kT\left(1 + e^{\dfrac{\mu - E}{kT}}\right)\left(\frac{1}{kT}\right)\left(\frac{\left(e^{\dfrac{\mu - E}{kT}}\right)}{\left(1 + e^{\dfrac{\mu - E}{kT}}\right)^2}\right) = \frac{1}{\left(e^{\dfrac{E - \mu}{kT}} + 1\right)} \tag{1.10}$$

We can now choose a specific chemical potential μ as a reference energy, and we call this the Fermi level, E_f. The Fermi energy is the energy at which the probability that a state is occupied is equal to ½.

$$n(E) = \frac{1}{\left(e^{\frac{E-E_f}{kT}} + 1\right)} \tag{1.11}$$

The statistics for electrons were developed independently by Enrico Fermi and Paul Dirac. Equation (1.11) is called the Fermi-Dirac distribution.

1.5 The Schrödinger Equation

Following the experimental verification of the de Broglie's proposal that electrons (and holes) have a characteristic wavelength, we can write down the wave function of a one-dimensional free electron as a general wave as follows:

$$\Psi(x, t) = Ae^{i(kx-\omega t)} = Ae^{ikx}e^{-i\omega t} \tag{1.12}$$

The first spatial derivative of this wavefunction gives:

$$\frac{d}{dx}\Psi(x, t) = ik\Psi(x, t) \tag{1.13}$$

Multiplying both sides by $-i\hbar$:

$$-i\hbar\frac{d}{dx}\Psi(x, t) = \hbar k\Psi(x, t) = p\Psi(x, t) \tag{1.14}$$

where p is the linear momentum associated with the wavefunction, following de Broglie's hypothesis. We define the momentum operator as:

$$p_{\text{op}} \equiv -i\hbar\frac{d}{dx} \equiv -i\hbar\nabla \tag{1.15}$$

Taking the 2nd derivative of the wavefunction gives:

$$\frac{d^2}{dx^2}\Psi(x, t) = -k^2\Psi(x, t) \tag{1.16}$$

Multiplying both sides by $\frac{-\hbar^2}{2m^*}$, where m^* is the effective mass of the electron (or hole)

$$-\frac{\hbar^2}{2m^*}\frac{d^2}{dx^2}\Psi(x, t) = \frac{p^2}{2m^*} = \frac{\hbar^2k^2}{2m^*}\Psi(x) = T\Psi(x, t) \tag{1.17}$$

where T is the kinetic energy associated with the electron.

It is left as an exercise to show that:

$$i\hbar\frac{d}{dt}\Psi(x,t) = E\Psi(x,t) \tag{1.18}$$

The principle of conservation of total energy states that

$$\text{Kinetic energy} + \text{Potential energy} = \text{Total energy}$$

Adding an external potential energy term $V(x,t)$ to (1.17) gives

$$-\frac{\hbar^2}{2m^*}\frac{d^2}{dx^2}\Psi(x,t) + V(x,t)\Psi(x,t) = E\Psi(x,t) = i\hbar\frac{d}{dt}\Psi(x,t) \tag{1.19}$$

Because this differential equation is separable in time and position, we can solve for the spatial dependence of the energies independently of the time evolution. This is easily verified by applying the wave function of (1.12) to the LHS of (1.19).

$$-\frac{\hbar^2}{2m^*}\frac{d^2}{dx^2}\Psi(x) + V(x)\Psi(x) = E\Psi(x) \tag{1.20}$$

A similar procedure can be used to solve for the time evolution of the wavefunction. The solutions of Schrödinger's equation give stationary states for which total energy is conserved.

1.6 De Broglie's First Idea

On the third page of his thesis, Louis de Broglie professed his belief that Planck's quantum relationship, $E = \hbar\omega$, applied not only to photons, but also to matter in general. Thus, a particle like an electron, with a rest-mass energy given by Einstein's relationship $E = m_0c^2$, would also have an innate frequency given by $\omega = \frac{m_0c^2}{\hbar}$. He wrote:

> The quantum relationship $(E = hf)$ would probably not have much meaning if its energy were distributed in a continuous way in space, and we have seen that it is certainly not the case. We can therefore propose a grand law of nature: each amount of energy associated with a mass m_0 is endowed with a periodical behavior of frequency f_0 such that:
>
> $$hf_0 = m_0c^2 \tag{1.21}$$
>
> where f_0 is of course measured in the system appropriate to that amount of energy.

This is a daring conjecture, based on little more than intuition and dimensional equivalence.

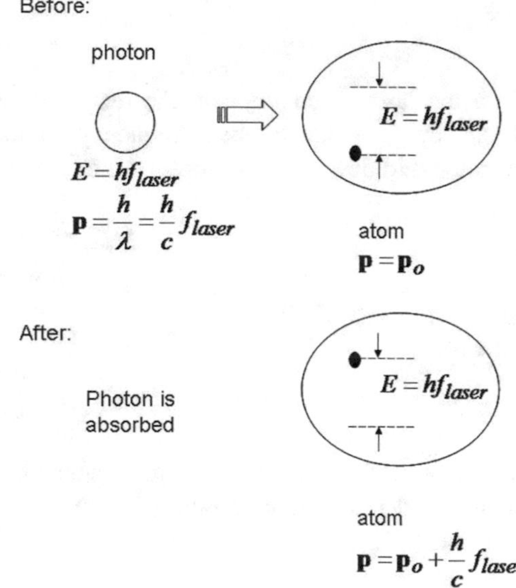

Fig. 1.5 Diagram of proposed experiment: a photon is emitted by a laser with frequency $= f_{\text{laser}}$ tuned to the energy difference between two levels in an atom. To conserve momentum, the atom recoils with momentum $= \frac{h}{c} f_{\text{laser}}$

Does a particle have a measurable frequency that is proportional to its mass? In the case of a non-relativistic electron, this frequency is about 1.3×10^{20} Hz. This frequency is more than 5 orders of magnitude too high to be measured directly using today's best instruments. Protons and atoms, of course, would have an even higher *de Broglie* frequency. However, it is possible to imagine an experiment that might be attempted using frequency-comb techniques developed by John Hall and Theodor Hänsch, and recognized by a Nobel Prize (2005). This is the subject of the following example.

Example 1.2 The following thought-experiment, diagrammed in Fig. 1.5, is based on an idea developed by Prof. Holger Müller (S-Y. Lan et al. 2013).

An atom is excited by a laser and absorbs a single photon with frequency $= f_{\text{Laser}}$. By conservation of momentum, the atom recoils with momentum $= m_0 V = \frac{h}{\lambda} = \frac{h}{c} f_{\text{Laser}}$

$$\text{The recoil energy} = hf_{\text{Recoil}} = \frac{1}{2} m_0 V^2 = \frac{1}{2} \frac{h^2 f_{\text{Laser}}^2}{m_0 c^2} \tag{1.22}$$

The recoil frequency would be, according to Planck's law ($E_{\text{Recoil}} = hf_{\text{Recoil}}$):

$$f_{\text{Recoil}} = \frac{hf_{\text{Laser}}^2}{2m_0 c^2} \tag{1.23}$$

Applying de Broglie's first idea: that each quantity of matter has a frequency determined by its mass: $f_{\text{Broglie}} = \frac{m_0 c^2}{h}$, we substitute this quantity in (1.22).

$$f_{\text{Recoil}} = \frac{f_{\text{Laser}}^2}{2 f_{\text{Broglie}}} \tag{1.24}$$

Both the laser frequency and the recoil frequency can be locked to a reference frequency f_{ref} generated by a frequency comb. This is the measurement technique invented and developed by John Hall and Theodor Hänsch.

$$f_{\text{Laser}} = n_1 f_{\text{ref}} \tag{1.25}$$

$$f_{\text{Recoil}} = n_2 f_{\text{ref}} \tag{1.26}$$

As a result:

$$f_{\text{Broglie}} = \frac{1}{2}\frac{n_1^2}{n_2} f_{\text{ref}} \tag{1.27}$$

Note that no fundamental constants appear in this expression. The de Broglie frequency is determined by a ratio of n_1 and n_2.

From (1.23), we can estimate

$$f_{\text{laser}} \approx 10^{15}\ \text{Hz}$$

$$f_{\text{Broglie}} \approx 10^{24}\ \text{Hz}$$

$$f_{\text{Recoil}} \approx 10^{6}\ \text{Hz}$$

The de Broglie hypothesis would be established if the recoil frequency can be measured. In turn, this would establish equivalence between mass and frequency. A recently published experiment has claimed such a result (S.-Y. Lan et al. 2013). However, this claim is contested by some experts in the field (L. Blanchet et al. 2013).

De Broglie received the Nobel Prize for his *second idea*, about the wavelength that can be attributed to matter. The Nobel award followed the experimental verification of this effect. The de Broglie wavelength is not an intrinsic property. Rather, it depends on the relative energy of the particle with respect to the observer. Thus, the wavelength of an electron in an electron microscope depends on its velocity with respect to the target it is probing. The de Broglie frequency, on the other hand, would be an intrinsic property of an atom or electron, at least in the non-relativistic regime.

The experiment of (S.-Y. Lan et al. 2013) addresses the equivalence between mass and frequency. For many years the unique standard for mass has been a block of platinum and iridium, located in Saint-Cloud, a suburb of Paris, France, with a mass that defined the kilogram. The precise determination of mass for any other object required a trip to The International Bureau of Weights and Measures (BIPM) in Saint-Cloud for a direct comparison, or using an instrument with a precision and accuracy directly traceable to this standard.

During the past decade, the BIPM has been working on a system of standards for mass, distance, time and other essential units of measurement that is related directly to fundamental constants, such as Planck's constant, for example, that would be independent of prototype objects like the block of Pt-Ir or a "standard meter". It completed an important phase of this work in May, 2019 with the publication of new definitions of SI (Système International) units, commonly referred to as "the metric system". In this revision, fundamental constants are defined as exact rational numbers. Then objects, such as the block of Pt-Ir can be assessed in terms of these definitions. The result is truly revolutionary. With this approach, any laboratory in the universe can have its own precise set of standards, without having to travel to Saint-Cloud. (NIST publication SP-330 2019). Kettlerle and Jamison have prepared an excellent discussion of the changes and implications of this new approach to the definition of SI units (Ketterle and Jamison 2020)

Referring to some of the constants that we have used so far,

$$\text{Planck's constant} : h \equiv 6.62607015 \times 10^{-34}\,\text{Joule} - \text{s}$$
$$\text{Speed of light} : c \equiv 2.99792458\,\text{m} - \text{s}^{-1}$$

Time-frequency: $f_{\text{Cs}} \equiv 9.192631770 \times 10^9\,\text{Hz}$, where f_{Cs} is the frequency of the hyperfine transition of the Cs-133 atom in its unperturbed ground state.

The BIPM has adopted the de Broglie conjecture (1.21) in order to define a standard for mass, based on the three constants as defined above.

$$\text{mass} : m \equiv \frac{h f_{\text{Cs}}}{c^2}$$

Thus, mass is determined in terms of the fundamental constants, h, c and f_{Cs}.

For example using (1.21), the mass of the photon (m_ϕ) emitted by the hyperfine transition in Cs-133 is $m_\phi \equiv 6.777 \times 10^{-41}$ kg, where we have not displayed all the digits that compose the exact expression for the mass.

Mass is conserved. When a Cs-133 atom is excited and absorbs a photon at the hyperfine frequency f_{Cs}, the mass of the atom should increase by $\Delta m = 6.777 \times 10^{-41}$ kg. (S-Y. Lan et al. 2013) have measured the difference in mass between the Cs-133 atom in the excited and ground states by using a Ramsey-Bordé atom interferometer. Their results show that such a difference occurs experimentally.

1.7 Summary

The electron is a fundamental particle with both wave and particle aspects. The Pauli exclusion principle says that no two electrons can occupy the same state (same place, same time, same state with the same magnetic moment), and we have derived the Fermi-Dirac statistical distribution, which gives the probability that a state can be occupied by an electron. The wavelength—momentum relationship that follows from the de Broglie's hypothesis leads naturally to the Schrödinger wave equation that carries important information concerning the energy and momentum of the electron.

1.8 Exercises

1.1 a. What is the wavelength of a 1 eV photon? Prove that a free electron cannot interact with a free photon conserving energy and momentum.
b. Calculate the wavelength of a 1 eV phonon in silicon. Comment on your result.

1.2 Derive the time-dependent Schrödinger equation:

$$i\hbar\frac{d}{dt}\Psi(x,t) = E\Psi(x,t)$$

Show that the wavefunction can be expressed:

$$\Psi(x,t) = Ae^{-i\frac{E}{\hbar}t}$$

where the allowed values of E are identical to those determined from the solutions of the time-independent Schrödinger equation

1.3 An electron is localized in a one-dimensional potential well of width $= L$ meters. Choose the value of the potential $V = 0$ at the bottom of the well and infinitely large outside the boundaries of the well. Fix one boundary of the well at $x = 0$ and the other boundary at $x = L$ as diagrammed in Fig. 1.6.

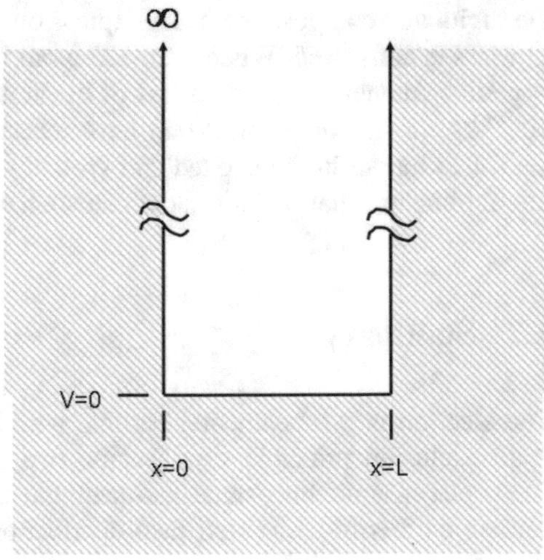

Fig. 1.6 1-dimensional infinite potential well

Assume that the electron wavefunction can be expressed: $\Psi(x) = A\sin(x) + B\cos(x)$

Show that the energies of the stable electron states are quantized and expressed by:

$$E_n = \frac{n^2\pi^2\hbar^2}{2L^2m}$$

1.4 Compare the Fermi distribution to the Boltzmann distribution.
1.5 Determine the frequency and vacuum wavelength of a photon having the mass equivalent energy of an electron at rest.

References

L. Blanchet, P. Wolf, C. Bord, S. Reynaud, C. Salomon, C. Cohen-Tannoudji, Equivalence principle and matter wave interferometry, July 2013. http://vietnam.in2p3.fr/2013/Cosmology/transparencies/Blanchet2.pdf

L. de Broglie, *Recherches sur la Théorie des Quanta* (1924). https://tel.archives-ouvertes.fr/tel-00006807/document

L. de Broglie, Nobel Prize (1929). http://www.nobelprize.org/nobel_prizes/physics/laureates/1929/

E. Fermi, Nobel Prize (1938). http://www.nobelprize.org/nobel_prizes/physics/laureates/1938/

T. Hänsch, Nobel Prize (2005). http://www.nobelprize.org/nobel_prizes/physics/laureates/2005/

W. Ketterle, A.O. Jamison, An atomic physics perspective on the kilogram's new definition, Phys. Today **73**(5), 33–38 (2020). https://physicstoday.scitation.org/doi/pdf/10.1063/PT.3.4472

S.-Y. Lan, P.-C. Kuan, B. Estey, D. English, J.M. Brown, M.A. Hohensee, H. Müller, A clock directly linking time to a particle's mass. Science **339**, 554–557. http://citeseerx.ist.psu.edu/viewdoc/download?doi=10.1.1.911.169&rep=rep1&type=pdf

Solvay Conference. http://en.wikipedia.org/wiki/Solvay_Conference

The International System of Units, D. B. Newell and E. Tiesinga (eds.), NIST publication no. SP-330 (Physical Measurement Laboratory National Institute of Standards and Technology Gaithersburg, 2019) https://nvlpubs.nist.gov/nistpubs/SpecialPublications/NIST.SP.330-2019.pdf

Chapter 2
Photons

Abstract The photon is an elementary particle characterized by its frequency and polarization. The photon displays both wave-like and particle-like behavior. A photon is the unit intensity of the electro-magnetic field. Once created, a photon is characterized by its energy ($E = hf$) which remains unchanged until the photon is destroyed. Maxwell's equations describe the electromagnetic fields in the classical regime, including propagation of photon beams as a function of frequency and polarization. Maxwell's equations represent a staggering accomplishment in physical theory, but they do not give a complete description of photon behavior. Two photons with the same quantum numbers (that is the same frequency and polarization) can (unlike electrons) occupy the same place at the same time. The probability that an energy state is occupied by a photon is given by the Bose-Einstein distribution. A single photon is an indivisible elementary particle. Single photons can be produced by spontaneous parametric down-conversion. This is an optical process that cannot by described using Maxwell's equations. When a single photon is incident upon a beamsplitter, its wavefunction is separated spatially into two parts, corresponding to reflection and transmission. When a measurement is made, the entire photon will be detected either as 100% transmitted or 100% reflected. This result is an example of the Copenhagen interpretation of quantum mechanics.

2.1 Introduction

A photon is the unit intensity of the electro-magnetic field. Once created, a photon is characterized by its energy ($E = hf$) which remains unchanged until the photon is destroyed. A photon has wavelike properties, and so it has a wavelength. While the energy or frequency is fixed, the wavelength of a photon can change depending on the index of refraction of the medium in which it is propagating. Photons, unlike electrons, do not interact with each other. They cannot exchange energy or momentum.

© Springer Nature Switzerland AG 2020

T. P. Pearsall, *Quantum Photonics*, Graduate Texts in Physics,
https://doi.org/10.1007/978-3-030-47325-9_2

Two photons with the same quantum numbers (that is the same frequency and polarization) can (unlike electrons) occupy the same place at the same time. In this situation, the photons are indistinguishable, and the respective wavefunctions of the photons can experience quantum interference.

The optical beamsplitter is an instrument, for example a simple glass slide, that divides a photon beam into two different paths by partial reflection and partial transmission of the beam. When a single photon is incident upon a beamsplitter, its wavefunction is spatially separated into two parts, corresponding to reflection and transmission. However, a photon is an indivisible elementary particle. When a measurement is made, the entire photon will be detected either as 100% transmitted or 100% reflected. In the time and space interval between the beamsplitter and the detector, the photon is delocalized. The wavefunction is present in both paths, and these two components belonging to a single photon can be made to interfere with each other.

In addition to reshaping the wavefunction of a single photon, the optical beamsplitter can be used to combine the wavefunctions of two separate photons. When these photons have the same properties of frequency and polarization, the beamsplitter, used in a combining mode will cause quantum interference effects to appear between the two photons.

These two effects: separation and combination are among the key elements used to implement important applications of quantum photonics: quantum computation and quantum communication.

2.1.1 Photon Properties

A photon is a quantum of excitation of the electro-magnetic field. The characteristics of photons are:

Energy is proportional to frequency: $E = hf = \hbar\omega$
Speed: $c = 2.98 \cdot 10^8$ m s^{-1} in vacuum
Mass: Because a photon has energy, it has non-zero mass. $m_{\text{photon}} = \frac{hf}{c^2}$
Wavelength: $\lambda = \frac{c}{nf}$, where n is the index of refraction
Spin (intrinsic angular momentum: $= +1$ (in units of $\hbar$) for left-hand circularly-polarized light and -1 for right-hand circularly polarized light.

There are two additional fundamental properties of photons.

- Any number of photons can occupy the same quantum state. This means that two or more photons can be indistinguishable. Quanta with this property are called bosons, and they obey Bose-Einstein statistics.
- Photons are not conserved. Photons can be created (emission) or annihilated (absorption). Photon emission or absorption assures energy conservation in photonic materials and devices.

2.1.2 *Brief History of Photons*

Time zero: The Big Bang

There is today significant experimental evidence for the reality of the Big Bang. According to this model, nearly 14 billion years ago, quantum fluctuations of the vacuum released a huge amount of energy over a space whose diameter was hundreds of millions of light years in less than 10^{-32} s (Lineweaver and Davis 2005). Following this explosion and subsequent expansion of space-time, the universe was composed of a soup of electrons, positrons, protons, neutrinos, and other fundamental particles, so dense that any photons emitted were reabsorbed. Some time afterward, when the universe had cooled sufficiently, protons and electrons could condense to form the first stable hydrogen atoms. Photons with energy insufficient to ionize these atoms could then propagate in space. This radiation is the oldest known photonic signal that can be measured directly. Measurement of this background radiation is an area of intense research today. Some important projects are the Cosmic Microwave Background Explorer (COBE) and the Wilkinson Microwave Anisotropy Probe (WMAP). Results from these projects have produced a compact timeline of the universe shown in Fig. 2.1.

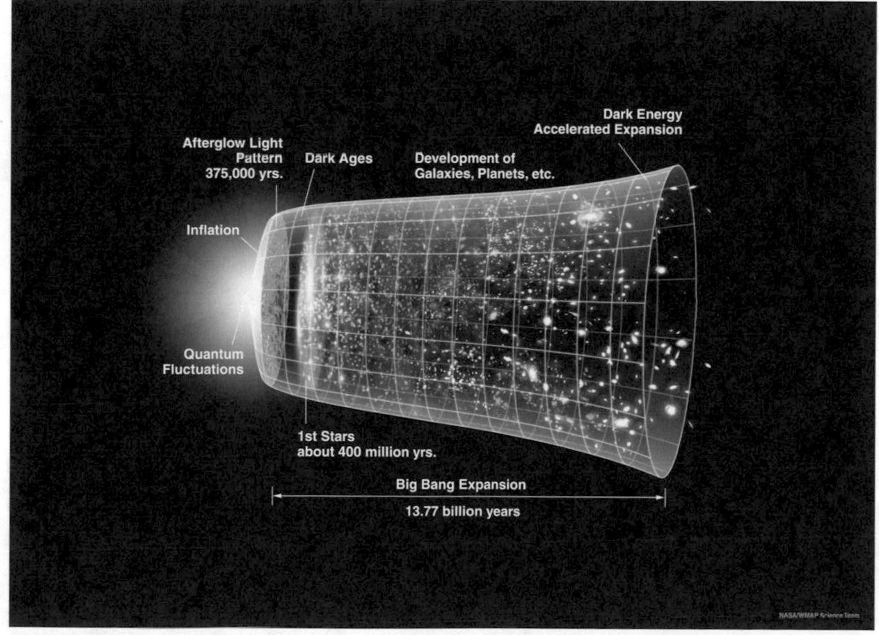

Fig. 2.1 *Credit* WMAP Wilkinson Microwave Anisotropy Map. Photons, unlike electrons, were evident from the beginning (http://map.gsfc.nasa.gov/media/060915/index.html public domain)

1665: Francesco-Maria Grimaldi carries out the first experiments on diffraction of light, inventing the word itself and publishes his results in his book **Physics and Mathematics of Light, Colors and Rainbows**, a significant work published two years after his death.

1675: Newton studies Grimaldi's work which serves as an inspiration for his study of light. Newton develops a corpuscular theory of light, but does not publish it, being aware of intense competition coming from the Paris Observatory.

1676: Ole Roemer, working at the Paris Observatory demonstrates that light has a finite speed, by measuring the variations in periodicity of the eclipse of Jupiter's moons. This periodicity oscillates as the earth orbits the sun, changing the distance between the earth and Jupiter. Roemer's first measurement of this speed was 2.2×10^8 m s^{-1}, about 26% less than today's established value.

1678: Christiaan Huygens, working at the Paris Observatory presents wave theory of light, based on experimental observations. However, aware of the contradictory Newtonian theory, he delays publication until 1690. His book becomes a scientific classic.

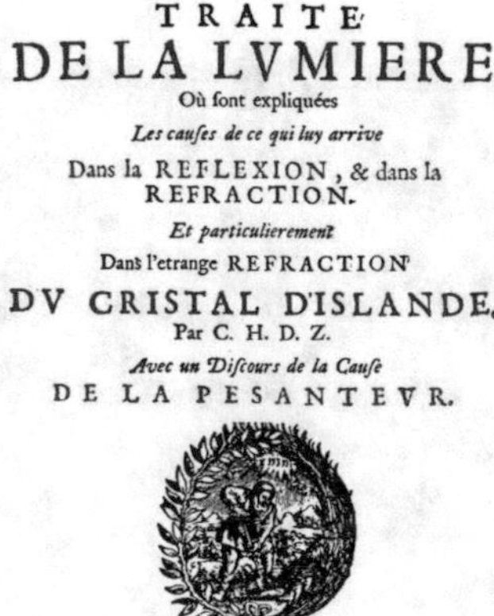

TRAITE
DE LA LVMIERE.
Où ſont expliquées
Les cauſes de ce qui luy arrive
Dans la REFLEXION, & dans la REFRACTION.
Et particulierement
Dans l'etrange REFRACTION
DV CRISTAL D'ISLANDE.
Par C. H. D. Z.
Avec un Diſcours de la Cauſe
DE LA PESANTEVR.

A LEIDE,
Chez PIERRE VANDER AA, Marchand Libraire.
MDCXC.

Fig. 2.2 Christiaan Huygens, title page of *Traité de la Lumière*, published in 1690

1704: Newton, seeking to have the last word, publishes his corpuscular theory, hoping to supplant Huygen's wave theory, and waiting until both Huygens and Robert Hooke, the major scientific advocates for the wave theory, were dead. For most of the eighteenth century, the Newton's reputation insures that his theory prevails.

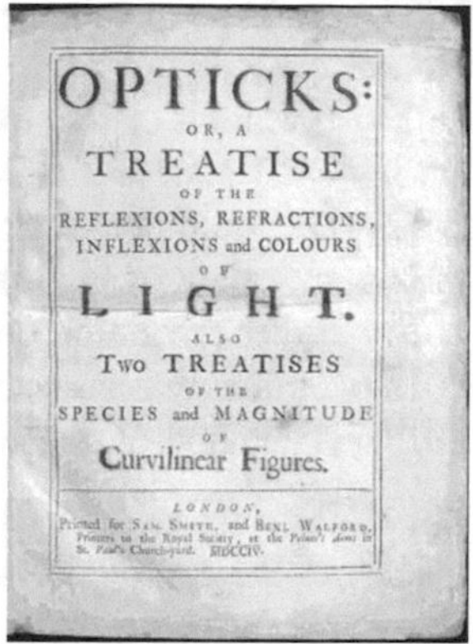

OPTICKS:
OR, A
TREATISE
OF THE
REFLEXIONS, REFRACTIONS,
INFLEXIONS and COLOURS
OF
LIGHT.
ALSO
Two TREATISES
OF THE
SPECIES and MAGNITUDE
OF
Curvilinear Figures.

LONDON,
Printed for SAM. SMITH, and BENJ. WALFORD,
Printers to the Royal Society, at the [illegible] Arms in
St. Paul's Church-yard. MDCCIV.

Fig. 2.3 Isaac Newton, title page of Opticks: or a Treatise on the Reflexions, Refractions, Inflexions and Colours of Light, published in 1704

1805: Thomas Young studies diffraction of light by a slit, experimental proof of the wave behavior of light.

1835: James Maxwell introduces Maxwell's equations which give a complete description of the propagation of light, uniting electricity and magnetism as aspects of the same phenomenon.

1884: Oliver Heaviside studies Maxwell's equations and starting from the 20 equations and 20 unknowns proposed by Maxwell, reduces them to the four equations we use today.

1901: Max Planck studies light radiated from objects as a function of their temperature. He concludes that light waves behave like particles with energy proportional to frequency.

Planck wrote in his autobiography (Planck 1948): *A new scientific truth does not triumph by convincing its opponents and making them see the light, but rather because its opponents eventually die, and a new generation grows up that is familiar with it.*

1905: Einstein studies the photoelectric effect, and argues that photo-ionization of an electron from a solid is made by a single photon having enough energy (proportional to its frequency) to free the electron from its bonded state.

1927: Paul Dirac invents quantum electrodynamics, introducing Maxwell's field equations into quantum mechanical calculations

1982: Quantum entanglement of photon pairs—theoretical conjecture dating from 1935 is tested by experiments devised by Alain Aspect that demonstrate the non-locality of quantum mechanical particles like photons.

2000–2018: After many decades of near neglect, Nobel Prize committees recognize the importance of photons and photonics ten times:

2000—The semiconductor double heterostructure laser.
2001—Observation of Bose-Einstein condensation using laser cooling.

2005—High precision measurement of time using laser frequency comb techniques.
2009—Invention of low-loss optical fibers.
2009—CCD photon imaging.
2012—Quantum non-destructive measurements of single photons and atoms.
2014—Invention of high-brightness solid-state blue LEDs and lasers.
2014—Super resolution optical microscopy.
2018—Optical tweezers and their application to biological systems.
2018—A method of generating high-intensity, ultra-short optical pulses.

2019: The International Bureau of Weights and Measures (BIPM) uses the frequency of the photon corresponding to the hyperfine transition of the Cs-133 atom in its unperturbed ground state $f_{\text{Cs}} \equiv 9.192631770 \times 10^9$ Hz to define basic unit of time: $\tau \equiv \frac{1}{f_{\text{Cs}}}$. The BIPM also used this frequency to define the basic unit of mass: $m \equiv \frac{hf_{\text{Cs}}}{c^2}$.

2.2 Bose Einstein Statistics

Following the work of Planck and Einstein, an electromagnetic wave is composed of photons. The energy of the wave is given by the energy of each photon: $\hbar\omega$ multiplied by the number of photons having that frequency, summed over all the frequency states:

$$E(\omega) = N(\omega)\hbar\omega$$

To determine the statistical distribution of photon states as a function of temperature, we use the same approach as we used in Sect. 1.4 (1.7 and following) to determine the Fermi distribution function.

Each state has an occupation probability P given by:

$$P = e^{\frac{\Omega + \mu N - E}{kT}} \tag{2.1}$$

N number of particles in the state in question
μ chemical potential
E energy of the state
K Boltzmann's constant
Ω is the grand potential:

$$\Omega(E) = -kT \ln\left(\sum_{\text{states}} e^{\frac{\mu N - E}{kT}}\right) = -kT \ln\left(\sum_{\text{states}} e^{\frac{\mu N - \varepsilon N}{kT}}\right), \tag{2.2}$$

where ε is the energy of a particle, (in this case the photon) making up that state.

Photons and electrons belong to entirely distinct classes of particles. Unlike the electron, any number of photons can occupy the same energy state:

$$\Omega(E) = -kT \ln\left(\sum_{\text{states}} e^{\frac{\mu N - \varepsilon N}{kT}}\right) = -kT \ln\left(\sum_{N=0}^{\infty} \left(e^{\frac{\mu - \hbar\omega}{kT}}\right)^N\right) = -kT \ln\left(1 - e^{\frac{\mu - \hbar\omega}{kT}}\right)^{-1} \tag{2.3}$$

where $\varepsilon = \hbar\omega$ is the energy of a photon.

As a result,

$$\Omega(\hbar\omega) = +kT \ln\left(1 - e^{\frac{\mu - \hbar\omega}{kT}}\right) \tag{2.4}$$

The occupation density n as a function of energy is given by

$$\langle n(\hbar\omega)\rangle = \frac{\partial \Omega}{\partial \mu} = -kT\left(-e^{\frac{\mu - \hbar\omega}{kT}}\right)\left(\frac{1}{kT}\right)\left(\frac{\left(1 - e^{\frac{\mu - \hbar\omega}{kT}}\right)}{\left(1 - e^{\frac{\mu - \hbar\omega}{kT}}\right)^2}\right) = \frac{1}{\left(e^{\frac{\hbar\omega - \mu}{kT}} - 1\right)} \tag{2.5}$$

A second fundamental difference between electrons and photons is particle conservation. In the low energy environment of photonics ($E < 10$ keV), mass is neither created nor destroyed, and electron number is conserved. Photons, on the other hand, can be created or destroyed, and there is no conservation requirement on the number of particles.

The chemical potential $= \mu$, measures the equilibrium condition of a system of particles. At equilibrium, μ is the same everywhere. Otherwise, the system is considered to be in a non-equilibrium state. Consider first the case of electrons. If two regions with differing chemical potential, or Fermi level, are brought together to form a system, then electrons will move from one region to the other until the chemical potentials are the same and equilibrium is restored. The displacement of electrons creates an electric field and thus an electric potential difference between the two regions. This is why a pn-junction has a built-in potential.

In the case of photons, where particle number is not conserved, if the chemical potential somewhere was higher than zero, photons would spontaneously disappear from that area until the chemical potential went back to zero; likewise if the chemical potential somewhere was less than zero, photons would spontaneously appear until the chemical potential went back to zero. When the number of particles is not conserved, as in the case of photons, we can set the chemical potential to zero everywhere. Making this substitution in (2.5) results in the Bose-Einstein distribution function:

$$n(\hbar\omega) = \frac{1}{\left(e^{\frac{\hbar\omega}{kT}} - 1\right)} \tag{2.6}$$

The spectral radiance of an ideal black body is given by Planck's law:

$$R(\omega, T) = \hbar\omega \frac{\omega^2}{2\pi^2 c^2} \frac{1}{\left(e^{\frac{\hbar\omega}{kT}} - 1\right)} \quad \text{in units of W Hz}^{-1}\ \text{sr}^{-1}\ \text{m}^{-2} \tag{2.7}$$

2.3 Propagation of Photons

The propagation of photons as waves is described by Maxwell's equations which model light as a fluid. The validity of these equations has been proved over a truly astonishing range of conditions. When the theory of special relativity was proposed by Einstein in 1905, it was shown that Maxwell's equations can be used without any modification to account for relativistic behavior. The scientific research of a number of scientists in the eighteenth and nineteenth centuries, synthesized by Maxwell as coupled electric and magnetic fields, and transformed by Heaviside into usable mathematical format, must rank among the greatest accomplishments of the human spirit.

2.3.1 Maxwell's Equations

$\mathbf{E}$ and $\mathbf{H}$ are the electric and magnetic vector fields
$\mathbf{D}$ is the electric displacement vector field
$\mathbf{B}$ is the magnetic induction vector field
ρ is the free charge density
ε is the dielectric constant
$\mathbf{J}$ is the current density

We will assume some simplifying conditions

If the materials are macroscopic and isotropic, then $\mathbf{D}(\mathbf{r}) = \varepsilon(\mathbf{r})\mathbf{E}(\mathbf{r})$
In dielectric materials, $\mu = 1$, so $\mathbf{B} = \mathbf{H}$
In lossless materials, $\mathbf{J} = 0$ and $\rho = 0$

$$\begin{aligned} &\nabla \cdot \mathbf{B} = 0 \\ &\nabla \cdot \mathbf{D} = 0 \\ &\nabla \times \mathbf{E} = -\frac{\partial \mathbf{B}}{\partial t} = -\mu\frac{\partial \mathbf{H}}{\partial t} \\ &\nabla \times \mathbf{H} - \frac{\partial \mathbf{D}}{\partial t} = \mathbf{J} = 0 \end{aligned} \tag{2.8}$$

Any vector field (strain is a good example) can be expressed in terms of two concepts.

Expansion, that is hydrostatic strain.
Rotation, that is shear strain.

The divergence of a field measures the linear expansion of a vector field at a given point. This is a scalar number and is expressed as the dot product of the gradient and the field vector: $\nabla \cdot \mathbf{E}(\mathbf{r})$.

The rotation is given by the cross-product of the gradient vector and the field vector. The rotation of the field is a vector, giving both the magnitude and the axis of the rotation.

If $\nabla \cdot \mathbf{B} = 0$ everywhere, that is, if there is not expansion or contraction, then only rotational components of the field remain, and so $\mathbf{B}$ can be expressed as a curl of a vector $\mathbf{A}$.

$$\mathbf{B} = \nabla \times \mathbf{A}$$

Maxwell's equations are quite general, and as a result, solutions differ depending on the geometry and properties of the medium in which light is propagating.

In quantum photonics we often seek to find a relationship between quantum fields at the input and the output of a 'device'. Many quantum photonic devices are produced as 2-dimensional planar structures, as a direct result of epitaxial crystal growth. The spacing between layers can be reduced in a quite controllable manner to quantum size dimensions; that is, for spacing dimensions similar to the wavelength of the particle-wave. Quantum effects can be observed in the 3rd dimension, perpendicular to the plane of crystal growth. It is convenient use a transfer matrix formalism that relates incident and reflected travelling field amplitudes.

Example 2.1 Propagation of light through a stack of materials having differing indices of refraction: an interference filter.

In this example, diagrammed in Fig. 2.4, a photon enters from the left, propagating in a dielectric medium Region 1 with $\varepsilon = \varepsilon_1\varepsilon_0$ and $\mu = \mu_0$. It then traverses Region 2 with $\varepsilon = \varepsilon_2\varepsilon_0$, having thickness a, and exits in Region 3 having the same parameters as Region 1. At each interface there is a finite probability for reflection and transmission. The objective of the example is to calculate the transmission coefficient $T = \frac{||F||^2}{||A||^2}$, in Region 3 as a function of photon frequency.

Solution:
The electromagnetic field is transverse to the propagation direction. Without loss of generality we can assume that $\mathbf{E}(z,t) = E_x(z,t)$. That is, the electric field is polarized in the x-direction for a wave propagating along the z-axis.

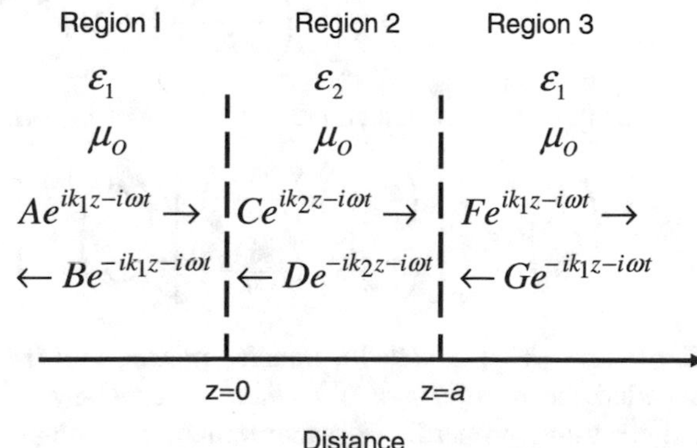

Fig. 2.4 Structure for Example 2.1, showing light propagation in the z-direction through a structure with various dielectric constants

Maxwell's equation for wave propagation is:

$$\frac{d^2}{dz^2}E(z,t) = \varepsilon\mu\frac{\partial^2}{\partial t^2}E(z,t) \tag{2.9}$$

A general expression for the solution to this equation is: $E(z,t) = E_0 e^{i(kz-\omega t)}$. In Region 1, we have:

$$E(z,t) = Ae^{i(kz-\omega t)} + Be^{i(-kz-\omega t)} \tag{2.10}$$

and similarly for Regions 2 and 3.

In this particular example, note that $G = 0$, because there is no wave coming from the right in Region 3. The wave vector $k = (\mu\varepsilon)\omega$ will differ in each region according to the values of μ and ε, while ω remains constant throughout the structure.

Boundary conditions, specific to the example are necessary to proceed further. These apply to the interfaces between adjacent regions:

(i) continuity of the electric field amplitude across the interface. This means that photons are neither destroyed or created as the wave crosses the interface.
(ii) continuity of the first derivative of the electric field across the interface. This means that the momentum of the photons is conserved as the wave crosses the interface.

For the interface at $z = 0$,

$$\begin{aligned} &\text{(i)} \quad A + B = C + D \\ &\text{(ii)} \quad k_1 A - k_1 B = k_2 C - k_2 D \end{aligned} \tag{2.11}$$

Solving for A and B in terms of C and D:

$$A = \frac{1}{2}\left[C\left(\frac{k_1 + k_2}{k_1}\right) + D\left(\frac{k_1 - k_2}{k_1}\right)\right]$$

and

$$B = \frac{1}{2}\left[C\left(\frac{k_1 - k_2}{k_1}\right) + D\left(\frac{k_1 + k_2}{k_1}\right)\right]$$

This result can be written conveniently in matrix form

$$\begin{pmatrix} A \\ B \end{pmatrix} = \frac{1}{2}\begin{bmatrix} \left(\frac{k_1+k_2}{k_1}\right) & \left(\frac{k_1-k_2}{k_1}\right) \\ \left(\frac{k_1-k_2}{k_1}\right) & \left(\frac{k_1+k_2}{k_1}\right) \end{bmatrix}\begin{pmatrix} C \\ D \end{pmatrix} = \begin{bmatrix} M_{ij} \end{bmatrix} \cdot \begin{pmatrix} C \\ D \end{pmatrix} \tag{2.12}$$

The matrix $[M_{ij}]$ is called a transfer matrix, and the procedure described below is called the *transfer-matrix method*. It can be adapted to calculate the optical and electronic properties of planar structures, both periodic and aperiodic, that are encountered in quantum photonic materials.

For the interface at $z = a$

$$\text{(i)}\quad Ce^{ik_2a} + De^{-ik_2a} = Fe^{ik_1a} + Ge^{-ik_1a}$$
$$\text{(ii)}\quad Ce^{ik_2a} + De^{-ik_2a} = Fe^{ik_1a} + Ge^{-ik_1a}$$

Solving for C and D in terms of F and G:

$$\begin{pmatrix} C \\ D \end{pmatrix} = \frac{1}{2}\begin{bmatrix} \left(\frac{k_1+k_2}{k_2}\right)e^{i(k_1-k_2)a} & \left(\frac{k_2-k_1}{k_2}\right)e^{-i(k_1+k_2)a} \\ \left(\frac{k_2-k_1}{k_2}\right)e^{i(k_1+k_2)a} & \left(\frac{k_1+k_2}{k_2}\right)e^{-i(k_1-k_2)a} \end{bmatrix}\begin{pmatrix} F \\ G \end{pmatrix} = \begin{bmatrix} N_{kl} \end{bmatrix}\cdot\begin{pmatrix} F \\ G \end{pmatrix} \tag{2.13}$$

These matrices can be combined:

$$\begin{pmatrix} A \\ B \end{pmatrix} = \begin{bmatrix} M_{ij} \end{bmatrix}\begin{pmatrix} C \\ D \end{pmatrix} = \begin{bmatrix} M_{ij} \end{bmatrix}\cdot\begin{bmatrix} N_{kl} \end{bmatrix}\begin{pmatrix} F \\ G \end{pmatrix} \tag{2.14}$$

This allows us to determine directly:

$A = (M_{11}N_{11} + M_{21}N_{12})F$, where we have set $G = 0$.

$$\left\|\frac{A}{F}\right\|^2 = \left(\frac{A}{F}\right)\cdot\left(\frac{A^*}{F^*}\right) = \frac{1}{4k_1^2k_2^2}\left(4k_1^2k_2^2\cos^2(k_2a) + \left(k_1^2+k_2^2\right)^2\sin^2(k_2a)\right)$$

The transmission coefficient T is expressed:

$$T \equiv \left\|\frac{F}{A}\right\|^2 = \frac{1}{\cos^2(k_2a) + \frac{1}{4}\left(\frac{k_1^2+k_2^2}{k_1k_2}\right)^2\sin^2(k_2a)} \tag{2.15}$$

Note that when $k_2a = n\pi$ where n is an integer, the transmission is unity, otherwise T < 1.

As an example, we consider visible light passing through a thin film of SiO_2 having an index of refraction of 1.5. We will assume the thickness of the film to be 2 μm. Equation (2.15) is evaluated as a function of photon frequency, because this remains constant in all regions, whereas the wavevector in Region 1 is different from that in Region 2.

The graph of the transmission coefficient versus frequency for this case is shown in Fig. 2.5.

The expected features are present: the transmission is periodic, although the minimum transmission remains about 85%. The periodicity depends directly on the film thickness, and it follows that transmission of photons away from the normal axis will increase the effective thickness of the film and therefore allow 100%

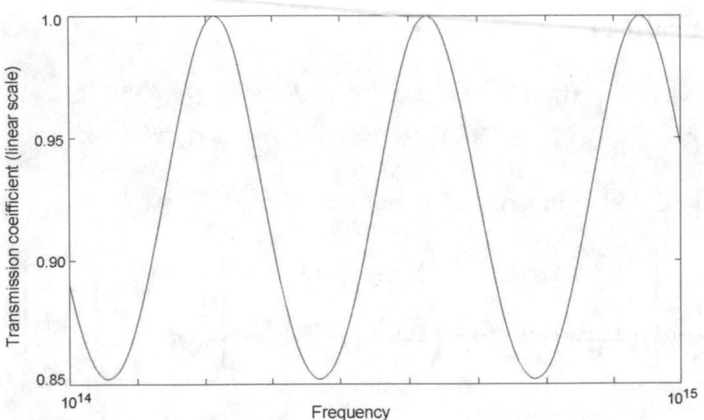

Fig. 2.5 Transmission of light through a thin film of SiO_2 at normal incidence

transmission at a different wavelength of incident light, creating a rainbow effect. Similar effects are observed if the film is not uniform in thickness across its plane, or if the film surface is not planar.

2.3.2 Reflection and Transmission: Fresnel's Equations

When a lightwave leaves one medium and enters another, a part of the beam is reflected, and a part is transmitted. In the absence of scattering or absorption at the interface, the sum of the intensities of the transmitted and reflected components is equal to the intensity of the incident beam: ($T^2 + R^2 = 1$, where T and R are respectively the transmission and reflection coefficient) The details of how the beam behaves will depend on the angle of incidence and polarization of the incident beam as well as the optical properties of the media on each side of the interface.

Maxwell's equations can be directly solved to obtain the amplitude and phase of the electric field relative to that of the incident beam resulting in Fresnel's equations. The reader is referred to any introductory text on electromagnetic behavior (for example, Hecht 1987) for the derivation.

The case of an electromagnetic wave incident on a boundary is diagrammed in Fig. 2.6. We assume that the materials on both sides of the boundary are dielectrics. The electric field vector is chosen to be perpendicular to the plane of incidence so that it lies parallel to the plane of the interface. Solving Maxwell's equations gives:

$$R = \left(\frac{\mathbf{E}_R}{\mathbf{E}_i}\right)_\perp = \frac{n_1 \cos\theta_1 - n_2 \cos\theta_2}{n_1 \cos\theta_1 + n_2 \cos\theta_2} = \frac{\left(\frac{n_1}{n_2}\right)\cos\theta_1 - \cos\theta_2}{\left(\frac{n_1}{n_2}\right)\cos\theta_1 + \cos\theta_2} \tag{2.16}$$

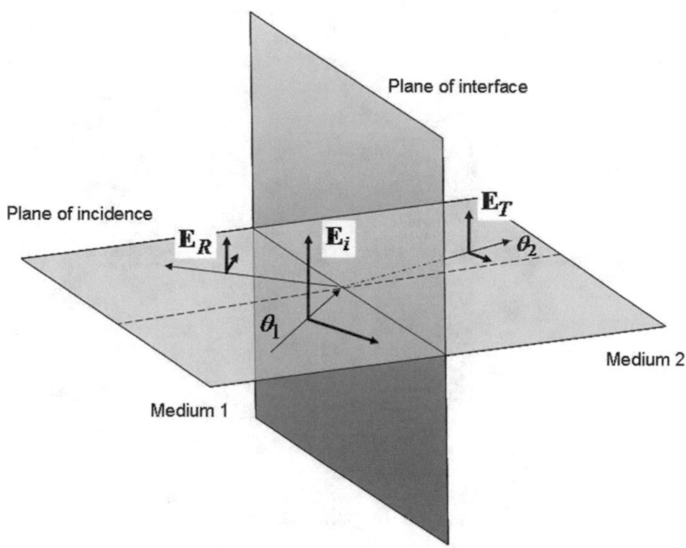

Fig. 2.6 When an electromagnetic wave is incident on a boundary between two materials, part of the beam is reflected and part is transmitted. Maxwell's equations require that the angle of incidence equal the angle of reflection. The plane of incidence is defined by the propagation vectors of the incident and reflected waves

and

$$T = \left(\frac{\mathbf{E}_T}{\mathbf{E}_i}\right)_\perp = \frac{2n_1 \cos\theta_1}{n_1 \cos\theta_1 + n_2 \cos\theta_2} \tag{2.17}$$

These are two of Fresnel's equations. The remaining two describe the situation where the electric field is parallel to the plane of incidence.

Equations (2.16) and (2.17) can be used to model the behavior of an optical beam splitter consisting of a simple glass plate. There are 2 cases to consider.

Case 1: $n_1 < n_2$
We could suppose that $n_1 = 1$ (air), and that $n_2 = 1.5$ (representing glass), for example. The incident beam strikes the glass surface from the air side. It is clear that $\left(\frac{\mathbf{E}_T}{\mathbf{E}_i}\right)_\perp$ is always real and positive. Thus the transmitted wave is in phase with the incident wave. On the other hand, if $n_1 < n_2$, then $\theta_1 > \theta_2$, and $\cos\theta_1 < \cos\theta_2$. Thus $\left(\frac{\mathbf{E}_R}{\mathbf{E}_i}\right)_\perp$ is negative, indicating a phase change or polarization change of π radians ($e^{i\pi} = -1$) of the reflected wave relative to the incident beam.

Case 2: $n_1 > n_2$
This case corresponds to the wave incident on the interface from the glass side. The transmitted wave is in phase with the incident wave as before. For the reflected wave, we note that $n_1 > n_2$, and that $\cos\theta_1 > \cos\theta_2$. $\left(\frac{\mathbf{E}_R}{\mathbf{E}_i}\right)_\perp$ is positive for all angles of incidence. Therefore the reflected wave is in phase with the incident wave.

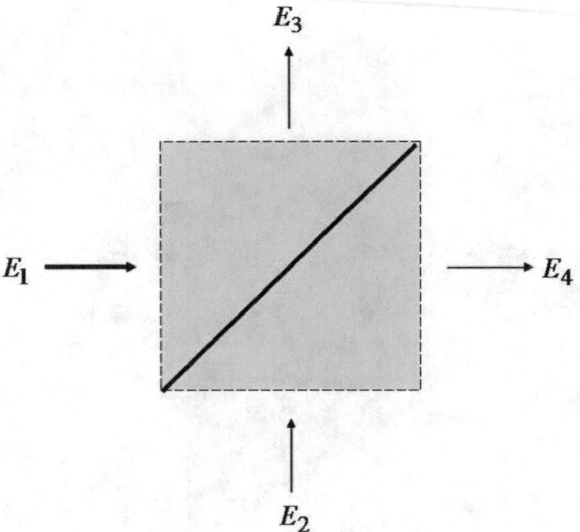

Fig. 2.7 A model of an optical beamsplitter as a device with two input ports (E_1 and E_2), and two output ports (E_3 and E_4)

The example in the previous section could serve as the basis for the analysis of a simple beam-splitter. We can make a description of such a beam splitter as a device with two inputs E_1 and E_2. The outputs are E_3 and E_4 as shown in Fig. 2.7.

We can write the outputs as a function of the inputs. The electric field amplitudes at the 2 inputs are E_1 and E_2, and the output field amplitudes are E_3 and E_4.

$$E_3 = (RE_1 + TE_2) \tag{2.18}$$
$$E_4 = (TE_1 + e^{i\pi} RE_2) \tag{2.19}$$

where the polarization reversal specified by Fresnel's equations is required in (2.19).

$$\begin{aligned} (E_3)^2 &= (RE_1)^2 + (TE_2)^2 + 2TRE_1E_2 \\ (E_4)^2 &= (TE_1)^2 + (RE_2)^2 - 2TRE_1E_2 \end{aligned} \tag{2.20}$$

and

$$E_3^2 + E_4^2 = (E_1)^2(T^2 + R^2) + (E_2)^2(T^2 + R^2) = E_1^2 + E_2^2, \tag{2.21}$$

conserving energy as required.

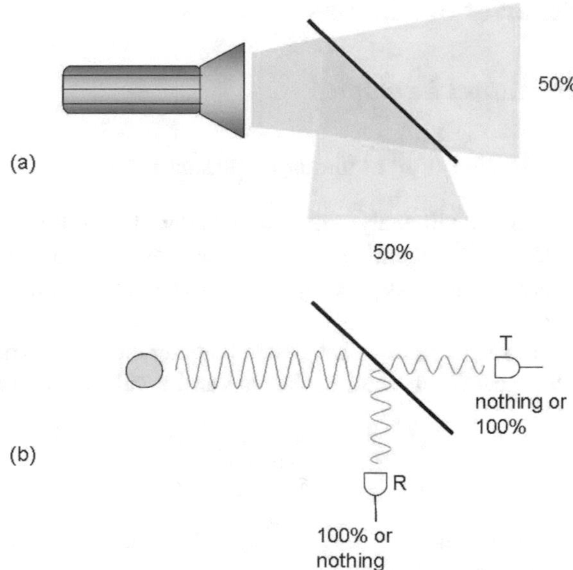

Fig. 2.8 A schematic view of photons passing through a beam splitter. **a** In the case of a beam of photons, Maxwell's equations give a precise account of how the intensity of the beam is divided. **b** Although a single photon cannot be divided in two, the wavefunction is present in both paths. The photon, however, can be detected only once. Therefore it must be detected either at T, showing that the photon propagated straight through the beam splitter, or at R showing that the photon was reflected. Experiments by Grangier et al. (1986) have demonstrated that the choice between transmission and reflection does not occur when the photon passes through the beam splitter, but rather when the photon is detected

2.3.3 Single Photon Behavior

In Fig. 2.8a, we show a beam splitter at work. A beam of light strikes a beam splitter, and the intensity of the beam is divided according to the properties of the beam splitter and the angle of incidence.

In place of the light beam, suppose that we could substitute a source containing only a single photon. This case is shown in Fig. 2.8b. A photon is an elementary particle and cannot be divided. Hence, it must either be entirely detected by the detector in the transmission path or entirely by the detector in the reflection path. However, the Schrödinger equation shows that the probability intensity is non-zero for both paths. According to the Copenhagen interpretation of quantum mechanics, detection of the photon shows which path the photon has taken, and the wavefunction in the path not taken collapses to zero.

How does the photon decide which route to take? No existing theory can explain this fundamental mystery of quantum photonics. The photon wavefunction is non-zero along both paths until detection occurs. We could say the photon "decides" which path it has taken only after it is detected. We continue this discussion in Sect. 2.4.3.

2.4 Wave Behavior of Photons

2.4.1 Computational Examples

Example 2.3 Propagation of light through a plasma layer.

Next, using (2.15) we will analyse the propagation of light through a plasma of electrons and protons. With equal numbers of electrons and protons, the plasma is electrically neutral. For simplicity, we will consider the problem in one dimension, as shown in Fig. 2.9.

A free electron in such a plasma is repelled by other electrons. The electric field experienced by an electron in such a plasma can be calculated from Maxwell's first equation:

$$\frac{d}{dz}\mathbf{E}(z) = \frac{\rho}{\varepsilon} = \left(\frac{N_e q}{\varepsilon}\right), \tag{2.22}$$

where q is the charge of one electron, and N_e is the electron density.

The force on the electron is expressed

$$\mathbf{F} = m_e\mathbf{a} = m_e\frac{d^2 f(z)}{dt^2} = Q\mathbf{E} = -\frac{N_e q^2}{\varepsilon}z \tag{2.23}$$

where $f(z)$ gives the position of the electron. Equation (2.17) is in the form of an oscillator

$$\frac{d^2 f(z)}{dt^2} = -\omega_p^2 z,$$

where

$$\omega_p \equiv \frac{N_e q^2}{m_e \varepsilon} \tag{2.24}$$

is defined as the plasma frequency.

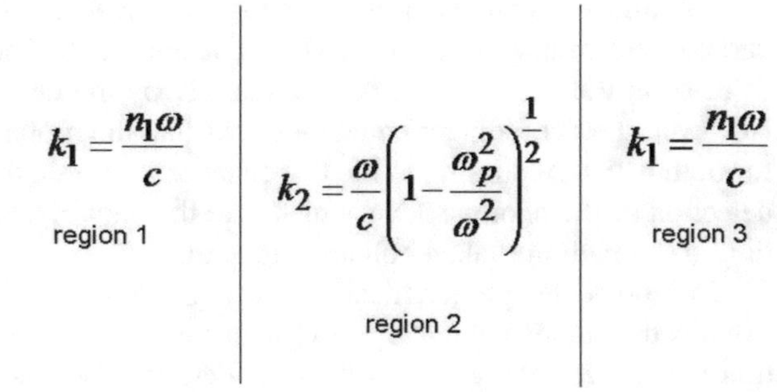

$$k_1 = \frac{n_1\omega}{c}$$

region 1

$$k_2 = \frac{\omega}{c}\left(1 - \frac{\omega_p^2}{\omega^2}\right)^{\frac{1}{2}}$$

region 2

$$k_1 = \frac{n_1\omega}{c}$$

region 3

Fig. 2.9 Schematic diagram showing the index of refraction of a planar plasma layer embedded in a dielectric region having an index of refraction $= n_1$

In this example, a photon wave travelling along the z-axis encounters a region of plasma.

Maxwell's equations are:

$$\nabla \times \mathbf{E} + \frac{\partial \mathbf{H}}{\partial t} = 0$$

$$\nabla \times \mathbf{H} - \frac{\partial \mathbf{E}}{\partial t} - \sigma \mathbf{E} = 0$$

where $\sigma = -i\frac{N_e q^2}{m\omega}$ is the conductivity of the plasma.

The fields can be expressed as waves,

$$\mathbf{E} = \mathbf{E_0} e^{i(kz-\omega t)}$$

Maxwell's equation for the electric field is:

$$\frac{\partial^2 \mathbf{E}}{\partial z^2} - \varepsilon\mu \frac{\partial^2 \mathbf{E}}{\partial t^2} - \sigma\mu \frac{\partial E}{\partial t} = 0$$

Substituting for **E** yields the dispersion relationship between ω and k:

$$k^2 = \mu\varepsilon\omega^2 - i\omega\sigma\mu$$

and

$$k^2 = \mu\varepsilon\omega^2 \left(1 - \frac{N_e q^2}{\varepsilon m \omega^2}\right) = \mu\varepsilon\omega^2 \left(1 - \frac{\omega_p^2}{\omega^2}\right) \tag{2.25}$$

Taking $\mu = \mu_0$ and $\varepsilon = \varepsilon_0$,

$$k^2 = \frac{\omega^2}{c^2}\left(1 - \frac{\omega_p^2}{\omega^2}\right) \tag{2.26}$$

When the wave frequency is less than the plasma frequency, (2.26) shows that the wave vector is imaginary (Fig. 2.10). Thus E-M wave propagation is strongly attenuated in this regime (Fig. 2.11).

Assume that Regions 1 and 3 are free space:

$$k_1 = k_3 = \frac{\omega}{c}$$

For the plasma layer

$$k_2 = \frac{\omega}{c}\left(1 - \frac{\omega_p^2}{\omega^2}\right)^{\frac{1}{2}} \tag{2.27}$$

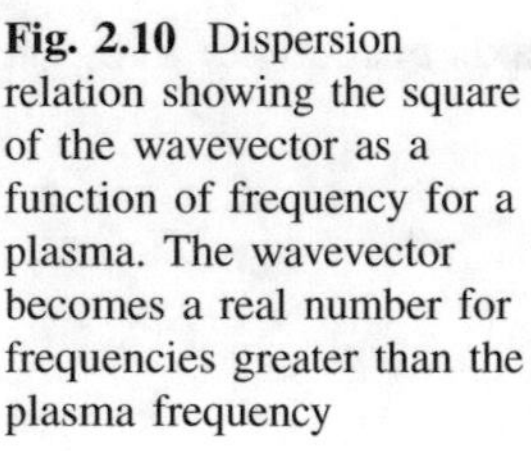

Fig. 2.10 Dispersion relation showing the square of the wavevector as a function of frequency for a plasma. The wavevector becomes a real number for frequencies greater than the plasma frequency

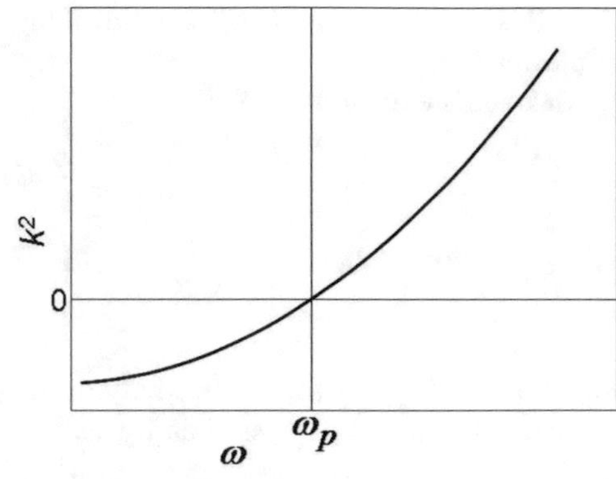

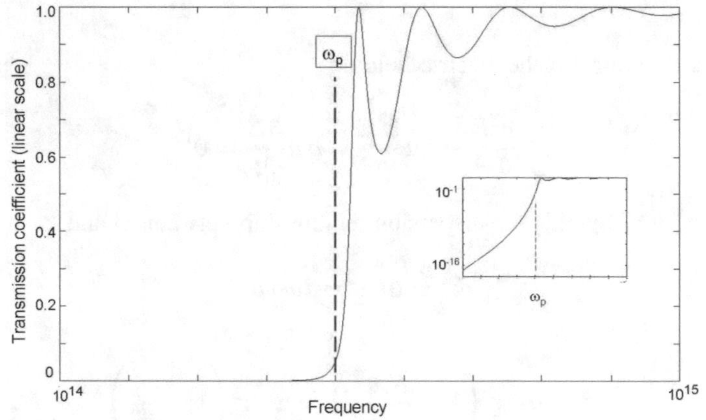

Fig. 2.11 Propagation of photons through a plasma sheath as a function of photon frequency. The *vertical dashed line* indicates the plasma frequency of 5×10^{14} Hz. The *inset* in the figure shows the same calculation with the transmission coefficient plotted on a logarithmic scale

The dependence of k_2^2 on frequency is shown in Fig. 2.10.

The dispersion relation is a non-local relationship between wavevector and frequency of wave-particles. Since momentum is directly proportional to wavevector, and energy is directly proportional to frequency, the dispersion relation can also be considered as the energy-momentum relationship for the wave-particle. Knowing the dispersion relation is often more important and informative than knowing the wavefunction.

In the low frequency regime, $\omega < \omega_p$ and the wavevector in Region 2 is imaginary. As a result the photon waves are strongly attenuated.

For incident waves having a frequency above the plasma frequency, the wavevector is real, and the conditions of the calculation resemble those we have just considered.

Despite these rather different situations, the transmission coefficient is given by (2.15) for the entire frequency range both below and above the plasma frequency.

In Fig. 2.11, we show the result of this calculation for a photon wave passing through a plasma sheath having a plasma frequency of 5×10^{14} Hz and a thickness of 5 μm.

At first view, the transmission coefficient is near 0 for frequencies below the plasma frequency, and rapidly increases above it, showing an oscillatory behavior. However, there is more to this result. It can clearly be seen that some transmission of photons is permitted through the barrier, even when the wavevector is imaginary. By examining the inset of Fig. 2.11, it can be seen that for **all** frequencies, there is transmission of photons through the barrier, although the attenuation becomes stronger as the frequency diminishes. This phenomenon: transmission of a photon through a region where the photon should not be allowed is called tunneling. Tunneling is a fundamental feature of wave behavior. All particle-waves, photons, electrons, even atoms can display tunnelling. Important quantum photonic devices depend on tunnelling to function properly.

2.4.2 Young's Double Slit Experiment

In 1805, Thomas Young demonstrated and interpreted the passage of a photon beam through a pair of slits in an opaque screen. His experiment showed the photon beam created a diffraction pattern associated with wave behavior of light. This work was revolutionary, overturning 100 years of belief in the Newtonian corpuscular theory, and rehabilitating the studies of Huygens.

In Fig. 2.12 we show a sketch of the experiment as devised by Young. A light beam passes through a narrow opening and is incident on a pair of slits, charac-

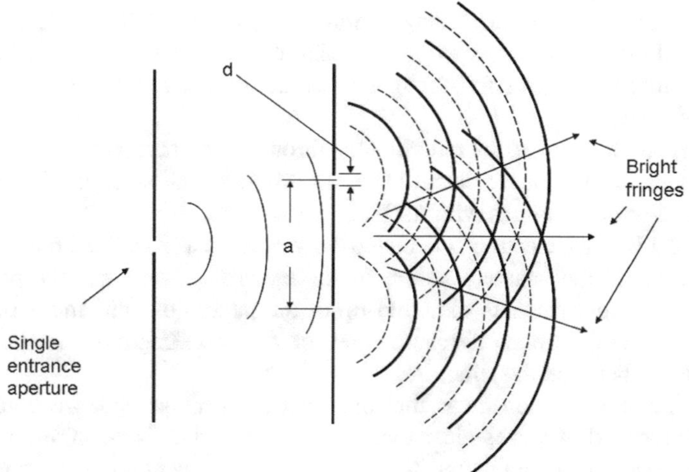

Fig. 2.12 Young's two-slit experiment of 1805 demonstrated wave-like behavior of photons. Each slit radiates light with circular wave fronts. The *arrows* show where the beams overlap in-phase, giving rise to maxima in the intensity In between, destructive interference reduces the intensity

terized by the slit width (d) and their separation (a). The passage through the slits gives rise to diffraction effects and under the right conditions, a pattern of light and dark fringes can be captured on a screen behind the slits.

Young sought to make the wave nature of light clear by demonstration. In order to optimise the diffraction effects, the light should be monochromatic and the phase front of the light should be the same at the entry of both slits. His critical innovation was the addition of a small opening in front of the two-slit passage in order to create a point-like source with spatial coherence.

A simple geometrical analysis relates the wavelength of light, the distance between the slits and the angle of divergence between the rays of maximum intensity: $n\lambda \approx d \sin(\alpha)$.

Young's experiment proves that photons have wave properties. However, it does not eliminate the possibility that photons also have particle properties. A century later these were shown by Planck and Einstein. The photon is now understood as being both a wave and a particle. Two centuries later, Young's two-slit experiment is a subject at the frontier of fundamental understanding of quantum behavior. Experimental instrumentation has reached the point where it is possible to detect individual photons. Sources can be chosen so that the passage of individual photons though the two-slit interferometer can be recorded at the image plane. Since only one photon is present in the apparatus at any time, multiple photons cannot interfere with each other. Thus, the interference pattern that builds up is the result of each photon interfering with itself. An elegant experiment has been carried out by Dykstra and colleagues at the University of Leiden, showing the detection of photons and the build-up of the interference pattern as a function of time.[1] A video of this experiment can be viewed at: https://www.youtube.com/watch?v=MbLzh1Y9POQ.

Electrons, like photons, exhibit wave and particle behavior simultaneously (Jönsson 1961, 1974). Akira Tonomura and colleagues from Hitachi Labs in Japan have carried out analogous experiments using electrons passing through a double slit etched into in a copper foil. Images taken at successive times from their 1989 paper[2] are shown in Fig. 2.13.

Since each electron passes individually through the arrangement of slits, these measurements show that the electron passes through both slits, and that the electron particle-wave interferes with itself.

In Fig. 2.13a, each electron is detected only once at well-defined point in space, a direct result of the quantum nature of the electron. Over time, the passage of more electrons contributes to the build-up of the pattern of light and dark fringes, a manifestation of wave interference, based on the wavelength of the electron and the separation between the slits.

This experiment demonstrates that the electron displays both wave and a particle behavior. It thus makes clear that the electron wave function interferes with itself. The electron wavefunction that represents the quantum is present in both slits, reflecting its delocalized nature. It is this delocalized nature that produces

[1]Dykstra et al. (2008), exter@physics.leidenuniv.nl.

[2]Tonomura et al. (1989), reproduced by permission.

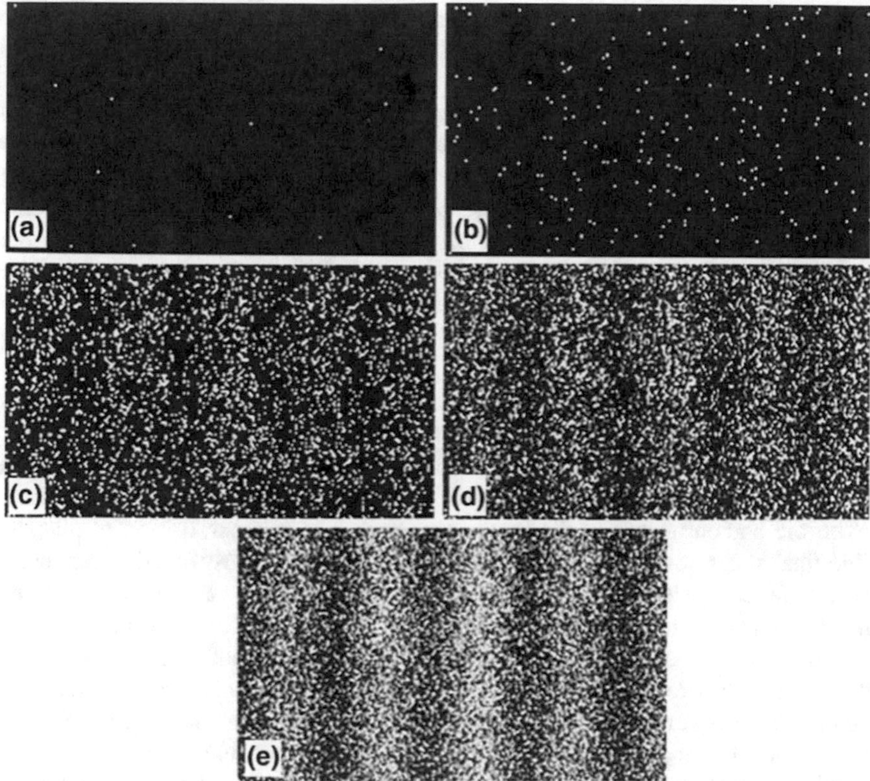

Fig. 2.13 Results of a double-slit-experiment performed by A. Tonomura showing the build-up of an interference pattern of single electrons. Numbers of electrons are 11 (**a**), 200 (**b**), 6000 (**c**), 40,000 (**d**), 140,000 (**e**). Reproduced from Tonomura et al. (1989) by permission of the American Physical Society

the interference shown in Fig. 2.13. It therefore makes little sense to ask through which slit the electron has passed.

In the classic Young's experiment where the source intensity is higher, each photon traverses the two-slit structure independent of all the others. Because the photon is a quantum, it can be detected on the other side at only one point. Thus, the process of detection forces the quantum to become localized at that point, so that its entire energy is given up to the detection apparatus in one interaction.

2.4.3 Single Photons

The complete understanding of measurements of photons in quantum systems is a work in progress. The double slit interference experiment reveals some important questions about the meaning of measurements, and the limits of what can be known and what must remain unknown in this simple experiment.

In a measurement, the photon (or an electron) can be measured at one location in space and time. The double-slit experiment shows that prior to measurement the photon can be delocalized, sampling two slits separated by a macroscopic distance. By observing the experiment in real time, we see that apparently identical photons wind up in different places as a result of traversing the same double-slit structure. We could think of the interference pattern as reflecting the probability distribution that describes where a photon quantum (electron) will be detected. This behavior might suggest that the impact of the photon (electron) on the double-slit structure introduces a measureable structured statistical uncertainty in the trajectory of the quantum.

The Copenhagen interpretation is a framework of principles that address this kind of question about the nature of quantum behavior prior to measurement. When applied to the double slit experiment, this interpretation would say that it is not possible to know how the photon passes through the double-slit structure. When the photon arrives at the image screen, it is in a superposition of all possible final states at the image plane. The amplitude of its wavefunction replicates the interference pattern that is observed after the arrival of many photons. When the photon is detected, by a sensor having a particular position fixed in space, the superposition of these states must collapse into the state that corresponds to the location of the detecting sensor. Thus, following the Copenhagen interpretation, the photon wave function remains delocalized up to the moment of detection.

Grangier, Roger and Aspect, published in 1986 an account of pioneering and elegant experiments that show the particle and wave behavior of a single photon in an interferometer like the double-slit experiment. The presentation below follows the treatment of this experiment by M. Beck[3] of Reed College.

1. Passage of a Photon Through a Beam Splitter

This event corresponds to the illustration in Fig. 2.8 earlier in this chapter. The solution of Maxwell's equations gives a transmission coefficient and a reflection coefficient. The situation is diagrammed in Fig. 2.14 (Bachor and Ralph 2004).

A single photon that traverses the beam splitter will either be transmitted or reflected. If the transmission detector T records a photon (= 1), then the reflection detector must record nothing (= 0). This result demonstrates anti-correlation. The calculation of the transmission and reflection coefficients of Example 2.1 indicates the relative numbers of counts at the transmission and reflection detectors if many experiments are conducted. In this description, we are supposing that the detectors are 100% efficient, that there are negligible losses along the trajectory of the photon.

We can quantify this result as follows:

Suppose that p_T = probability that a photon is detected only at T, p_R = probability that a photon is detected only at R, and that p_{TR} = probability that a photon

[3]Beck (2012).

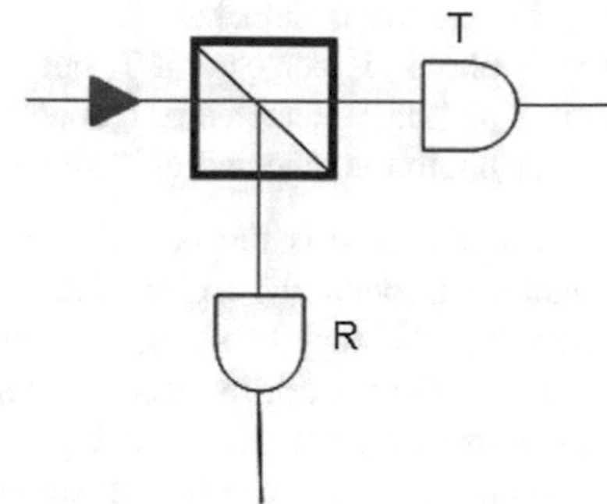

Fig. 2.14 A single photon incident on an ideal beam splitter is transmitted either 100% to T or reflected 100% to R

is detected at both T and R at the same time. We can express the degree of correlation, the correlation coefficient $g^{(2)}$ by the ratio:

$$g^{(2)}(\tau = 0) = \frac{(p_{\mathrm{TR}})}{(p_{\mathrm{T}})(p_{\mathrm{R}})} \tag{2.28}$$

where τ is the time delay between photon detection at T and R.

$g^{(2)}(0) = 0$ for a single photon, because $p_{\mathrm{TR}} = 0$.

If we allow the time delay to be long enough, eventually $p_{\mathrm{TR}} > 0$. For very long times $p_{\mathrm{TR}} \to (p_{\mathrm{T}})(p_{\mathrm{R}})$, and $g^{(2)}(\tau \to \infty) = 1$.

For a beam of classical lightwaves incident on the beam splitter, we measure intensities:

$$I_{\mathrm{T}} = \mathrm{T}I_i$$
$$I_{\mathrm{R}} = \mathrm{R}I_i$$
and
$$\mathrm{T} + \mathrm{R} = 1$$

In this case $g^{(2)}(0) = \frac{\langle I_{\mathrm{T}} I_{\mathrm{R}} \rangle}{\langle I_{\mathrm{T}} \rangle \langle I_{\mathrm{R}} \rangle} = \frac{\langle I_i^2 \rangle}{\langle I_i \rangle^2}$, and its value is determined by the statistics of the incoming beam.

The Cauchy-Schwarz inequality, however, specifies that $\langle I_i^2 \rangle \geq \langle I_i \rangle^2$, so in the classical case:

$$g^{(2)}(0) \geq 1 \tag{2.29}$$

Thus, the single photon case can be distinguished from the classical case in a measurement.

A Note on the Losses and Non-Ideality

In a real experiment, single-photon detection is not 100%. In addition, the detector has a dark count rate. As a result, when a photon passes through the beam splitter and is incident on the detectors, there are four possible results:

(a) No photon is detected
(b) A photon is detected at T, but not at R
(c) A photon is detected at R, but not at T
(d) A photon is detected at T and at R.

Result **b** or **c** is the behavior we would expect from the passage of a single quantum through the experimental set-up. Result **a** is the result of the limited quantum efficiency of single photon detectors. A typical value for this efficiency is 60%. Result **d** occurs because the time the detection window is open occasionally permits the presence of two photons. Both the dark count rate and the detector quantum efficiency can be measured, and this information can be used to correct the statistics of the raw data on many measurements. This short discussion emphasizes that while the ideal single quantum experiment is straightforward to describe, obtaining meaningful results from a real experiment requires careful analysis to account for non-idealities of which this is an important example.

Patrick Bronner of the Department of Physics at the Universtät Erlangen has prepared this experiment on-line (http://www.didaktik.physik.uni-erlangen.de/quantumlab/english/index.html see "photon existence experiment"), and the reader can follow the set-up, the taking and analysis of the data. (Quantum Lab 2009)

2. Generation of Single Photons

Beta Barium Borate or BBO, is a nonlinear optical crystal. BBO's unique combination of properties makes it an excellent choice for second and higher order harmonic generation, frequency mixing, broadly tunable sources, and parametric down-conversion applications. In the following experiment, BBO is used for parametric frequency down-conversion of a single photon of frequency ω_0 into two photons of frequency $\frac{\omega_0}{2}$. Energy is of course conserved. Parametric frequency down-conversion is treated in more detail in this book in Chaps. 5 and 9.

Conservation of momentum (referred to as phase-matching for photon beams) requires that the two lower frequency photons be emitted at an angle of $\pm\theta$ each side of the primary beam. This feature enables the easy separation of the photons into two measurement circuits as shown in Figs. 2.15 and 2.16. When a photon is detected by the gating detector G, then we know with certainty that a single photon has been launched into the single photon experiment.

Experiment no. 1: Passage through a single beam splitter: measurement of photon quantum behavior (see Fig. 2.16).

This experiment is quite close to Young's double-slit experiment. There are two paths for the photon: transmission or reflection. By putting a detector in each arm we can determine with certainty along which path the photon travels. A single photon can be detected only once, and it must generate a count either at T or at R, but not both. Thus we expect to see anticorrelation of detection events confirming 2 results: a single photon in the interferometer and particle-like behavior.

Results of this experiment reported by Thorn et al. (2004) give $g^{(2)}(\tau = 0) = 0.0177 \pm 0.0026 \ll 1$. Quantum behavior is clearly established. The result does not show perfect anticorrelation, because of less than perfect detection efficiency, for example, and other non-idealities in the experiment.

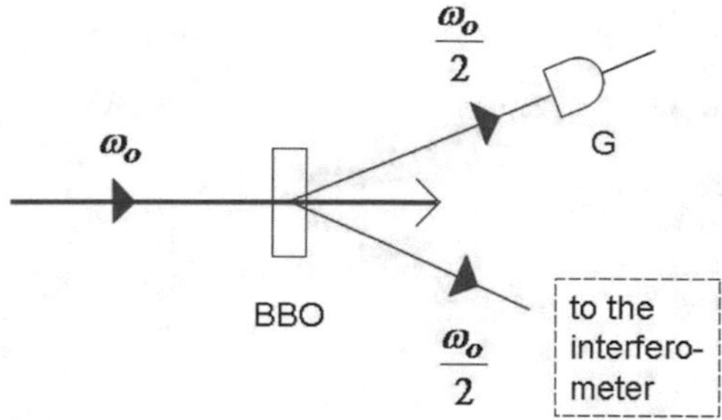

Fig. 2.15 Non-linear materials, in this case BBO, can be used to generate 2 photons each having half the frequency of the incoming beam ω_0. These photons are emitted in at a well-defined angle relative to the incoming beam (in this case about 3°) allowing for easy separation from each other and from the incoming beam

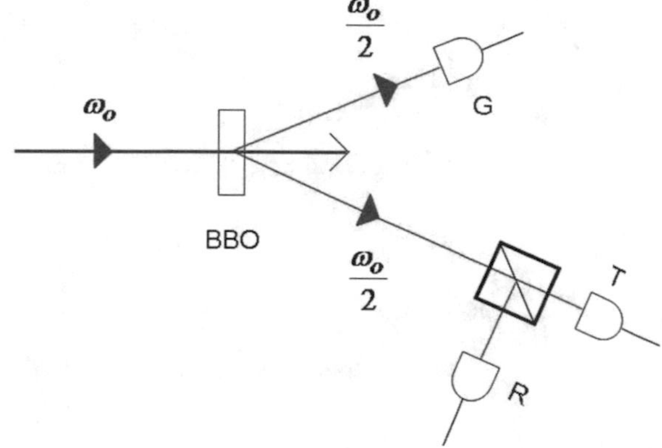

Fig. 2.16 Detection of a single photon by the gating detector G confirms that a second single photon will enter the beam splitter and result in a signal at either R or T, but not both. The results of this experiment show this anticorrelation effect which is a non-classical quantum result. Data from this experiment can be analyzed in Exercise 2.8

Experiment no. 2: Passage through 2 beam splitters: measurement of single photon interference and photon wave behavior.

In this experiment we insert a second beamsplitter and two mirrors, creating a two parallel propagation paths (called a Mach-Zehnder interferometer) as shown in Fig. 2.17.

When a photon is detected at G, this heralds the presence of an identical photon in the Mach-Zehnder interferometer. By varying the position of one of the mirrors as shown, the path length of one arm can be changed relative to the other. When the path difference is equal to a multiple of the wavelength of the light the photon interferes constructively with itself at the second beamsplitter. When the path length is changed by an additional $\frac{\lambda}{2}$, the interference is destructive. Grangier et al. (1986) first performed this experiment using photon pairs generated by fluorescence from an atomic beam of calcium using 2-photon excitation, before down-conversion using BBO was recognized as a superior source. The results of Grangier et al. are shown in Fig. 2.18.

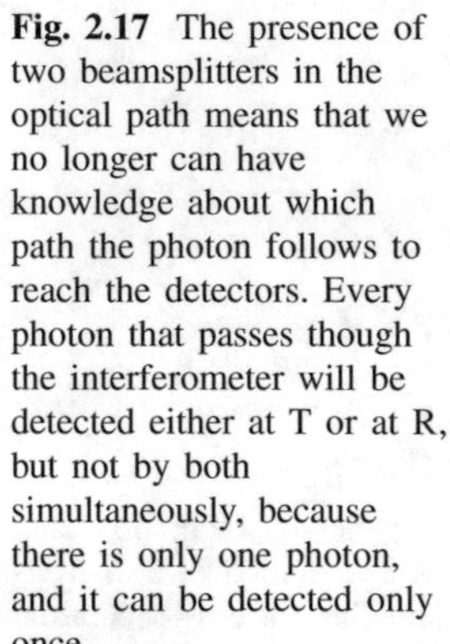

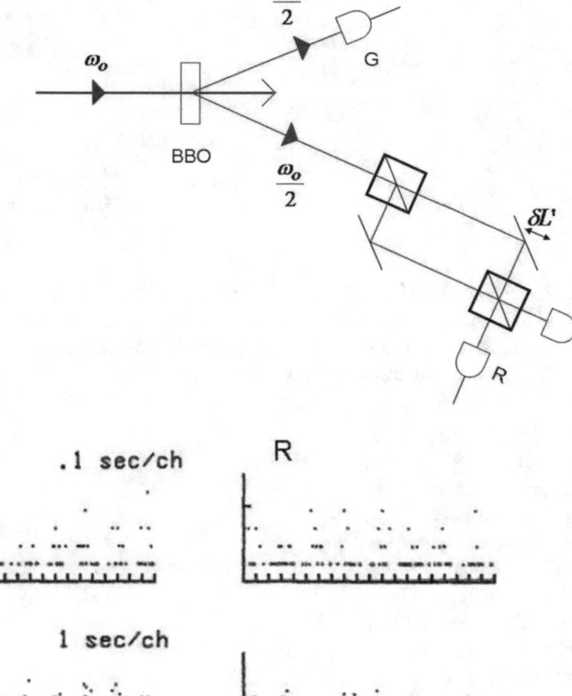

Fig. 2.17 The presence of two beamsplitters in the optical path means that we no longer can have knowledge about which path the photon follows to reach the detectors. Every photon that passes though the interferometer will be detected either at T or at R, but not by both simultaneously, because there is only one photon, and it can be detected only once

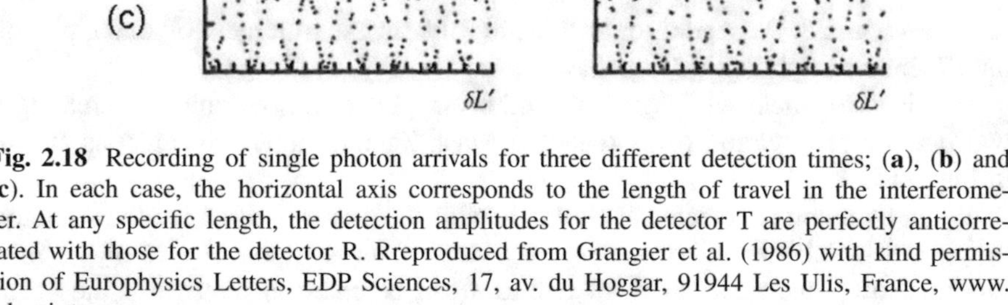

Fig. 2.18 Recording of single photon arrivals for three different detection times; (**a**), (**b**) and (**c**). In each case, the horizontal axis corresponds to the length of travel in the interferometer. At any specific length, the detection amplitudes for the detector T are perfectly anticorrelated with those for the detector R. Rreproduced from Grangier et al. (1986) with kind permission of Europhysics Letters, EDP Sciences, 17, av. du Hoggar, 91944 Les Ulis, France, www.edpsciences.org

The plots in Fig. 2.18 show the number of photon counts as a function of the change in the interferometer path length $\delta L'$. The groups (a), (b), and (c) correspond to increasing integration times. In Fig. 2.18a the integration time is short enough that the recordings of individual photons detected at T or at R are clearly seen. In Fig. 2.18c, the integration time is long enough that the clear pattern, corresponding to wavelength interference, is seen. As expected, when the detector T receives a maximum signal, detector R is at a minimum and vice versa.

In the intermediate regions where the signals at T and R are comparable, the single photon has similar probabilities for transmission to T or to R. It is important to remember that this cannot correspond to the classical interpretation that the intensity of the wave is divided by the beam splitter, because a single photon cannot be detected more than once. In the single photon case, all the intensity is collected by either T or R. This behavior is summarized by the necessity of wavefunction collapse upon measurement which is an important component of the Copenhagen interpretation of quantum mechanics.

2.4.4 The Copenhagen Interpretation of Quantum Measurements

The Copenhagen interpretation is a set of guidelines for understanding and interpreting experiments on quanta, and for connecting theoretical calculations with experiment. It was developed mainly by Bohr and Heisenberg in the first half of the twentieth century. For a number of years following in the twentieth century, these guidelines provided a useful framework for communication among physicists, acting like the "holy tablets" passed down by the "founding fathers" of the quantum revolution. However, there is as yet no theory to explain the physics of wavefunction collapse.

The progress in instrumentation during the last two decades now enables "table-top" experiments that measure the behavior of single quanta. The group of Jan-Peter Meyn at the Universität Erlangen-Nürnburg has created on-line versions of key experiments in quantum photonics. The user can pilot the measurement, record and analyze data (http://www.didaktik.physik.uni-erlangen.de/quantumlab/english/index.html). In these circumstances, physicists around the world can make measurements on the most basic of all elementary particles: the photon. This is an area of very exciting research, the results of which are addressing questions of what constitutes knowledge and reality.

Current experimental results in this field are causing an examination of the Copenhagen interpretation. One objective is to replace the Copenhagen "rules" with physical theory that explains observed behavior. We will not attempt to review the directions of current thought here, but the reader is directed to the Wikipedia page on the Copenhagen interpretation (http://en.wikipedia.org/wiki/Copenhagen_interpretation) which gives an idea of the range of the current debate.

2.5 Quantum Entanglement

Measurements of more than one photon can be divided into two categories. In the classic case photons behave independently. Photons that pass through the double-slit apparatus, for example, usually do so independently, and do not interfere with each other. Each photon does interfere with itself. Such photons are called separable quanta. If we can write the states of two such quanta as 3-dimensional vectors: $|A\rangle = (a_1|i_1\rangle \;\; a_2|i_2\rangle \;\; a_3|i_3\rangle)$ and $|B\rangle = (b_1|j_1\rangle \;\; b_2|j_2\rangle \;\; b_3|j_3\rangle)$, then the combined state of $|A\rangle$ and $|B\rangle$ is written as a tensor. If this can be accomplished by a straightforward multiplication:

$$|\Psi\rangle = |A\rangle \otimes |B\rangle = \begin{pmatrix} a_1b_1|i_1, j_1\rangle & a_1b_2|i_1, j_2\rangle & a_1b_3|i_1, j_3\rangle \\ a_2b_1|i_2, j_1\rangle & a_2b_2|i_2, j_2\rangle & a_2b_3|i_2, j_3\rangle \\ a_3b_1|i_3, j_1\rangle & a_3b_2|i_3, j_2\rangle & a_3b_3|i_3, j_3\rangle \end{pmatrix} \tag{2.30}$$

then the constituent wavefunctions can be measured independently and are separable.

However, this is only one of the ways to write the matrix. In general the basis states i and j are not equivalent and the result of the combination is different. If the combinations are made in any other way, then the wavefunctions cannot be measured independently, and the wavefunctions are said to be entangled.

$$|\Psi\rangle = \sum_{ij} c_{ij}|i\rangle \otimes |j\rangle = \begin{pmatrix} c_{11} & c_{12} & c_{13} \\ c_{21} & c_{22} & c_{23} \\ c_{31} & c_{23} & c_{33} \end{pmatrix} \tag{2.31}$$

The case of photon polarization is important because this observable is used to characterize photon quantum effects.

We can express the polarization state of each photon as combination of parallel and perpendicular states:

$$|A\rangle = a_1|\uparrow\rangle + a_2|\rightarrow\rangle \text{ and } |B\rangle = b_1|\uparrow\rangle + b_2|\rightarrow\rangle$$

A combined but separable wavefunction is written:

$$|\Psi\rangle = \begin{pmatrix} c_{11}|\uparrow\uparrow\rangle & c_{12}|\uparrow\rightarrow\rangle \\ c_{21}|\rightarrow\uparrow\rangle & c_{22}|\rightarrow\rightarrow\rangle \end{pmatrix} = (c_{11}|\uparrow\uparrow\rangle + c_{12}|\uparrow\rightarrow\rangle + c_{21}|\rightarrow\uparrow\rangle + c_{22}|\rightarrow\rightarrow\rangle)$$

On the other hand, the combined wavefunction, with zero net polarization:

$$|\Phi\rangle = \frac{1}{\sqrt{2}}(|\uparrow, \uparrow\rangle + |\rightarrow\rightarrow\rangle) \tag{2.32}$$

is entangled, because it cannot be expressed as the product matrix shown in (2.30).

When two (or more) quanta are entangled, then they behave as a single particle with respect to measurement of the entangled property. Thus the constituent quanta are correlated. A measurement on one of the constituent quanta becomes a simultaneous measurement on all the constituent quanta, even though they may appear to be spatially separated (Bell 1971). Entanglement is closely related to the macroscopic concept of entropy.

As an example, parametric down conversion creates a pair of entangled photons with zero net polarization. Prior to a measurement, the polarization of each photon is unknown. However, the net polarization of the pair is known. A measurement of polarization on one of the photons must indicate either parallel or perpendicular orientation of the polarization. As a result, the other unmeasured photon must assume the opposite orientation in order to conserve the feature that the pair has zero net polarization (Yin et al. 2013). Such an event represents a transition from the quantum mechanical world where particles may exist in a superposition of possible states to the classical world where particles must assume specific values.

2.6 Summary

The photon is an elementary particle characterized by its frequency and polarization. The photon displays both wave-like and particle-like behavior. Maxwell's equations describe the electromagnetic fields in the classical regime, including propagation of photon beams as a function of frequency and polarization. Maxwell's equations represent a staggering accomplishment in physical theory, but they do not give a complete description of photon behavior. It was subsequently shown by Planck that the energy carried by a photon beam radiated from a body heated to a temperature T could be properly explained only if each photon were carrying a quantum of energy proportional to its frequency. Maxwell's equations are often not well adapted for describing the behavior of the photon as an individual quantum.

Single photon behavior can be studied by using spontaneous parametric down conversion to transform one photon into a pair of photons, each of which having one-half the frequency of its parent. (Spontaneous parametric down-conversion is a good example of photon behavior that cannot be accounted for by Maxwell's equations.) A basic tool for studying single photons is the beamsplitter. The beamsplitter can be analyzed in the classical regime using Maxwell's equations, but this analytical approach fails when faced with the propagation of a single photon through a beamsplitter apparatus. This failure, however, does not prevent the single photon from propagating through the beamsplitter. The photon wavefunction is seen to be present in both exit paths of the beam splitter, and when reflecting mirrors are used to bring the paths to overlap, interference, that is, self-interference is recorded with results that echo Thomas Young's demonstration of two-slit interference in the early nineteenth century. This and related experiments show that the photon is a particle with wavelike behavior, and that as a single "particle" it can be present over an extended macroscopic region of space. This kind of non-local behavior of the photon is a fundamental building block of important applications like quantum cryptography and quantum computing.

2.7 Exercises

2.1 Using MATLAB, calculate the transmission coefficient of a photon beam through a plate of glass (SiO_2) using the parameters and following the procedure given in Example 2.1. Plot the dependence of the transmission coefficient as a function of wavelength and as a function of frequency. Identify the wavelength above which oscillations in the transmission coefficient no longer occur.

2.2 Using Maxwell's equations, derive the expression for the transmission coefficient and the reflection coefficient for an electromagnetic wave passing from one dielectric material to another with the electric field polarized parallel to the plane of incidence. These expressions are Fresnel's 3rd and 4th equations.

2.3 Derive the analytical expression for photon propagation through a plasma layer when the wavevector is an imaginary number ($k_2 = i\kappa_2$ where κ_2 is real).

Show that:

$$T \equiv \left\| \frac{F}{A} \right\|^2 = \frac{1}{\cosh^2(\kappa_2 a) + \frac{1}{4}\left(\frac{k_1^2 - \kappa_2^2}{k_1\kappa_2}\right)^2 \sinh^2(\kappa_2 a)}$$

2.4 A general interface between two planar regions having indices of refraction n_1 and n_2 is shown in Fig. 2.19:
Show that the general expression for the transfer matrix that summarizes the boundary conditions at the interface at z = Z is:

$$\begin{bmatrix} M_{ij} \end{bmatrix} = \frac{1}{2}\begin{bmatrix} (1+R)e^{iZB} & (1-R)e^{-iZA} \\ (1-R)e^{iZA} & (1+R)e^{-iZB} \end{bmatrix}$$

where

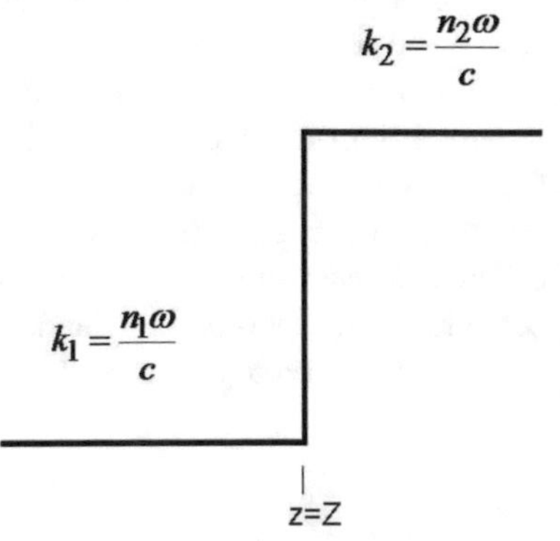

Fig. 2.19 Diagram of index of refraction versus distance for Exercise 2.4

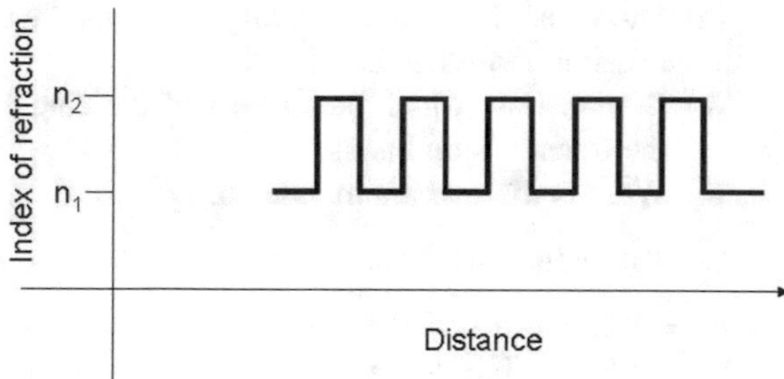

Fig. 2.20 Model structure for a five-layer planar dielectric stack

$$R = \frac{k_2}{k_1}$$
$$A = k_1 + k_2$$
$$B = k_2 - k_1$$

k_1 is the wavevector on the left-hand side and k_2 is the wavevector on the right side of the interface, and Z is the coordinate of the interface.

2.5 Compute the transmission coefficient for photons passing through a grating formed by 5 planar layers of alternating index of refraction, making use of the expression in Exercise 2.4 for the transfer matrix (Fig. 2.20).
Evaluate your computing routine using:

$n_1 = 1.5$
$n_2 = 1.9$.
Thickness of each region = 500 nm.
Over a frequency range: 10^{14} Hz $< \omega < 10^{15}$ Hz.

Make a plot of the transmission and reflection coefficients as a function of free-space wavelength.
Hint: Write a subroutine that determines the transfer matrix for *two* adjacent interfaces of the structure.

2.6 Calculate the transmission coefficient as a function of free-space wavelength for photons passing through a dielectric stack of 11 planar layers of alternating index of refraction, making use of the calculation routine developed in Exercise 2.4. Refer to Fig. 2.21.

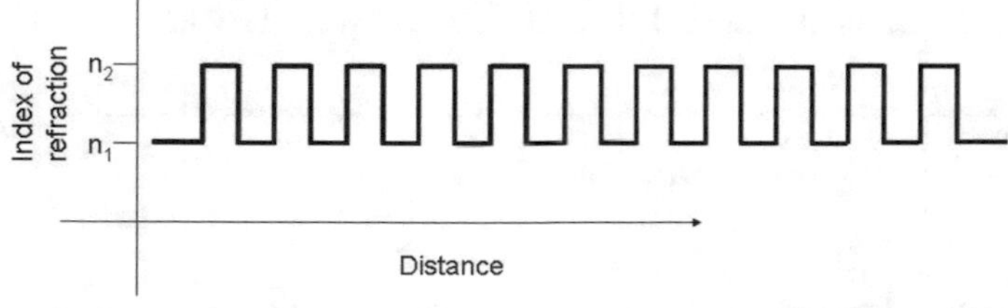

Fig. 2.21 The 11-period structure. The index of refraction is changed periodically, simulating a multi-layer dielectric filter

(a) Show that the effect of adding additional layers is to create bands of transmission and reflection.
(b) Demonstrate which parameters of the calculation determine the width of the transmission bands.
(c) What is the maximum value of reflectivity that you can achieve?

Evaluate your simulation using

$n_1 = 1.5$, width = 500 nm
$n_2 = 3.0$, width = 500 nm
$500\,\text{nm} < \lambda_{\text{free-space}} < 1500$ nm.

Such a structure is called a photonic crystal, because it is periodic structure, like a crystal, but with a period that is similar to the wavelength of light that propagates through the structure.

2.7 By introducing a defect in the structure of Fig. 2.21 we can change its optical properties. There are different types of defects:

Vacancy: a layer can be left out of the stack
Impurity: a layer can have a different index of refraction
Phase: a layer can be placed at a non-periodic location.

Each of these defects can be modelled in a straightforward way using the transfer-matrix method. In this example we show the effect of introducing a phase defect. We advance the position of the 6th layer in the stack by 30% of a full period as shown in Fig. 2.22.
Referring to Fig. 2.20, introduce a defect in the 11 period structure by enlarging the lower index layer in the center of the periodic structure.
Using the same parameters as those for Exercise 2.5, substitute a $\frac{\lambda}{4}$ phase defect for the 6th period. In a simple form, this could be a section 250 nm in length with $n = n_1$

(a) Calculate the transmission coefficient as a function of free-space wavelength and compare to your result in Exercise 2.4.

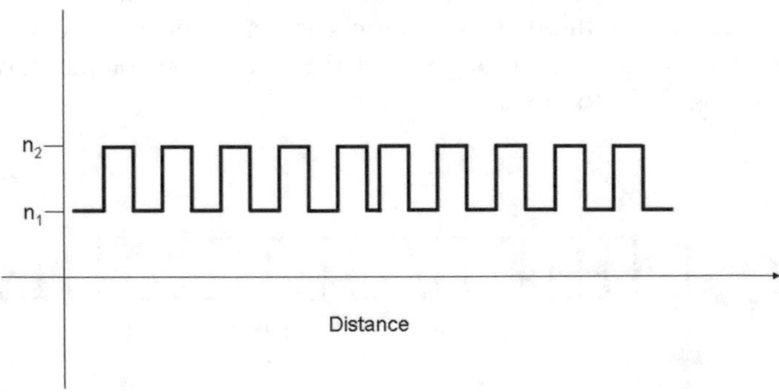

Fig. 2.22 Periodic dielectric contrast grating. The sixth period has been replaced by a phase defect

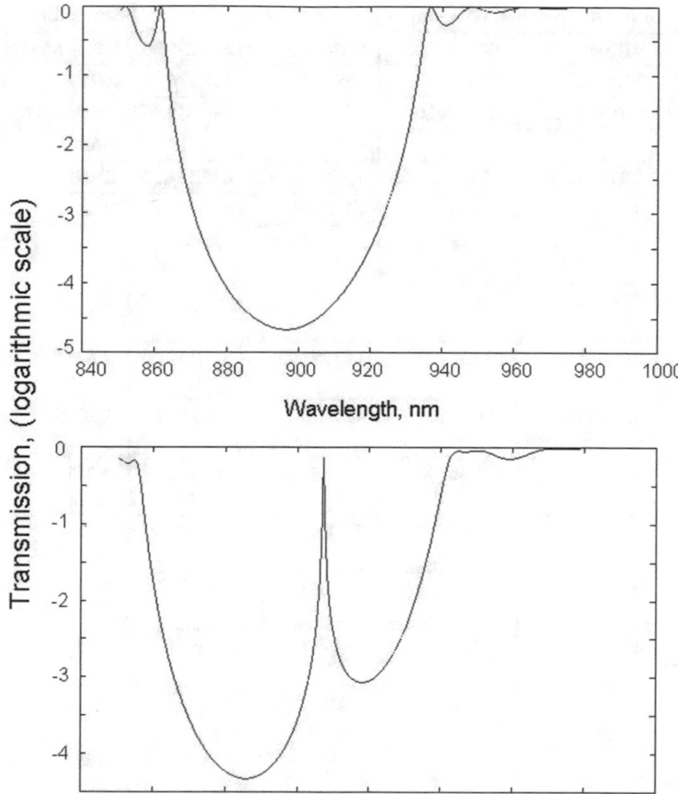

Fig. 2.23 Sample simulation of defect in a dielectric stack. This result shows the effect of a quarter-period phase defect inserted in a dielectric stack. The defect creates a highly transmitting, narrow bandwidth mode in the center of a highly reflecting ($R > 0.9999$) band

(b) Show that the presence of a wider layer introduces a region where transmission can occur in the middle of a region where transmission is normally forbidden. (This is called a defect mode, and is used to achieve single-mode operation of lasers that use photonic crystal reflectors as mirrors.) (see Fig. 2.23 for an example simulation).

2.8 Some data taken from the experiment "Existence Photon" on the Quantum Lab website and described in Fig. 2.16 are summarized in Table 2.1. Detector D1 refers to the transmitted photon and detector D2 refers to the reflected photon. T refers to the trigger which heralds the arrival of a single photon at the beam splitter. These data are taken for a series of laser power used to generate photon pairs by spontaneous parametric down conversion

(a) Complete the table by calculating the second-order correlation coefficient $g^2(0)$ for the measurements with trigger and without trigger for laser power = 500 and 50 μW.

Table 2.1 Data for the passage of a single photon passing through a beam splitter

Counts D1	971,650.700	781,706.300	578,977.667	384,096.367	193,297.533	96,938.467
Error D1	1928.581	1295.531	958.530	577.874	409.237	324.212
Counts D2	960,647.500	764,864.200	560,194.467	369,650.367	185,951.100	93,124.300
Error D2	1455.367	964.423	608.047	604.612	516.928	338.532
Counts T	2,270,792.800	1,793,066.333	1,302,454.000	856,532.200	429,910.333	214,817.333
Error T	4131.687	2195.611	1307.544	1099.080	715.257	479.135
Counts D1 and T	221,116.333	178,162.100	131,663.400	87,088.800	43,596.467	21,692.967
Error D1 and T	713.366	491.231	454.170	265.738	200.273	145.135
Counts D2 and T	221,075.567	176,539.133	129,278.733	85,011.100	42,624.267	21,126.767
Error D2 and T	628.030	508.930	414.670	317.893	220.700	134.916
Counts D1 and D2	971.133	632.467	344.633	149.967	36.267	10.067
Error D1 and D2	29.586	24.787	20.830	15.721	6.236	3.552
Counts D1 and D2 and T	391.900	256.467	141.033	59.500	15.133	4.033
Error D1 and D2 and T	20.444	15.885	12.971	9.149	3.875	2.251
$g^2(0)$ without trigger		1.024	1.028	1.022	0.976	
Error $g^2(0)$ without trigger	0.031	0.040	0.062	0.107	0.168	0.381
$g^2(0)$ with trigger		0.015	0.011	0.007	0.004	
Error $g^2(0)$ with trigger	0.001	0.001	0.001	0.001	0.001	0.001
Laser-Power (μW)	500.000	400.000	300.000	200.000	100.000	50.000
Measure-Time (s)	60.000	60.000	60.000	60.000	60.000	60.000
Number of measurements	30.000	30.000	30.000	30.000	30.000	30.000

The experiment set-up is shown in Fig. 2.16. D1 corresponds to transmission, while D2 corresponds to reflection. There is a "trigger" detector T that heralds the presence of a single photon at the beam splitter. These data are extracted from the website: **Quantum Lab**, http://www.didaktik.physik.uni-erlangen.de/quantumlab/english/index.htm, "Existence Photon" and used with the kind permission of Prof. Jan-Peter Meyn

(b) Comment on the relationship between the determination of $g^2(0)$ and laser power that you observe in the two cases of trigger and without trigger.

References

H.-A. Bachor, T.C. Ralph, *Guide to Experiments in Quantum Optics*, 2nd edn. (Wiley-VCH Verlag, Weinheim, 2004). ISBN 13: 978-3-52-740393-6

M. Beck, *Quantum Mechanics (Theory and Experiments)* (Oxford University Press, New York, 2012). ISBN 978-19-979812-4

J.S. Bell, Introduction to the hidden variable question, in *Proceedings, International School of Physics*, '*Enrico Fermi*', (Academic Press, New York, 1971) pp. 171–181 (Chapter X in *Speakable and Unspeakable in Quantum Mechanics*, Cambridge University Press, Cambridge, 1987). ISBN 0-521-33495-0

D. Dykstra, S. Busch, W. Peters, M. van Exter, Leiden University CCD (2008). https://www.youtube.com/watch?v=MbLzh1Y9POQ

P. Grangier, G. Roger, A. Aspect, Experimental evidence for a photon anticorrelation effect on a beam splitter: a new light in single photon interferences. Europhys. Lett **1**, 173–179 (1986). https://courses.physics.illinois.edu/phys513/sp2016/reading/week1/GrangierSinglePhoton1986.pdf

E. Hecht, *Optics*, 2nd edn. (Addison-Wesley, Reading, 1987). ISBN 0-201-11609-X

C. Jönsson, Elektroneninterferenzen an mehreren künstlich hergestellten Feinspalten (Electron interference in a fabricated metal-slit grating). Zeitschrift für Physik **161**, 454–474 (1961). https://link.springer.com/article/10.1007/BF01342460

C. Jönsson, Electron diffraction at multiple slits. Am. J. Phys. **42**, 4–11 (1974). http://materias.df.uba.ar/f4Aa2012c2/files/2012/08/multiple_slit.pdf

C.H. Lineweaver, T.M. Davis, Misconceptions about the Big Bang. Sci. Am. **36–45** (2005). https://www.researchgate.net/profile/Charley_Lineweaver/publication/7879288_Misconceptions_about_the_Big_Bang/links/0c96052ce5d7551b95000000/Misconceptions-about-the-Big-Bang.pdf

Max Planck, *A Scientific Autobiography and Other Papers*, (New York, Philosophical Library, 1949) ISBN 0-8065-3075-8

Quantum Lab. http://www.didaktik.physik.uni-eriangcn.dc/quantumlab/english/index.htm. An interactive presentation of quantum photon experiments. Project leader: Jan-Peter Meyn, Physikalisches Institut–Didaktik der Physik, Universität Erlangen-Nürnberg (2009)

J.J. Thorn, M.S. Neel, V.W. Donato, G.S. Bergreen, R.E. Davies, M. Beck, Observing the quantum behavior of light in an undergraduate laboratory. Am. J. Phys. **72**, 1210–1219 (2004). https://digitalcommons.usu.edu/cgi/viewcontent.cgi?article=1797&context=psc_facpub

A. Tonomura, J. Endo, T. Matsuda, T. Kawasaki, H. Ezawa, Demonstration of single-electron build-up of an interference pattern. Am. J. Phys. **57**, 117–120 (1989). http://www.theo-physik.uni-kiel.de/~bonitz/D/vorles_13ws/tonomura_AJP000117.pdf

J. Yin, Y. Cao, H.-L. Yong, J.-G. Ren, H. Liang, S.-K. Liao, F. Zhou, C. Liu, Y.-P. Wu, G.-S. Pan, L. Li, N.-L. Liu, Q. Zhang, C.-Z. Peng, J.-W. Pan, Bounding the speed of spooky action at a distance. Phys. Rev. Lett. **110**, 260407 (2013). https://arxiv.org/abs/1303.0614

Chapter 3
Free Electron Behavior in Semiconductor Heterostructures

Abstract Solutions to the Schrödinger wave equation are used to study the allowed states of electrons and holes in the presence of a one-dimensional potential. In this analysis, we treat the electron as a wave with a single wavevector. We study the transit of an electron-wave through a series of potential barriers using the transfer-matrix method to determine the transmission coefficient as a function of kinetic energy. The Schrödinger equation shows that an electron can be transported through a potential barrier when its kinetic energy is not sufficient to surmount the potential barrier. This tunneling effect is an important contributor to reverse breakdown in p-n junction diodes. Tunneling behavior is more prominent in the transport of electrons through a sequence of potential barriers. As the number of barriers increases, the range of allowed energies for tunneling broadens into a transmission band. These bands are directly related to the electronic bandstructure of crystalline materials. Particle-like behavior of electrons can be described by constructing a wavepacket. The electron wavefunction consists of a sum of different wavevectors leading to localization in space and time of the electron wavefunction. This representation is needed to address transit time between two points in space.

3.1 Introduction

Crystalline semiconductor materials are typically grown as a series of epitaxial layers on a plane, single crystal substrate. The thickness of the layers can be controlled accurately from a nanometer to several micrometers. While charge carriers: electrons and holes, are largely free to move in the plane of these layers, transport in third dimension is modified strongly by differences in the band gaps of the layers composing the structure, leading to charge carrier confinement, tunneling, and local acceleration among some of the more interesting effects. Charge carrier confinement is the essential component of the silicon MOSFET, the room-temperature semiconductor injection laser and the quantum Hall effect.

In this chapter, we will study the allowed states of electrons and holes in the presence of one-dimensional potentials corresponding to the different bandgaps in

© Springer Nature Switzerland AG 2020

T. P. Pearsall, *Quantum Photonics*, Graduate Texts in Physics,
https://doi.org/10.1007/978-3-030-47325-9_3

the structures under consideration. In many instances we will treat the electron as a wave with a single wavevector. Solutions to the Schrödinger wave equation are quite accurate, to the point that they are now used as the tool of choice for design and analysis of semiconductor devices that depend on energy-band heterostructures for their operation. This analysis is called bandstructure engineering.

Particle-like behavior of electrons can be described by constructing a wavepacket. The electron wavefunction consists of a sum of different wavevectors leading to localization in space and time of the electron wavefunction. This representation is needed to address transit time between two points in space.

3.2 The Schrödinger Equation

Following the experimental verification of de Broglie's proposal that electrons (and holes) have a characteristic wavelength, we can write down the wave function of a one-dimensional free electron as follows:

$$\Psi(x) = A\mathrm{e}^{ikx} \tag{3.1}$$

The time-independent Schrödinger equation was presented in (1.17)

$$-\frac{\hbar^2}{2m^*}\frac{d^2}{dx^2}\Psi(x) + V(x)\Psi(x) = E\Psi(x) \tag{3.2}$$

Schrödinger's equation is a statement of conservation of total energy:

$$\text{Kinetic energy } + \text{ Potential energy } = \text{Total energy}$$

The solutions of Schrödinger's equation give the stationary states for which total energy is conserved.

3.3 Quantized Energy Levels for Electrons in an Infinitely Deep Potential Well

3.3.1 Potential Well with Infinitely High Boundaries

A simple point of departure is the case of energy levels for electrons confined in a potential well with infinite potential barriers on each side (Fig. 3.1):

The general solution to (3.2) in each region has the form of two counter-propagating plane waves:

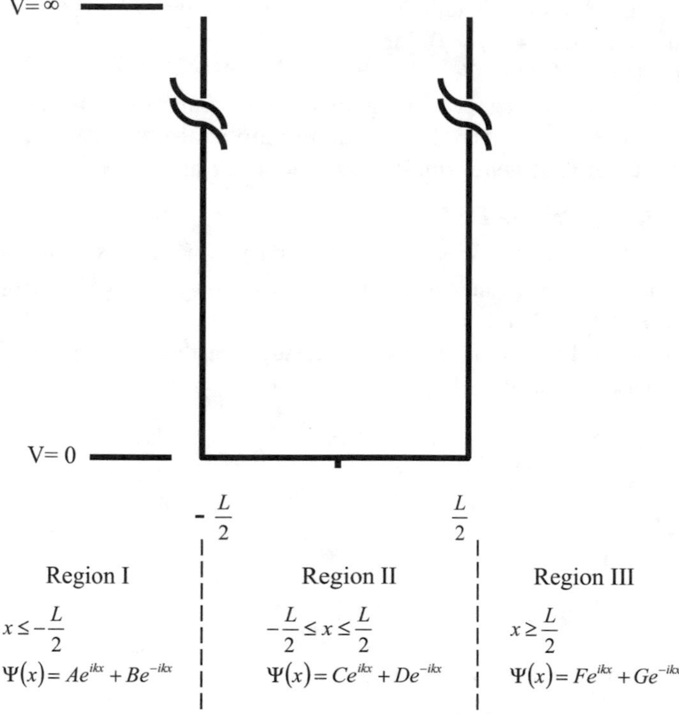

Fig. 3.1 Particle energy versus real-space distance in one dimension. The width of the well is given by L. The energy at the bottom of the well is set to 0 eV. The solution to the Schrödinger equation is a free-electron wavefunction, The particle wavevector k is imaginary in the barrier region and a real number in the well region

$$\Psi(x) = Ae^{ikx} + Be^{-ikx}, \quad \text{where} \quad k = \sqrt{\frac{2m^*}{\hbar^2}(E - V)} \tag{3.3}$$

The constants A and B are determined by application of boundary conditions.

3.3.2 *Boundary Conditions*

The boundary conditions are an essential input to the physics of the problem. We will treat them region by region.

Region I: $\Psi(x) = Ae^{ikx} + Be^{-ikx}$
In region I, V is a positive real number, so k is purely imaginary. This means that the wavefunction consists of 2 terms, one that grows exponentially and the other which decays exponentially. The exponentially growing term is unphysical, so the coefficient A = 0.

Because the value of V is infinite, the remaining term is 0 everywhere, and B remains undetermined. $\Psi(x) = 0$, for all $x \leq -\frac{L}{2}$.
Region III $\Psi(x) = Fe^{ikx} + Ge^{-ikx}$
The boundary condition applied to region III is similar to that for region I. The coefficient F is set to zero and the remaining term is also 0 because the value of V is infinite. G remains undetermined. $\Psi(x) = 0$, for all $x \geq \frac{L}{2}$.

Region II $\Psi(x) = Ce^{ikx} + De^{-ikx}$
In region II, V = 0, and $E = \frac{\hbar^2 k^2}{2m^*}$. The boundary condition is continuity of the wavefunction across the boundaries. This means $\Psi\left(-\frac{L}{2}\right) = \Psi\left(\frac{L}{2}\right) = 0$ and requires that $kL = n\pi$, n = 1, 2, 3...

It then follows that $C = D$ in order to assure that the wavefunction is zero at the boundaries of the well.

Solution:

For $x \leq -\frac{L}{2}$, $x \geq \frac{L}{2}$: $\Psi(x) = 0$
For $-\frac{L}{2} < x < \frac{L}{2}$,
k is quantized:

$$k = \frac{n\pi}{L} \tag{3.4}$$

The quantized energy levels are given by

$$E_n = \frac{n^2\pi^2\hbar^2}{2L^2 m^*}. \tag{3.5}$$

The lowest-lying solution, for n = 1 is $\Psi(x) = C\cos\left(\frac{\pi}{L}x\right)$.

The value for C can be determined by normalizing the wavefunction to unity in the well:

$$\int_{-\frac{L}{2}}^{\frac{L}{2}} C^2\cos^2\left(\frac{\pi}{L}x\right)dx = 1, \quad C = \sqrt{\frac{2}{L}}. \tag{3.6}$$

For n = 2, the wave function is $\Psi(x) = \sqrt{\frac{2}{L}}\sin\left(\frac{2\pi}{L}x\right)$.
For $x \geq \frac{L}{2}$, $\Psi(x) = 0$.

In general the wavefunctions are:

$$\Psi(x) = \begin{cases} \sqrt{\frac{2}{L}}\cos\left(\frac{n\pi}{L}x\right), & \text{n} = 1, 3, 5, \ldots \\ \sqrt{\frac{2}{L}}\sin\left(\frac{n\pi}{L}x\right), & \text{n} = 2, 4, 6, \ldots \end{cases} \tag{3.7a, b}$$

The first three quantized energy levels are shown in Fig. 3.2. These are calculated using (3.5), with a well width of 10 nm and using an effective mass of 0.066 m_0, which is that of an electron in the conduction band of GaAs.

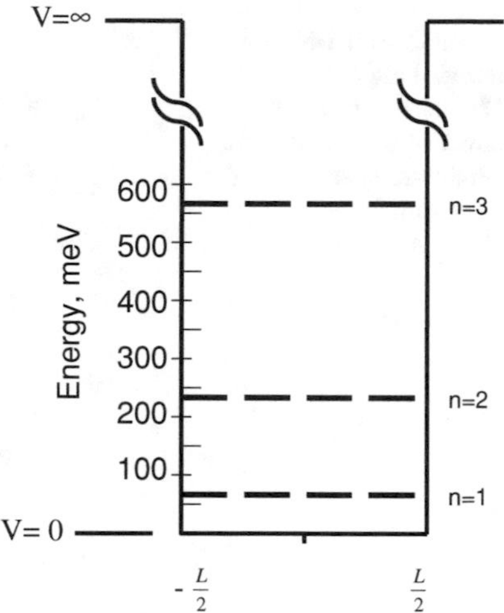

Fig. 3.2 Quantized energy levels in a well having a width of 10 nm and infinitely high potential barriers. The electron effective mass = 0.066 m_0

3.4 Electrons Confined by a Square Potential Well with Finite Barriers

The previous analysis of the square potential well is helpful in understanding some general behavior of confined electrons in semiconductor heterostructures. However, real heterostructure devices must be designed with finite barriers. Both the analysis and the results have significant differences compared to the infinite potential model. In addition, the finite potential barrier model that we present here gives a precise and quantitative account of electronic energy levels in important semiconductor devices such as lasers and modulators.

As illustrated in Fig. 3.3, the potential barrier is defined:

$$V(x) = \begin{cases} V_0, & \text{for } |x| \leq \frac{L}{2} \\ 0, & \text{for } |x| \leq \frac{L}{2} \end{cases} \tag{3.8}$$

Schrödinger's equation is:

$$-\frac{\hbar^2}{2m^*}\frac{d^2}{dx^2}\Psi(x) + V(x)\Psi(x) = E_n\Psi(x) \tag{3.9}$$

where $m^* = m_b^*$ in the barrier material, such as $Al_xGa_{1\text{-}x}As$, and $m^* = m_w^*$ in the well material, such as GaAs.

We are principally interested in the range of electron energies $E < V_0$ where electrons are confined by the potential well. As before, we can write the solution to (3.9) in the three different regions.

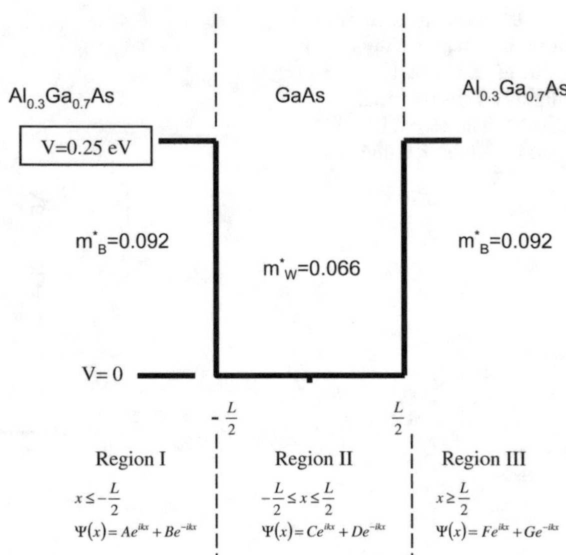

Fig. 3.3 Diagram of a potential well with finite barrier height. The parameters for electron effective mass and barrier height are specific for the heterostructure: $GaAs/Al_{0.3}Ga_{0.7}As$

3.4.1 Boundary Conditions

There are 2 boundary conditions which are necessary to assure that the physics of the situation is respected:

(i) continuity of the wavefunction across the boundary.
(ii) continuity of the electric current across the boundary.

This second condition arises because electrons are not confined entirely to the well, due to the finite nature of the confining potential. Continuity of current is a physical requirement that assures conservation of charge throughout the heterostructure.

The current carried by one electron is:

$$I = -qv = -q\frac{m^*v}{m^*} = -q\frac{\hbar k}{m^*} = iq\frac{\hbar}{m^*}\frac{d}{dx}\Psi(x) \tag{3.10}$$

The boundary condition that conserves electric current is therefore:

$$\frac{1}{m_W^*}\frac{d}{dx}\Psi_{\text{region II}}(x)_{x=\frac{L}{2}} = \frac{1}{m_B^*}\frac{d}{dx}\Psi_{\text{region III}}(x)_{x=\frac{L}{2}}$$

D.J. BenDaniel and C.B. Duke proposed and explained this requirement in a landmark publication. Nearly twenty years later, G. Bastard showed the importance of these boundary conditions for the physics of semiconductor heterostructures (BenDaniel and Duke 1966; Bastard 1981).

Applying these conditions at $x = \frac{L}{2}$,

$$Ce^{ik\frac{L}{2}} + De^{-ik\frac{L}{2}} = Fe^{-\kappa\frac{L}{2}}$$

and

$$\frac{ik}{m_W^*}\left(Ce^{ik\frac{L}{2}} - De^{-ik\frac{L}{2}}\right) = -\frac{\kappa}{m_B^*}Fe^{-\kappa\frac{L}{2}}.$$

These results can be combined:

$$\frac{ik}{m_W^*}\left(Ce^{ikL} - D\right) = -\frac{\kappa}{m_B^*}\left(Ce^{ikL} + D\right), \tag{3.11}$$

giving a homogeneous linear equation for the unknowns C and D.

$$Ce^{ikL}\left(\kappa m_W^* + ikm_B^*\right) + D\left(\kappa m_W^* - ikm_B^*\right) = 0$$

Applying the same procedure at the boundary $x = -\frac{L}{2}$ gives:

$$Ce^{-ikL}\left(\kappa m_W^* - ikm_B^*\right) + D\left(\kappa m_W^* + ikm_B^*\right) = 0 \tag{3.12}$$

A solution to (3.11) and (3.12) exists if and only if the determinant of coefficients is equal to zero:

$$2i(\kappa m_W^*)^2 \sin(kL) + 4ikm_B^*\kappa m_W^* \cos(kL) - 2i(km_B^*)^2 \sin(kL) = 0$$

$$\cot(kL) = \frac{\left(km_B^*\right)^2 - \left(\kappa m_W^*\right)^2}{2km_B^*\kappa m_W^*} \tag{3.13}$$

Equation (3.13) gives the energy of the quantized states in the potential well. It can be solved by numerical iteration to any degree of accuracy desired. It is helpful to view the overall situation graphically as shown in Fig. 3.4. As is often the case, solving for the wavefunction is of secondary interest.

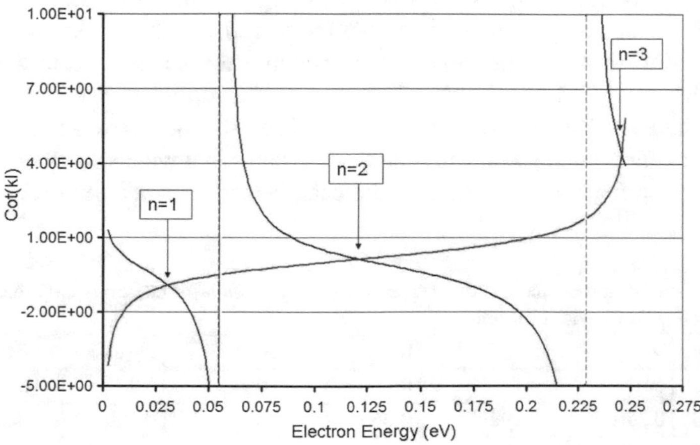

Fig. 3.4 Plot of cotangent (kl) (in *blue*) versus $\frac{(km_B^*)^2-(\kappa m_W^*)^2}{2km_B^*\kappa m_W^*}$ (in *red*). This calculation is based on a well width of 10 nm and parameters for the well and barrier regions based on GaAs and $Al_{0.3}Ga_{0.7}As$ respectively. For the details, refer to Sect. 3.4.2

3.4.2 Discussion of Results for Quantization of Energy Levels in a Square Potential Well

The results for the case of an electron confined in an infinite potential well are intuitive. Since we treat the electron as a wave, then it follows that the allowed states will be those for which an integral number of wavelengths can fit between the geometrical boundaries of the potential well. The mathematics of this model system corresponds to the case of sound waves in an organ pipe or the vibrations of a guitar string. The energy of a free electron is proportional to the square of its momentum, and from the de Broglie relation, this implies that the energy is proportional to the inverse square of the wavelength: $(\hbar k)^2 = \left(\frac{n\pi}{L}\right)^2$. Thus, the quantized energy levels increase in proportion to $n^2 = 1{::}4{::}9$, etc. The solution to the equations for the model system of the square potential well of infinite height is straightforward. The results, however, do not give a good approximation to those for electron confinement by real potential wells, such as those used to make semiconductor lasers.

In a second case, we have considered electron states confined by a square potential well of finite height. A significant feature of this system is the penetration of the electron states into the barrier. One obvious consequence is that the physical dimension of confinement is larger than the physical dimension of the potential well. This causes the quantized energies to be lower than those found for the case of the infinite potential well. A second important result is that the penetration of the electron wave function in the barrier region requires boundary conditions that properly reflect the physics of the situation.

In Table 3.1, we summarize the calculation of the energy of the confined states in the two cases, for a potential well having a width of 10 nm.

Note that the quantized energies for the case of the infinite well follow the sequence: E_0, $4E_0$, $9E_0$. For the finite well the series is; E_0, $4E_0$, $8.2E_0$. This illustrates that the highest energy state is not as strongly confined by the potential barrier. Indeed, it lies only a few meV below the barrier height. Weaker confinement means that the wave function extends farther into the barrier region than is the case for the n = 1 and n = 2 states. This lowers the confinement energy.

The finite barrier model presented here has been shown to give accurate quantitative results for both electron and hole energy states in a wide variety of semiconductor heterostructures. By measuring the energies, one can determine the physical

Table 3.1 Comparison of the first three quantized energy levels for electrons in GaAs, confined in a potential well of width 10 nm

	V_0	m^*_B	m^*_W	E_1 (eV)	E_2(eV)	E_3 (eV)
Infinite well	∞	–	0.066	0.056	0.223	0.502
Finite well	0.25 eV	0.092	0.066	0.03	0.121	0.245

In the case of an infinitely high potential barrier, the quantized energies are significantly larger than the example of a finite barrier height based on $Al_{0.3}Ga_{0.7}As$ as the barrier material width of the well, and this calculated width can be compared to direct measurement by electron microscopy. The agreement between calculation and experiment confirms that electronic states are free-electron-like and that the boundary conditions used to solve Schrödinger's equation are correct.

3.4.3 Energy Levels in a Quantum Well in the Presence of an Electric Field: The Quantum-Confined Stark Effect

The application of an electric field (Stark effect) changes the relative energy separation between adjacent quantum levels (Fig. 3.5). The energy shift can be calculated using perturbation theory.

The Hamiltonian of the perturbation is expressed:

$$\mathcal{H}' = q\mathcal{E}\mathbf{x} \tag{3.14}$$

We will consider the modification to the lowest-lying (n = 1) level. The first order change is expressed

$$\Delta E_1 = \langle \Psi_1 | q\mathcal{E}x | \Psi_1 \rangle \equiv 0 \text{ by a simple symmetry argument}$$

Thus, the perturbation to the energy levels must be calculated in the 2nd order

$$\Delta E_2 = \sum_{j \neq 1} \frac{|\langle \Psi_j | q\mathcal{E}x | \Psi_1 \rangle|^2}{(E_1 - E_j)} \approx \frac{|\langle \Psi_2 | q\mathcal{E}x | \Psi_1 \rangle|^2}{(E_1 - E_2)} \tag{3.15}$$

where we have simplified the calculation by keeping only the first term of the sum, that is: $j = 2$. There are good reasons for this:

1. A quantum well in photonic materials rarely has more than 3 levels.
2. The magnitude of subsequent terms will be reduced by the larger energy level difference.
3. The matrix element is zero by symmetry whenever j is an odd number. (This is demonstrated in Sect. 7.5.2)

We can evaluate (3.15) in the case of the infinitely deep quantum well using the wavefunctions obtained in (3.7a, b), in order to estimate the magnitude of the effect.

$$\Delta E_2 \approx -\frac{2q^2\mathcal{E}^2}{L(E_2 - E_1)} \left(\int_{-\frac{L}{2}}^{\frac{L}{2}} x \cos\left(\frac{\pi x}{L}\right) \sin\left(\frac{2\pi x}{L}\right) dx \right)^2 = -24 \left(\frac{2}{3\pi}\right)^6 \frac{q^2\mathcal{E}^2 m^* L^4}{\hbar^2} \tag{3.16}$$

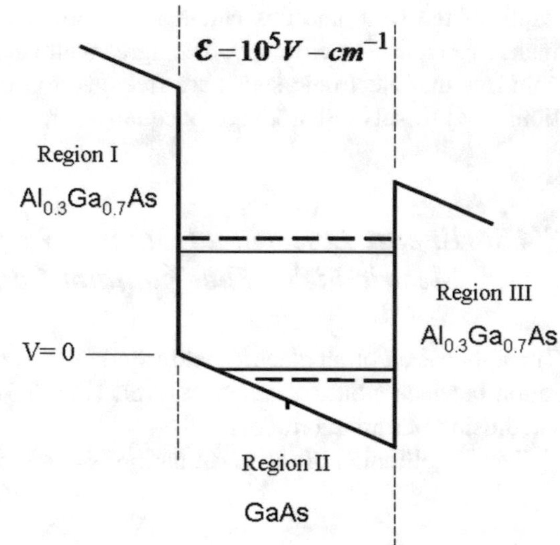

Fig. 3.5 Schematic diagram of a GaAs/AlGaAs quantum well in an electric field. An electric field of $\mathcal{E} = 10^5$ V cm^{-1} results in a voltage drop of 0.1 V across a 10 nm quantum well. The unperturbed quantized energy levels are shown as *solid lines*. These are positioned in the quantum well with the unperturbed energies (E = 0.03 eV and E = 0.121 eV) at the center of the well. The Stark effect pushes these levels apart, and the perturbed energies are shown as *dashed lines*

For a quantum well of GaAs having a width of 10 nm, the shift in energy is about 2 meV in an electric field of 10^5 V cm^{-1} and is shown in Fig. 3.5.

3.5 Quantum-Mechanical Tunneling

In this section we will investigate the transmission of an electron or other wavelike particle through a potential barrier of finite magnitude. This situation is nearly the reverse of that discussed in the previous section. Wavelike behavior permits particle transmission through a barrier even though its total energy is less than that of the potential. This phenomenon has no classical analog. Friedrich Hund was the first to take notice of tunneling in 1927 when he was calculating the ground state of the double-well potential. Its first application was a mathematical explanation for alpha decay, which was done in 1928 by George Gamow and independently by Ronald Gurney and Edward Condon. They simultaneously solved the Schrödinger equation for a model nuclear potential and derived a relationship between the half-life of the particle and the energy of emission that depended directly on the mathematical probability of tunneling.

The study of semiconductors and the development of transistors and diodes led to the acceptance of electron tunneling in solids by 1957. Leo Esaki demonstrated tunneling in pn-junction diodes for which he received the Nobel Prize in Physics in 1973.

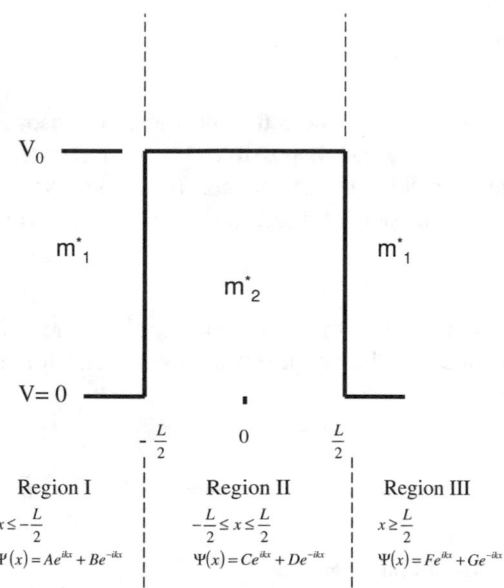

Fig. 3.6 Particle transmission in the presence of a potential barrier of width L and height V_0

Tunneling of particles is an important event in many areas of science and technology, including black holes, elementary particle decay, superconductivity, radar, chemical reactions, and integrated circuits to name just a few. The scanning tunneling microscope created a new window of surface science allowing direct imaging in real space with atomic resolution. Gerd Binnig and Heinrich Rohrer were awarded the 1986 Nobel Prize for demonstrating this capability. Recently, tunneling has been harnessed and controlled to produce a new kind of semiconductor laser, based on tunneling injection of electrons into a square potential well. These devices are the subject of Chap. 8. The diagram of a one dimensional potential barrier is shown in Fig. 3.6.

In this analysis, we suppose that a particle-wave is incident on the barrier from the left. In the classical case, the particle is reflected with 100% probability if its kinetic energy is less than the barrier height V_0. It will be transmitted with 100% probability if its kinetic energy is greater than V_0. In the quantum mechanical case there will be both reflection and transmission of the particle-wave at all energies.

The reflection coefficient is the probability that the particle is reflected:

$$R \equiv \left\| \frac{B}{A} \right\|^2 \tag{3.17}$$

where A and B are the amplitudes of the incident and reflected wavefunction respectively as shown in Fig. 3.6. The transmission coefficient is the probability that a particle incident in region I is transmitted to the right in region III (Fig. 3.7):

$$T \equiv \left\| \frac{F}{A} \right\|^2 \tag{3.18}$$

The particle in the geometry of Fig. 3.6 is nowhere confined. As a result, a continuous range of particle energies is possible. We are interested to learn about the conditions for tunneling through the barrier. To do this we need to develop an expression for the transmission coefficient. As before we divide the analysis into three regions.

Region I
The particle wave is incident from the left. In region I there will be both the incident wave traveling to the right, and a reflected wave traveling to the left:

$$\Psi(x) = Ae^{ikx} + Be^{-kx} \tag{3.19}$$

The particle effective mass is m_1^*. The wavevector $k = \frac{1}{\hbar}\sqrt{2m_1^* E}$.

Region III
The situation is identical that of Region I

$$\Psi(x) = Fe^{ikx} + Ge^{-ikx} \tag{3.20}$$

Note that the term Ge^{-ikx} represents a traveling wave incident from the right and moving to the left. In this tunneling model we could define $G \equiv 0$. However, there are some advantages to carry this term to the end of the analysis in order to develop a general solution that can be used with more complicated structures.

Region II
The wavefunction can be expressed in the same way

$$\Psi(x) = Ce^{ikx} + De^{-ikx} \tag{3.21}$$

In this region, $k = \frac{1}{\hbar}\sqrt{2m_2^* (E - V_0)}$.

In the range of energies where tunneling occurs, k is a pure imaginary, so we define:

$$\kappa = ik$$

Substituting in (3.18)

$$\Psi(x) = Ce^{-\kappa x} + De^{\kappa x} \tag{3.22}$$

We use the same two boundary conditions which are necessary to assure that the physics of the situation is respected:

(i) continuity of the wavefunction across the boundary.
(ii) continuity of the electric current across the boundary.

We apply the boundary conditions at each interface. For example, at the interface between regions II and III:

$$\Psi_{\text{region II}}(x)_{x=\frac{L}{2}} = \Psi_{\text{region III}}(x)_{x=\frac{L}{2}} \tag{3.23}$$

and

$$\frac{1}{m_2^*}\frac{d}{dx}\Psi_{\text{region II}}(x)_{x=\frac{L}{2}} = \frac{1}{m_1^*}\frac{d}{dx}\Psi_{\text{region III}}(x)_{x=\frac{L}{2}} \tag{3.24}$$

In a similar fashion, we apply the boundary conditions at the interface at $x = -\frac{L}{2}$.

$$Ae^{-ik\frac{L}{2}} + Be^{ik\frac{L}{2}} = Ce^{\kappa\frac{L}{2}} + De^{-\kappa\frac{L}{2}} \tag{3.25}$$

and

$$A\frac{ik}{m_1^*}e^{-ik\frac{L}{2}} - B\frac{ik}{m_1^*}e^{ik\frac{L}{2}} = -C\frac{\kappa}{m_2^*}e^{\kappa\frac{L}{2}} + D\frac{\kappa}{m_2^*}e^{-\kappa\frac{L}{2}} \tag{3.26}$$

Next, we solve for A and B:

$$A = \frac{1}{2}e^{\kappa\frac{L}{2}+ik\frac{L}{2}}\left[1 + \frac{i\kappa m_1^*}{k_2 m_2^*}\right]C + \frac{1}{2}e^{-\kappa\frac{L}{2}+ik\frac{L}{2}}\left[1 - \frac{im_1^*\kappa}{m_2^*k}\right]D$$

and

$$B = \frac{1}{2}e^{\kappa\frac{L}{2}-ik\frac{L}{2}}\left[1 - \frac{i\kappa m_1^*}{k_2 m_2^*}\right]C + \frac{1}{2}e^{-\kappa\frac{L}{2}ik\frac{L}{2}}\left[1 + \frac{im_1^*\kappa}{m_2^*k}\right]D$$

This is convenient to write in matrix format:

$$\begin{pmatrix} A \\ \\ B \end{pmatrix} = \begin{pmatrix} \frac{1}{2}e^{\kappa\frac{L}{2}+ik\frac{L}{2}}\left[1 + \frac{i\kappa m_1^*}{k_2 m_2^*}\right] & \frac{1}{2}e^{-\kappa\frac{L}{2}+ik\frac{L}{2}}\left[1 - \frac{im_1^*\kappa}{m_2^*k}\right] \\ \\ \frac{1}{2}e^{\kappa\frac{L}{2}-ik\frac{L}{2}}\left[1 - \frac{i\kappa m_1^*}{k_2 m_2^*}\right] & \frac{1}{2}e^{-\kappa\frac{L}{2}ik\frac{L}{2}}\left[1 + \frac{im_1^*\kappa}{m_2^*k}\right] \end{pmatrix} \begin{pmatrix} C \\ \\ D \end{pmatrix} \tag{3.27}$$

Applying the same boundary conditions at the next interface gives:

$$Ce^{-\kappa\frac{L}{2}} + De^{\kappa\frac{L}{2}} = Fe^{ik\frac{L}{2}} + Ge^{-ik\frac{L}{2}} \tag{3.28}$$

and

$$-C\frac{\kappa}{m_2^*}e^{-\kappa\frac{L}{2}} + D\frac{\kappa}{m_2^*}e^{\kappa\frac{L}{2}} = F\frac{ik}{m_1^*}e^{ik\frac{L}{2}} - G\frac{ik}{m_1^*}e^{-ik\frac{L}{2}} \tag{3.29}$$

Solving for C and D, we obtain:

$$C = \frac{1}{2}e^{\kappa\frac{L}{2}+ik\frac{L}{2}}\left[1 - \frac{ikm_2^*}{\kappa m_1^*}\right]F + \frac{1}{2}e^{\kappa\frac{L}{2}-ik\frac{L}{2}}\left[1 + \frac{ikm_2^*}{\kappa m_1^*}\right]G$$

and

$$D = \frac{1}{2}e^{-\kappa\frac{L}{2}+ik\frac{L}{2}}\left[1 + \frac{ikm_2^*}{\kappa m_1^*}\right]F + \frac{1}{2}e^{-\kappa\frac{L}{2}-ik\frac{L}{2}}\left[1 - \frac{ikm_2^*}{\kappa m_1^*}\right]G$$

In matrix format:

$$\begin{pmatrix} C \\ \\ D \end{pmatrix} = \begin{pmatrix} \frac{1}{2}e^{\kappa\frac{L}{2}+ik\frac{L}{2}}\left[1-\frac{ikm_2^*}{\kappa m_1^*}\right] & \frac{1}{2}e^{\kappa\frac{L}{2}-ik\frac{L}{2}}\left[1+\frac{ikm_2^*}{\kappa m_1^*}\right] \\ \frac{1}{2}e^{-\kappa\frac{L}{2}+ik\frac{L}{2}}\left[1+\frac{ikm_2^*}{\kappa m_1^*}\right] & \frac{1}{2}e^{-\kappa\frac{L}{2}-ik\frac{L}{2}}\left[1-\frac{ikm_2^*}{\kappa m_1^*}\right] \end{pmatrix} \begin{pmatrix} F \\ \\ G \end{pmatrix} \tag{3.30}$$

Substituting (3.27) in (3.24), gives the relationship between the incident and transmitted wavefunctions, which we can write in symbolic form:

$$\begin{aligned} \begin{pmatrix} A \\ B \end{pmatrix} &= \begin{pmatrix} M_1 & M_2 \\ M_3 & M_4 \end{pmatrix} \begin{pmatrix} M_5 & M_6 \\ M_7 & M_8 \end{pmatrix} \begin{pmatrix} F \\ G \end{pmatrix} \\ &= \begin{pmatrix} M_1M_5 + M_2M_7 & M_1M_6 + M_2M_8 \\ M_3M_5 + M_4M_7 & M_3M_6 + M_4M_8 \end{pmatrix} \begin{pmatrix} F \\ G \end{pmatrix} \end{aligned} \tag{3.31}$$

In (3.28), there is one term that allows the calculation of the transmission coefficient:

$$A = (M_1M_5 + M_2M_7)F + (M_1M_6 + M_2M_8)G = (M_1M_5 + M_2M_7)F, \text{ since } G = 0.$$

Therefore,

$$\frac{F}{A} = \left(\frac{1}{M_1M_5 + M_2M_7}\right) = \frac{1}{\cosh(\kappa L) - \frac{i}{2}\left(\frac{\kappa m_1^*}{km_2^*} - \frac{km_2^*}{\kappa m_1^*}\right)\sinh(\kappa L)} \tag{3.32}$$

$$T = \left\|\frac{F}{A}\right\|^2 = \frac{1}{\cosh^2(\kappa L) + \left(\frac{1}{4}\right)\left(\frac{\kappa m_1^*}{km_2^*} - \frac{km_2^*}{\kappa m_1^*}\right)^2 \sinh^2(\kappa L)} \tag{3.33}$$

Note that (3.32) is easily adapted to the case where $E > V_0$. The variable κ is replaced by ik_2 and the transmission coefficient can be calculated as shown here.

This approach can be extended in a straightforward way to structures having a series of potential barriers with arbitrary material properties, barrier height and position. The passage through each interface is treated by a 2×2 matrix that summarizes the boundary conditions.

To appreciate the quantum mechanical behavior of the particle, we will plot out the transmission coefficient as a function of incident particle energy (see Fig. 3.7). For the examples that follow, the barrier width is 10 nm and the barrier height is 0.25 eV.

Case I $m_1^* = m_2^* = 9 \times 10^{-31}$ kg

The first observation is that transmission through the barrier is negligibly small except where the incident kinetic energy of the particle approaches that of the barrier potential. Even in this case, the transmission coefficient does not exceed

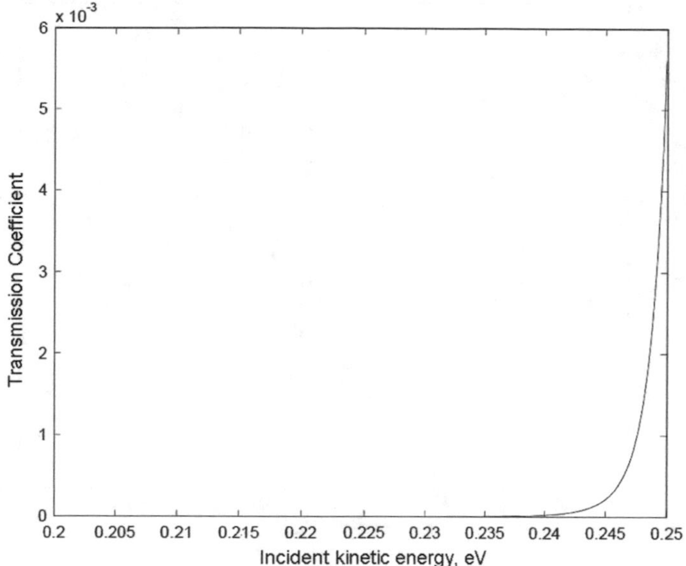

Fig. 3.7 Transmission coefficient versus incident energy, for a free-electron incident on a potential barrier of 0.25 eV. The barrier width is 10 nm. The electron mass is taken as the free-electron rest mass throughout. The transmission coefficient remains less than 1% even as its energy exceeds the barrier height

6×10^{-3}. The transmission coefficient for a wider spectrum of energies both below and above the barrier potential is shown in Fig. 3.8. This result is analogous to that shown in Fig. 2.11 for photon transmission through a plasma layer.

In the classical case, the transmission coefficient = 0 for energies less than the barrier potential energy. The transmission coefficient is 1 for energies greater than the barrier potential. The analysis of particle behavior in the quantum regime is different. For energies less than the barrier potential, the transmission coefficient is indeed small compared to unity, but as we have shown in Fig. 3.7, it is not zero, and its value is a function of incident energy. For incident energies greater than the barrier height, the envelope of the transmission coefficient tends toward unity, while the coefficient itself displays a series of pronounced resonances. These resonances occur at energies that correspond to the cases where an integral number of de Broglie wavelengths of the particle-wave adds up to the barrier width (Fig. 3.8).

Case II $m_1^* \neq m_2^*$.

In this example, $m_1^* = 9 \times 10^{-31}$ kg. $m_2^* = 0.066 m_1^*$. All other parameters remain the same (Fig. 3.9).

In this case we can see that the transmission coefficient has increased by nearly two orders of magnitude when the incident energy is close to the barrier height. The magnitude of the transmission coefficient is sensitive both to the ratio of the effective masses as well as their absolute values.

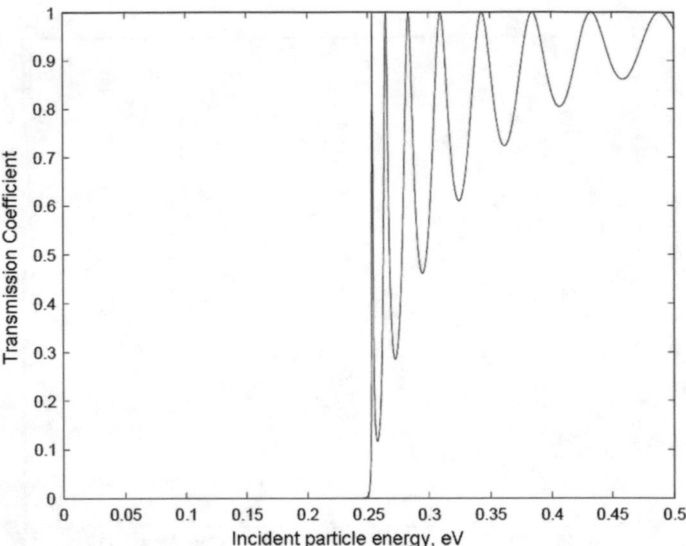

Fig. 3.8 Transmission coefficient versus incident particle energy using the same conditions as in Fig. 3.7. The transmission coefficient reaches unity for particle energies above the barrier potential, but only at discrete energies, showing that the barrier reflects the particle with significant efficiency, even when the incident energy of the particle is several times larger than the barrier height

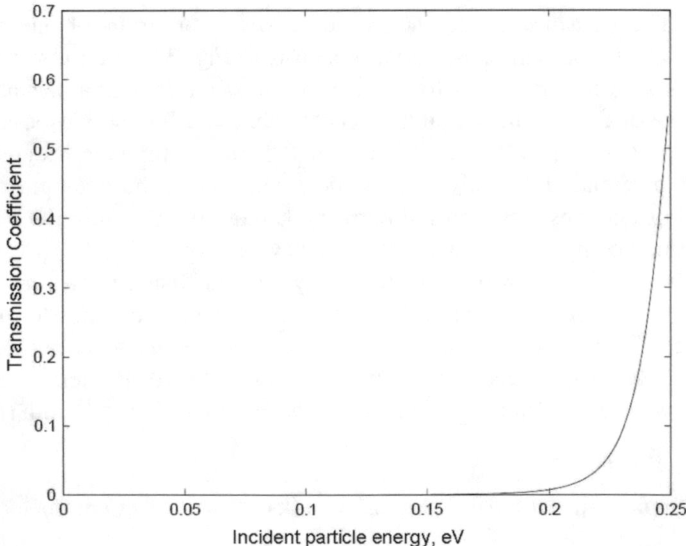

Fig. 3.9 Transmission coefficient versus incident particle energy. In this case the effective mass of the particle in the barrier region is taken to be 0.066. The result is that the barrier effectiveness is significantly reduced, and the transmission coefficient of the particle is close to unity as its incident energy approaches the barrier height

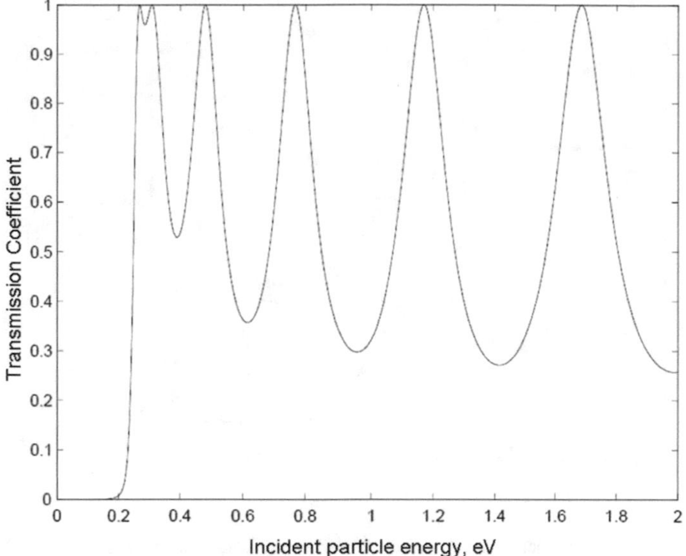

Fig. 3.10 Transmission coefficient versus incident particle energy using the same conditions as in Fig. 3.9. The transmission coefficient reaches unity for particle energies above the barrier potential, but only at discrete energies. However, the transmission coefficient does not approach unity at high energies, as in Fig. 3.8. Instead, the difference in effective masses inside and outside the barrier results in a significant reflection of the particle wavefunction, even in the classical energy regime

As before, we look at the overall behavior for incident energies both below and above the barrier height.

In Fig. 3.10, the behavior of the transmission coefficient for energies above the barrier height is quite different from that shown in Fig. 3.8. There are still the pronounced resonances that occur when the particle wavelength is an integral multiple of the barrier width. However, the transmission coefficient no longer evolves asymptotically toward unity at large energies. The barrier acts like a strong scattering centre at all non-resonant energies. Scattering occurs because of the discontinuity of the wavevector between region II and the surrounding regions. The difference in effective masses adds to the strength of scattering through the term $\left(\frac{\kappa m_1^*}{k m_2^*} - \frac{k m_2^*}{\kappa m_1^*}\right)^2$ in (3.29) which reflects the boundary condition of continuity of current.

3.6 Tunneling in pn-Junction Diodes

Tunneling is an important phenomenon in semiconductor devices like transistors and pn-junction diodes. In this section we will develop a model for tunneling in reverse-biased pn-junctions.

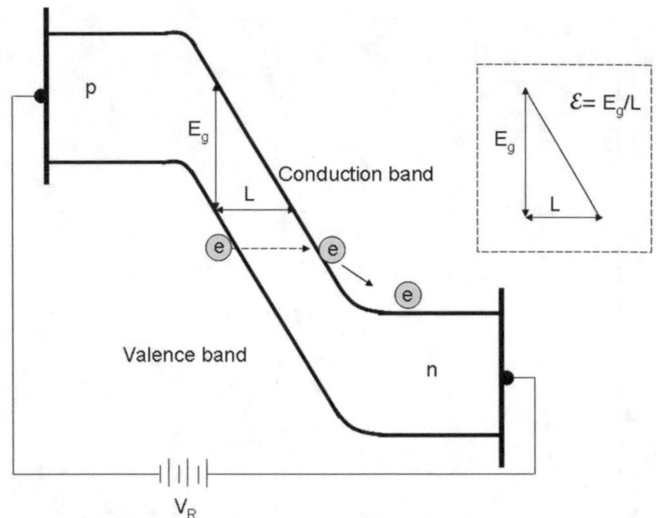

Fig. 3.11 A schematic energy level diagram for a pn-junction in reverse bias. An electron may tunnel from the valence band to the conduction band at constant potential. The tunneling barrier shown in the *inset* is *triangular* in shape

In Fig. 3.11, we show a schematic diagram of a pn-junction, plotting potential energy versus distance. In the model that follows, electrons in the valence band are confined in a potential well. The barrier height is the bonding energy (energy gap). By tunneling into the conduction band, a valence electron leaves the bonding state and transitions to the conduction band where it is free to carry current. All electrons in the valence band are candidates for tunneling. When a pn junction is reverse-biased, the conduction band and valence band can have the same potential energy, separated by a distance L as shown in the diagram. At any fixed point in space, the valence band and conduction band are separated by the bandgap energy. The equivalent electric field is $\mathcal{E} = \frac{E_g/q}{L}$. When tunneling occurs, an electron makes a transition from the valence band, across the forbidden gap to the conduction band at constant energy. Then it contributes to the current density in the conduction band according to the definition:

$$J = nqv_{\text{sat}}, \tag{3.34}$$

where q is the charge on the electron, and v_{sat} is the saturation velocity for electrons in the conduction band (for common semiconductors, this is typically $v_{\text{sat}} \approx 1 \times 10^5$ m s^{-1}). The electron density in the conduction band is expressed: $n = N_V T$, where N_V is the electron density in the valence band and T is the tunneling transmission coefficient.

Electrons in the conduction band contribute to the reverse leakage current. These electrons can gain additional energy from the electric field and initiate impact ionization, generating additional carriers, increasing the dark current and eventually leading to avalanche breakdown. Tunneling and impact ionization each make substantial contributions to the dark current near the reverse breakdown voltage. The relative contribution depends principally on the bandgap energy. Impact ionization dominates in materials like silicon and GaAs which have a bandgap greater than 1 eV. Tunneling dominates in materials with a lower bandgap, like InAs. ($E_G = 0.4$ eV at room temperature). An intermediate case, germanium, with a bandgap of 0.6 eV, is an example of a material in which tunneling and impact ionization make similar contributions to the dark current.

In the treatment which follows we will use the result that we have obtained for the tunneling transmission coefficient to characterize the tunneling current. This approach requires some approximations. We are not aiming for quantitative agreement with experiment, but rather an understanding of how the tunneling current varies as a function of applied bias voltage, and how the tunneling current changes with regard to parameters such as the fundamental bandgap energy and the effective mass of electrons. Some important approximations are discussed below.

(a) Tunneling barrier
The tunneling barrier as shown in Fig. 3.11 is triangular, whereas we have carried out a calculation for a rectangular barrier. Obviously the triangular barrier will allow more current to pass. A more precise calculation can be made to take full account of the actual barrier geometry by using the WKB method to handle the boundary conditions (Kane 1959).

(b) Effective mass
Electrons in the valence band of a semiconductor like Ge or GaAs are present in two bands, each of which has its own characteristic effective mass. In the conduction band, the electron mass is different still. The effective mass in the fundamental gap region cannot be measured directly. This poses a problem for how to assign the mass in regions I, II and III of the simple one-dimensional model for free-electron tunneling. In our approximation, we choose the same mass throughout the structure, while choosing the effective mass for electrons in Ge to be greater than that for GaAs.

The tunneling current is:

$$J_{\mathrm{Tunn}} = nqv_{\mathrm{sat}},$$

where

$$n = N_V T \tag{3.35}$$

$v_{\mathrm{sat}} = 1.2 \times 10^7$ cm s^{-1}. This is the saturated electron velocity in electric fields greater than 5×10^3 V cm^{-1}.

$N_V = 4$ valence electrons/atom $\times 8$ atoms/unit-cell/unit-cell volume $= 1.7 \times 10^{23}$ cm^{-3}

and

$$T = \frac{1}{\cosh^2(\kappa L) + \left(\frac{1}{4}\right)\left(\frac{\kappa m_1^*}{k m_2^*} - \frac{k m_2^*}{\kappa m_1^*}\right)^2 \sinh^2(\kappa L)}$$

Using parameters appropriate for GaAs.

The tunneling barrier is shown in the inset of Fig. 3.11. The electrons of interest lie at the valence band edge, and have only thermal energy

$$E = \frac{3}{2}k_B T. = \frac{\hbar^2 k^2}{2m^*} \tag{3.36}$$

The barrier height $V_0 = E_g = 1.43$ eV. The wavevector κ in the forbidden gap is:

$$\kappa = \frac{1}{\hbar}\sqrt{2m^*\left(E_g - k_B T\right)} \text{ and } |\kappa| = 1.4 \times 10^9 \text{ m}^{-1} \gg |k| \tag{3.37}$$

The width of the barrier L can be determined from the electric field:

$$\mathcal{E} = \frac{E_g/q}{L} = \sqrt{\frac{2qN_D V}{\epsilon\epsilon_0}}, \quad \text{where } V = V_{\text{Applied}} + \phi_{\text{built-in}} \tag{3.38}$$

$L = \frac{E_G}{\sqrt{\frac{2qN_D V}{\epsilon\epsilon_0}}} = 2.7 \times 10^{-1}$ m, for GaAs, with $N_D = 10^{16}$ cm^{-3} and a reverse bias of V = −1 V.

To evaluate the transmission coefficient, we set $m_1^* = m_2^* = m^*$, and we note that $\kappa L \gg 1$. The transmission coefficient can be simplified to:

$$T \cong 16e^{-2\kappa L}\left(\frac{k\kappa}{k^2+\kappa^2}\right)^2 \cong 16\left[\frac{k}{\kappa}\right]^2 e^{-2\kappa L} \tag{3.39}$$

$$\kappa L = \frac{E_G}{\hbar}\sqrt{\frac{2m^* E_G \epsilon\epsilon_0}{2qN_D V}} = C_0 V^{-\frac{1}{2}} \tag{3.40}$$

and

$$\left(\frac{k}{\kappa}\right)^2 \cong \frac{3k_B T}{2E_g}, \quad \text{where } T \text{ is the temperature} \tag{3.41}$$

This results in:

$$T \cong 16\left(\frac{3k_B T}{2E_g}\right) e^{-2C_0 V^{-\frac{1}{2}}} \tag{3.42}$$

and

$$J_{\text{tunneling}} \cong 16 N_V q v_{\text{sat}}\left(\frac{3k_B T}{2E_g}\right) e^{-2C_0 V^{-\frac{1}{2}}} \text{ amps}-\text{cm}^{-2}, \tag{3.43}$$

yielding the tunneling current density in terms of the reverse bias voltage. It is crucial to note that the applied voltage does not "ionize" or otherwise energize electrons in the valence band. The effect of increasing the reverse bias voltage is to decrease the physical width between the valence band and conduction band, increasing the probability of tunneling.

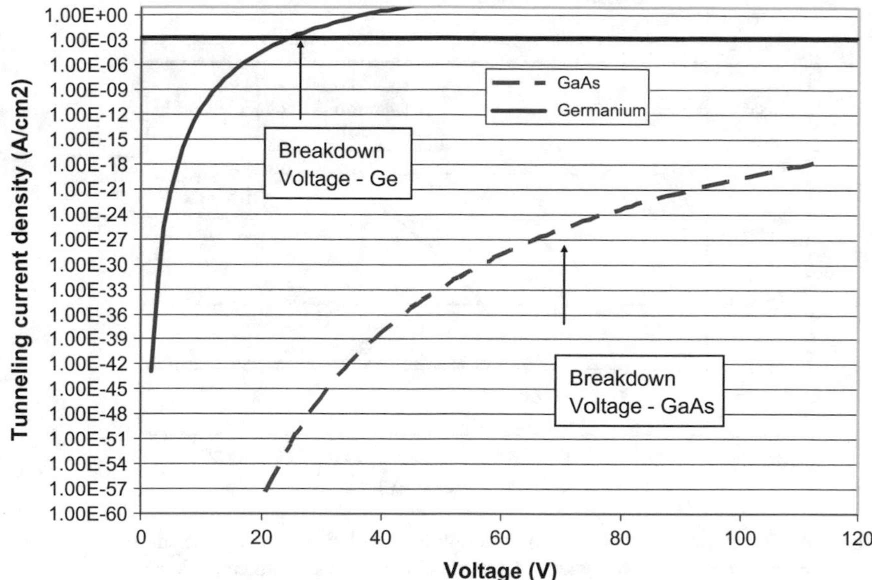

Fig. 3.12 Comparison of tunneling current density in Ge and GaAs pn diodes. Near the breakdown voltage for each diode, the tunneling current density in Ge is more that 20 orders of magnitude larger than in GaAs. The heavy line at 10^{-3} A cm^{-2} is a level indicative of the reverse breakdown regime

This model is evaluated for the cases of Ge and GaAs in Fig. 3.12, where a one-sided pn junction is assumed having a doping level of $N_D = 1 \times 10^{16}$ cm^{-3}. We indicate by the vertical arrows the measured breakdown voltages for both Ge and GaAs having this level of doping. The principal result is that the tunneling current density in GaAs is many orders of magnitude smaller than in Ge, due to the difference in the energy gap, which is the barrier for tunneling. In addition, it can be seen that electron tunneling occurs at all levels of applied bias.

From a practical viewpoint, one can associate a current density $J_R > 10^{-3}$ A cm^{-2} with the regime of reverse breakdown. Using this assumption, the tunneling current in Ge appears to be an important contribution to the breakdown regime. In the case of GaAs, the calculated tunneling current density lies well below the breakdown regime when the breakdown voltage is applied. It follows that tunneling is not a major contributor to reverse breakdown current in GaAs, while it is a major contribution to the breakdown current in Ge pn diodes.

The carrier concentration N_D can have an important effect on the tunneling current because it can be varied over several magnitudes. In Fig. 3.13 we give an example of this effect for a Ge p$^+$n diode, with 2 different levels of doping on the n-side.

Tunneling is a pure quantum-mechanical effect that accompanies the wave-like behavior of particles: (electrons, photons, protons, and even atoms).

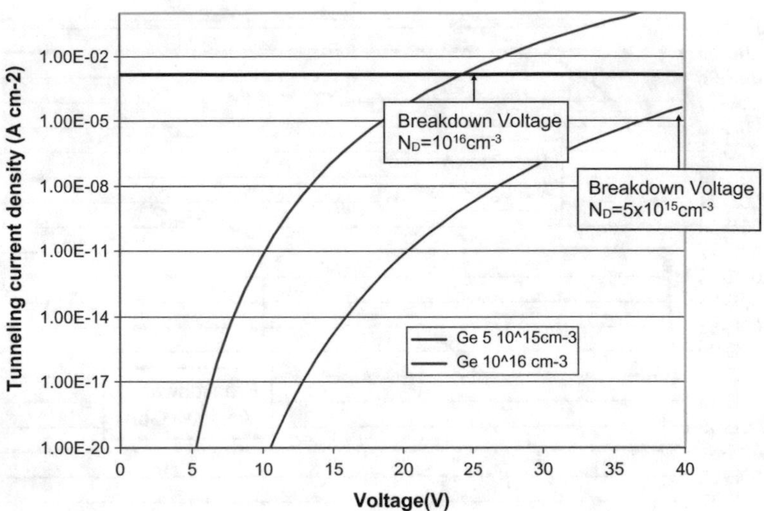

Fig. 3.13 The carrier concentration on the lightly-doped side of the pn diode determines the electric field and therefore the width of the tunneling region. Reducing this carrier concentration lowers the electric field at a given voltage. In this example for a Ge pn diode, lowering the carrier concentration by 50% reduces the tunneling current density by one order of magnitude near the breakdown voltage. This dependence is important because it is the only recourse available to designers for reducing leakage by tunneling in a device of a specific material composition

The tunneling probability increases:

as the kinetic energy of the incident particle approaches the barrier potential,
as the physical thickness of the barrier decreases, and
as the effective mass of the tunneling particle decreases.

Tunneling is a source of leakage current in pn diodes and thus in transistors, when conditions reduce the distance between conduction and valence bands. Reducing tunneling current in devices with nanometer dimensions is not straightforward. In the case of leakage current in diodes and transistors, we have shown that one option is to reduce the free carrier concentration at the junction, which serves to widen the tunneling barrier for a given reverse bias voltage. As a result, the tunneling current density near breakdown is also reduced, meaning that the competing mechanism, impact ionization, may make the major contribution to the reverse current. Putting several barriers in series might seem like a possible solution, but in the next section, we will see that this is not at all the case.

3.7 Tunneling in the Presence of Multiple Barriers

Sequential tunneling in the presence of multiple barriers is treated using the same approach as that for a single barrier. A one-dimensional model of a two-barrier environment is shown in Fig. 3.14.

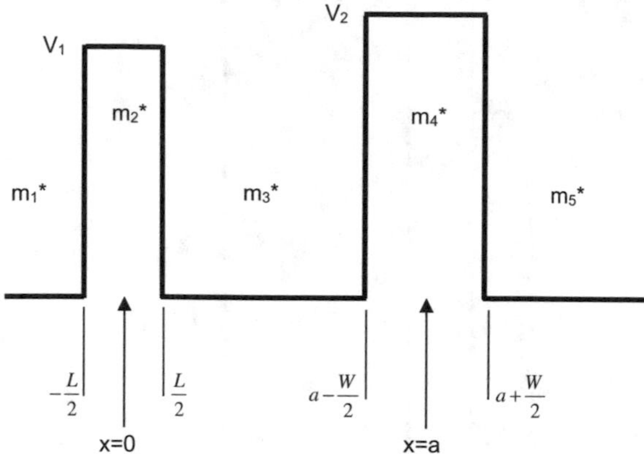

Fig. 3.14 Schematic model for a double potential barrier in one dimension. A boundary condition is determined for each interface. Note that positioning of each barrier, its width and magnitude are arbitrary

The objective is to develop an expression for the transmission coefficient. Using (3.21)–(3.27), we can immediately write out the solution by imposing the boundary conditions at each interface:

$$\begin{pmatrix} A \\ B \end{pmatrix} = \begin{pmatrix} \frac{1}{2} e^{\kappa \frac{L}{2} + ik \frac{L}{2}} \left[1 + \frac{i\kappa m_1^*}{k_2 m_2^*} \right] & \frac{1}{2} e^{-\kappa \frac{L}{2} + ik \frac{L}{2}} \left[1 - \frac{i m_1^* \kappa}{m_2^* k} \right] \\ \frac{1}{2} e^{\kappa \frac{L}{2} - ik \frac{L}{2}} \left[1 - \frac{i\kappa m_1^*}{k_2 m_2^*} \right] & \frac{1}{2} e^{-\kappa \frac{L}{2} - ik \frac{L}{2}} \left[1 + \frac{i m_1^* \kappa}{m_2^* k} \right] \end{pmatrix} \begin{pmatrix} C \\ D \end{pmatrix} = \begin{pmatrix} M_1 & M_2 \\ M_3 & M_4 \end{pmatrix} \begin{pmatrix} C \\ D \end{pmatrix}$$
$$\begin{pmatrix} C \\ D \end{pmatrix} = \begin{pmatrix} M_5 & M_6 \\ M_7 & M_8 \end{pmatrix} \begin{pmatrix} F \\ G \end{pmatrix}$$
$$\begin{pmatrix} F \\ G \end{pmatrix} = \begin{pmatrix} M_9 & M_{10} \\ M_{11} & M_{12} \end{pmatrix} \begin{pmatrix} H \\ I \end{pmatrix}$$
$$\begin{pmatrix} H \\ I \end{pmatrix} = \begin{pmatrix} M_{13} & M_{14} \\ M_{15} & M_{16} \end{pmatrix} \begin{pmatrix} J \\ K \end{pmatrix} \tag{3.44}$$

It is evident that this procedure can be extended easily to a large number of barriers of arbitrary height and position.

The transmission coefficient is determined by multiplying these matrix equations, giving

$$\begin{pmatrix} A \\ B \end{pmatrix} = \begin{pmatrix} M_{17} & M_{18} \\ M_{19} & M_{20} \end{pmatrix} \begin{pmatrix} J \\ K \end{pmatrix}$$

and since K is usually set to zero,

$$T = \left\| \frac{J}{A} \right\|^2 = \left\| \frac{1}{M_{17}} \right\|^2 \tag{3.45}$$

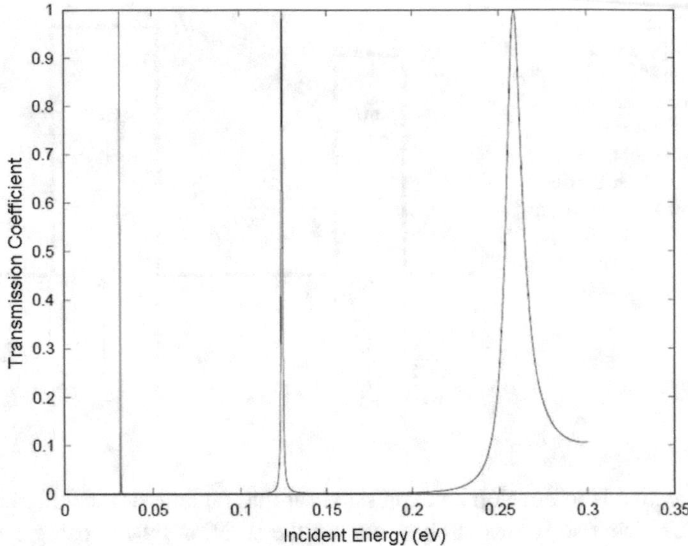

Fig. 3.15 Transmission coefficient for a double barrier structure. Unlike the single barrier example, tunneling with 100% transmission is possible at specific resonant energies well below the barrier potential. Resonant tunneling is a dramatic manifestation of the wavelike properties of quantum particles like electrons

In Fig. 3.15, we show the result of this calculation for a double-barrier of height 0.25 eV and a width of 5 nm for each barrier. The distance between barriers is 10 nm. The particle effective mass in the barrier region is 0.092 m_0 and the effective mass outside is 0.066 m_0. The results show that transmission by tunneling is seen at energies well below the barrier height. At specific resonant energies, the transmission coefficient is unity, indicating that the potential barrier is entirely transparent at these energies.

Resonant tunneling has been studied theoretically since the 1950s. In the general absence of computers, the first approximate solution was achieved using the WKB method (Duke 1969). The transfer matrix method developed here is more accurate, and solutions are easily implemented using MATLAB or similar computing environments (see Harrison 2006 and Wartak 2013). Chang, Esaki and Tsu first measured resonant tunneling in a GaAs/AlGaAs heterostructure having parameters similar to those used for the results of Fig. 3.15 (Chang et al. 1974). Resonant tunneling is used to enhance performance of transistors and as a means of injecting carriers into quantum well structures at energies below the barrier height of the well. This type of injection is key to the operation of the quantum cascade laser, for example.

It is instructive to display these results using a logarithmic scale and this is shown in Fig. 3.16. There is tunneling transmission at all energies, although the transmission coefficient remains far below unity with the exception of the two

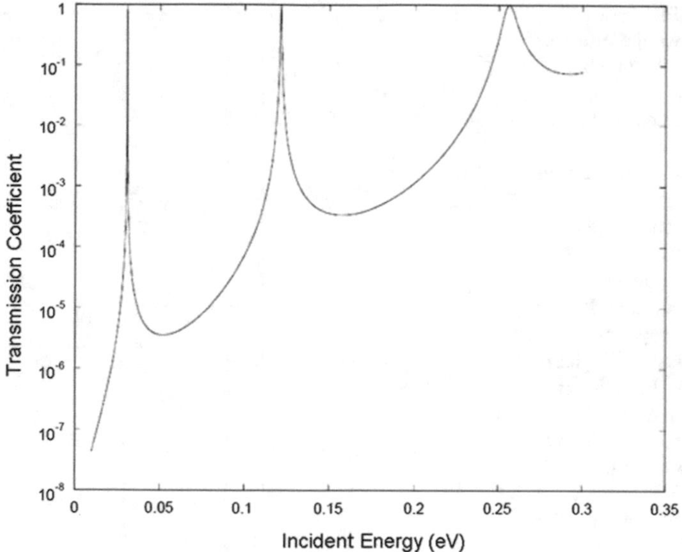

Fig. 3.16 Transmission coefficient versus incident energy for a double barrier structure using the same parameters as those for Fig. 3.15. Using a log scale reveals that the transmission coefficient increases gradually with incident energy, just as in the case of tunneling through a single barrier. However, there are in addition specific resonant energies that are closely related to the energies of the confined states in a potential well (see Fig. 3.4)

resonant energies at 0.035 and 0.124 eV. Note that these energies are quite close to those calculated earlier for the finite potential well having the same width and effective mass ratio. This is an important underlying principle of resonant tunneling: quantum mechanical particles can tunnel with near unity transmission coefficient through a multiple barrier structure at specific energies. These energies correspond to the quantised levels of a finite potential well having the same parameters of potential depth, width, and effective masses in the well and the barrier.

The transfer matrix method can be extended in a straightforward way to calculate tunneling transmission in a heterostructure containing many periods. The principal result is that the narrow resonances seen in Figs. 3.15 and 3.16 broaden into energy bands (Tsu and Esaki 1973). The mathematics of the calculation are similar to those used for computing the optical pass bands for interference filters, with the changes in refractive index playing an analogous role to changes in barrier height used in tunneling.

In Chap. 2, we examined the wave behavior of photons. In one example we considered the one-dimensional propagation of photons through a plasma layer barrier, (see Sect. 2.4.1). Photon propagation through a region which is lossy, because the photon wavevector is imaginary, is nonetheless allowed. The attenuation of the wavefunction amplitude depends on the energy of the photon and the thickness of the plasma region. The mathematics that describe penetration of

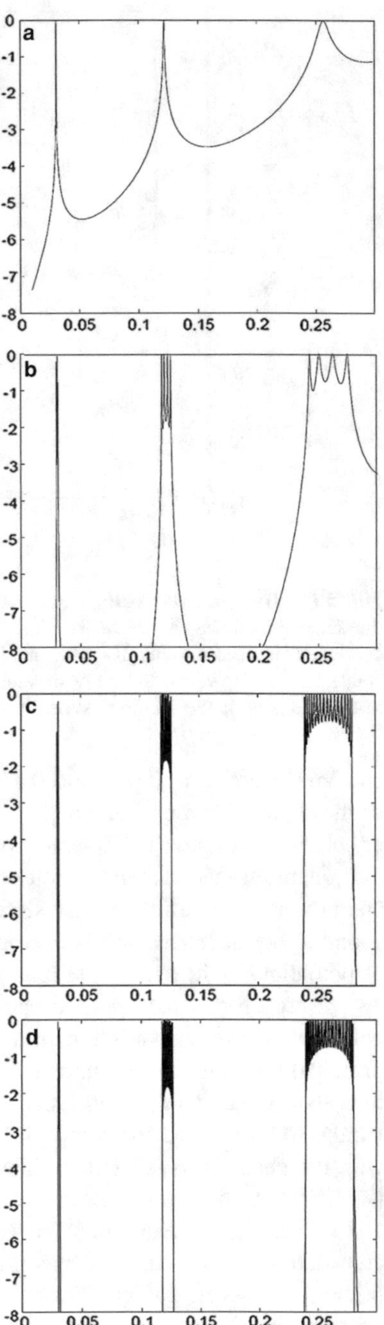

Fig. 3.17 This series of images shows the effect of augmenting the number of barriers in a resonant tunneling structure. In each figure we show the transmission coefficient on the vertical axis as a function of energy (eV). Each calculation uses the following parameters. Barrier potential = 0.25 eV, Barrier width = 5 nm, Effective mass in the barrier = 0.092 m_0. Effective mass outside the barrier = 0.066 m_0. Space between the barriers = 10 nm. **a** 2 barriers, **b** 5 barriers, **c** 20 barriers, **d** 50 barriers. The principal result is that adding more barriers does not impede the transmission by tunneling. Quite the reverse is true. Augmenting the number of barriers enlarges the energy region where transmission may occur with high efficiency. The calculations show that the resonant transmission window does not change its energy as the number of periods is augmented. Instead, the resonant transmission region widens to form an energy band. The energy width of each band is already well-defined after 5 barriers. Adding additional barriers increases the density of resonant states, without changing the width of the energy band

a photon through such a region are identical to those of an electron tunneling through a potential barrier discussed above.

In Chap. 2, we also discussed the propagation of photons through a structure where the index of refraction is modulated between two or more values. This is a different situation, in which the photon wave vector is always a real quantity. Such structures, whether periodic or not, give rise to interference, and the transmission coefficient shows regions of transmission and reflection. The exercises of Chap. 2 show that structures having an extended period variation of index of refraction lead to frequency bands of high transmission, separated by regions of high reflection. These results are supported by mathematical analysis that is similar, but not identical to the analysis that has produced the results shown in Fig. 3.17. The feature that is responsible for this similarity is the shared wave nature of photons and matter, proposed by Louis de Broglie and demonstrated by Clint Davisson and Lester Germer.

It is clear that the resonant tunneling calculation can be used to simulate electron energy levels in a one-dimensional periodic structure, i.e., a crystal. At the same time the transfer matrix method can also be used to simulate electron energy levels in a non-periodic structure, with varying barrier heights and non-periodic position. Calculations based on these structures also show that energy bands develop, although the resonances are not so perfect as in the case of the periodic structure.

Next, it is instructive to turn the transmission coefficient shown in Fig. 3.17 on its side. The transmission coefficient characterises the superlattice structure over its extent in real space. In Fig. 3.18, we can see immediately that there are bands of energy where the transmission coefficient is close to unity. We call these minibands. In between, there are regions where the transmission coefficient is close to zero, and we call these minigaps.

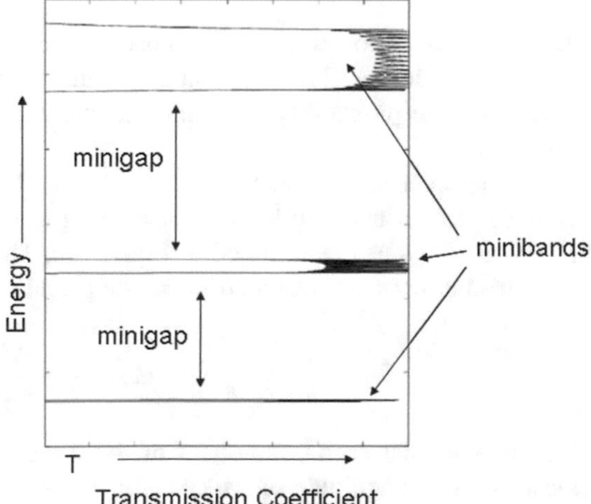

Fig. 3.18 The tunneling coefficient in a superlattice structure gives rise to energy minibands and minigaps. The minibands are usually referred to as subbands. This additional energy level structure, created by the geometry of the sample, can be exploited to make both photon detectors and photon emitters, using inter-subband electronic transitions

3.8 Tunneling Time

Tunneling is a fundamental manifestation of the Heisenberg uncertainty principle and quantum mechanics. In the previous sections we have studied the stable states of particle-waves in environments of varying potential energies. In each case we assign the particle a single well-defined wave-vector and proceed to evaluate the time-independent Schrödinger equation. Since a measurement must find the particle either on one side or the other, there is a finite probability of finding that the particle has tunneled through the barrier. We can examine the steady-state tunneling probabilities based on the solution of the time-independent Schrödinger equation probabilities in a straightforward way. However, this approach leaves a key question unanswered: how long does it take the particle to tunnel from one side to the other?

3.8.1 Wavepackets

So far we have treated the wavefunction as a free-particle wave, with a single k-vector. This is a convenient mathematically for solving the Schrödinger equation. However, the Heisenberg uncertainty principle requires that a particle characterized by a single k-vector (ie. $\Delta k \to 0$) be delocalized spatially ($\Delta x \to \infty$). The "transit-time" of such a wavefunction does not make physical sense.

As a first step toward understanding the tunneling transit-time, we modify the simple free-particle wavefunction into a spatial wavepacket consisting of a sum of such wavefunctions weighted by an appropriate envelope function.

$$\Phi(x) = \sum_n a_n \Psi_n(x) \tag{3.46}$$

The envelope function is typically chosen to be a probability distribution such as a Poisson function or Gaussian function. This permits the normalisation of $\Phi(x)$, assuring that the probability of finding the particle somewhere in a defined spatial range is unity.

A wavepacket is distributed in space, shown by the diagram in Fig. 3.19a. It must also be distributed in k-space, according to the Heisenberg uncertainty principle $\Delta x \Delta p \geq \hbar$, as diagrammed in Fig. 3.19b. The group velocity of each of the waves making up the wavefunction can be calculated from the relationship:

$$\vec{v}_G = \frac{\partial \omega}{\partial k} = \frac{1}{\hbar}\frac{\partial}{\partial k}E(k) = \frac{\partial}{\partial k}\frac{\hbar k^2}{2m} = \frac{\hbar \mathbf{k}}{m} = \frac{\mathbf{p}}{m} \tag{3.47}$$

which implies that each component of the wavepacket has a group velocity that depends on the magnitude of its k-vector. If the wavepacket has minimum uncer-

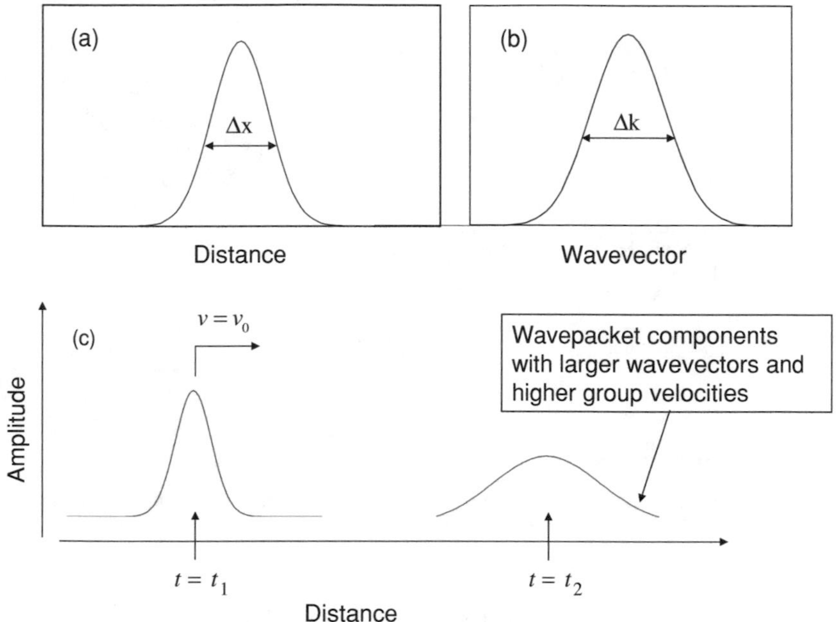

Fig. 3.19 **a** An example wavepacket with a Gaussian spatial distribution. $\Delta x \Delta \hbar k > h$. **b** The momentum ($\mathbf{p} = \hbar \mathbf{k}$) distribution of the wavepacket in **a**. **c** Larger wavevectors have higher group velocity. This causes the wavepacket to spread out spatially, even though the wavevector distribution remains constant

tainty at time $t = t_0$, that is: $\Delta x \Delta p = \hbar$, then at some later time, in the absence of any external interaction, the spatial extent of the wavepacket will have spread out, while the distribution in k-space remains constant, (conserving total momentum), and thus $\Delta x \Delta p > \hbar$ (Fig. 3.19c). An important result to retain is that higher k-vector components of the wavepacket have a larger group velocity than lower k-vector components.

3.8.2 Detection of a Wavepacket Representing a Single Particle

A single quantum particle such as an electron can be detected. In order to measure the transit time of an electron that may tunnel through a barrier, the electron wavefunction must be localized spatially. This means that the electron wavefunction must be a wavepacket composed of a distribution of **k**-vectors. To measure the time of flight of the electron, we refer to the schematic diagram of a possible experiment shown in Fig. 3.20. We have considerably simplified the details so that the one-dimensional electron wavepacket is launched toward the right toward an

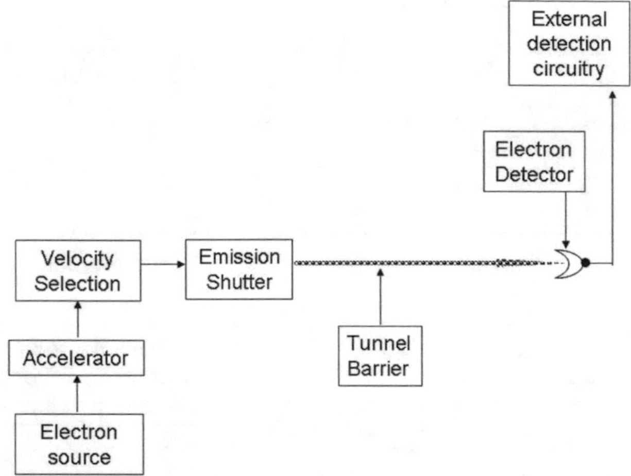

Fig. 3.20 Schematic diagram of an experiment to measure electron transit time in the presence and the absence of a tunneling barrier

electron detector placed at a specific distance from the emission shutter. At time $t = 0$, a gate is opened allowing the electron wavepacket to propagate to the right. What would the observer of the detector record?

The detector is capable of detecting an electron and causing a charge to circulate in the detector electronics resulting in a signal. When the detector captures an electron, the spatial extent of the wavepacket collapses to the location of the detector. The probability of finding the electron at the detector is unity, and zero everywhere else. The signal amplitude at the detector is 1 when an electron is detected; otherwise it is zero.

If we perform many such transmission experiments and superimpose the results, we see that that the detected arrival times are spread out, with the greatest frequency centered around the classical expected value: $\langle t \rangle = \frac{X_d}{v_p}$.

In the second group of experiments, we put a single potential barrier in the path of the electron wavepacket. The wavepacket is launched with a kinetic energy $\left(= \frac{mv_p^2}{2}\right)$ that is less than the barrier height. From the investigation of tunneling through a single barrier, we know that the transmission coefficient is less than unity, and in most cases far less than unity. The transmission coefficient increases exponentially as a function of the wavevector magnitude. In the wavepacket, these are the components with higher group velocity, and they are distributed at the leading edge of the wavepacket.

During the impact of the wavepacket on the barrier, the barrier acts as a passive filter, enabling transmission of the larger wavevector components and reflecting the smaller wavevector components. Thus the collision reshapes the wavepacket amplitude from its original form into a new profile. Focussing for a moment on the tunneling phenomenon, the highest wavevector components of the wavepacket

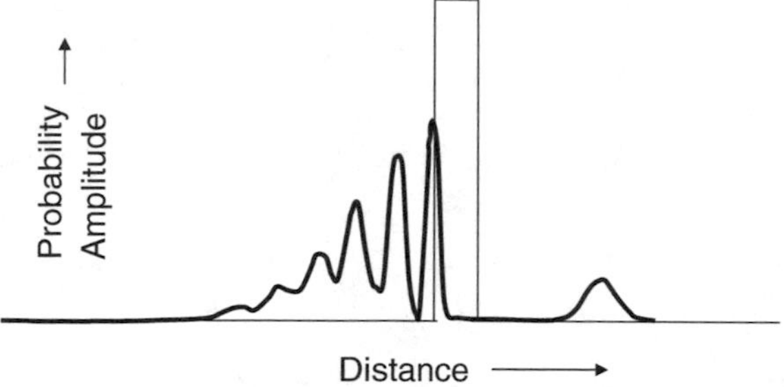

Fig. 3.21 When a particle represented by a wavepacket is incident on a barrier, the wavepacket is dramatically reshaped into 2 different spatial regions on either side of the barrier. On the *left-hand side*, the incident portion of the wavepacket interferes with the reflected portion. On the *right-hand side* a smaller peak represents the part of the wavepacket that has tunnelled through the barrier. When a measurement takes place, the particle is localized either on one side of the barrier or on the other

are clustered in the leading edge of the wavepacket. The probability that some of these components of the wavepacket will be spatially localized on the right hand side of the barrier will depend on both the pre-collision probability density of these elements and their corresponding transmission coefficients. Following the leading edge are wavepacket components with lower values of wavevector. While the probability density of these components is higher, their transmission coefficients are orders of magnitude lower. The result is that a small peak is formed in the wavepacket on the right-hand side of the tunneling barrier. This peak may be fully formed before the principal maximum of the wavepacket is incident one the barrier. Such a collision is shown schematically in Fig. 3.20. At this point it is important to remember that the entire wavepacket, on both sides of the barrier, represents a single and indivisible electron (Fig. 3.21).

The observer running the experiment at the detector has a chart recorder that marks the detection of an electron as a function of time. In the discussion that follows, we will assume that the observer records the results of many experiments on the same sheet of paper, thus accumulating results. Under the conditions of the experiment, the expected arrival time of the peak of the free-electron wavepacket occurs at $\langle t \rangle = 200\tau$, where τ is a unit of time. In Fig. 3.22 we show the results of an experiment in which a free-electron wavepacket is launched along an unimpeded pathway toward the detector. In experiment 1, shown in Fig. 3.22a, the electron is detected at $t = 191.5\tau$, slightly ahead of the expected arrival time. The following experiments show results following 6, 13, 21, and 50 repetitions. It is apparent that the measured arrival times are clustered around the expected arrival time $\langle t \rangle = 200\tau$. The electrons arrive one at a time, and for each event, the signal amplitude is unity at the time of detection, and zero elsewhere. Each such detection represents the transmission of information in the full sense of the theory of special relativity. If the electrons are launched at equal time intervals, then the density of points in Fig. 3.22e is proportional to the detection rate of electrons at the detector.

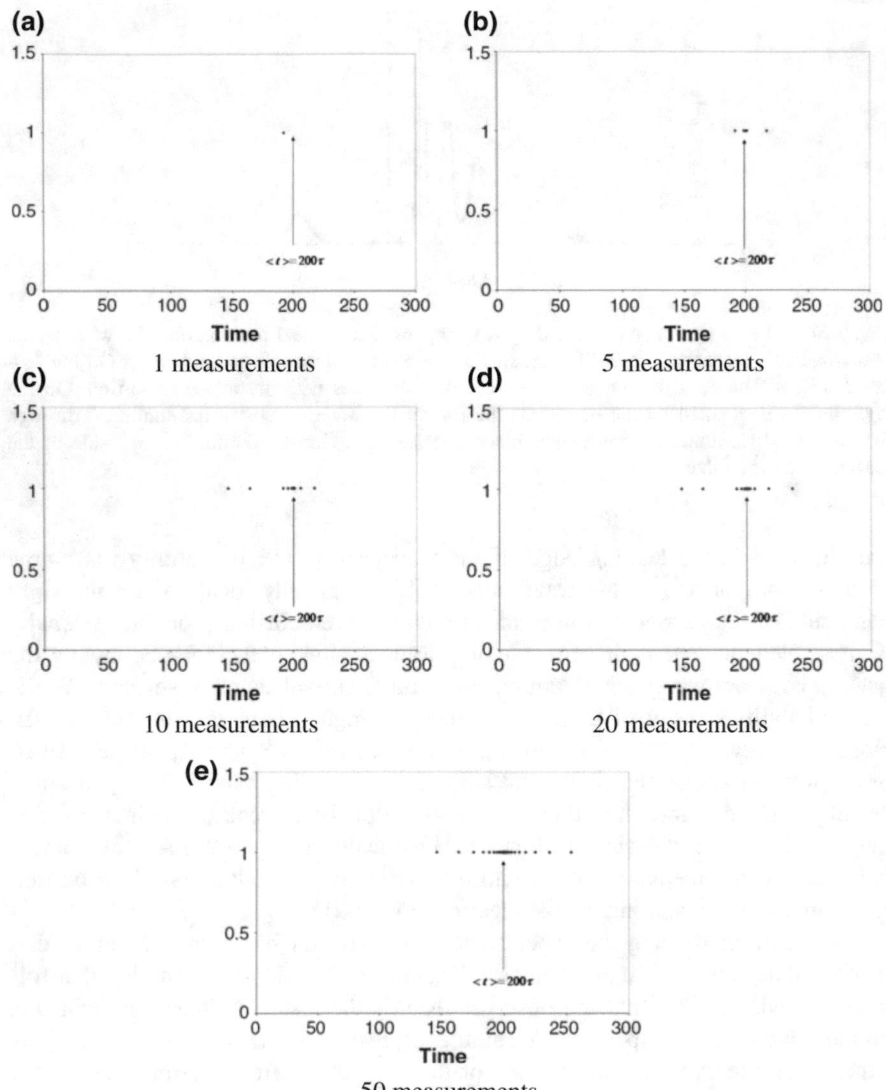

Fig. 3.22 a–e A series of measurements to detect the arrival time of an electron wavepacket emitted at a constant rate from an electron source with a well-defined group velocity. Detection is a statistical event based on the quantum mechanical behavior of electrons

By superimposing the free-electron wavepacket on the results in Fig. 3.22e, it can be seen in Fig. 3.23 that there is a one-to-one correspondence between the **magnitude** of the wavepacket and the **detection rate** of electrons at the detector.

When a tunneling barrier is placed in the propagation path, we can see whether the barrier retards or advances the arrival time. Having measured a given detection

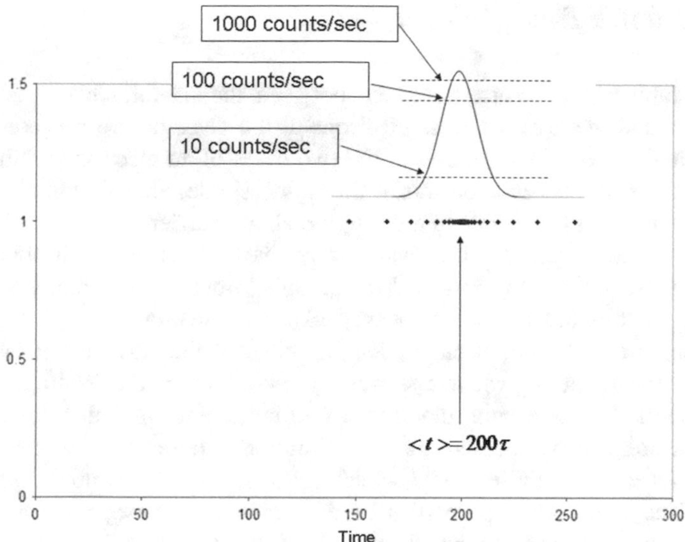

Fig. 3.23 The *inset* shows schematically a calibration of the magnitude of the wavepacket in terms of the arrival rate of electrons at the detector

rate at a particular transit time for electrons that have tunneled through a barrier, one can then ask what transit time corresponds to the same detection rate for an unimpeded wavepacket. Is it shorter or longer? This procedure enables a comparison of the relative arrival times of electrons at a detector. A hypothetical experimental procedure might include the following steps.

(i) Fix the position of the detector.
(ii) Determine the detection rate in the absence of a signal. Suppose that this rate is 1 count per second.
(iii) Choose a particle detection rate higher than the background. Suppose that a rate of 10 counts per second is chosen.
(iv) Launch a stream of electrons, one by one toward the detector. Choose a launch rate much larger than the detection rate. Suppose that a rate of 10^6 launches per second is chosen.
(v) Measure the time required for the detection rate to reach 10 counts per second.
(vi) Place a tunneling barrier in the path of the electron stream and repeat step v.
(vii) Compare the times in steps v and vi.

The paper by Eric Fossum (Fossum 2013) presents an example of an analogous method used to detect and analyze single photons in state-of-the-art quantum image sensors.

3.8.3 Relative Time of Flight

Having established the correspondence between the magnitude of the electron wavepacket and the arrival rate of electrons at the detector, we can compare the time required to detect an electron in the two cases of an electron transit with no barrier in the path and electron transit through a single, abrupt tunneling barrier. Consider first the case of tunneling through a single barrier.

First we need to choose an arrival rate r_0 that is well above the background arrival rate. Next the rate must be low enough to detect the small peak that is formed in the forward part of the wavepacket by tunneling through the barrier. The measurement is focused on the leading edge of the wavepacket. No matter how much the tunneling event reshapes the wavepacket, the leading edge will always appear the same, with the arrival rate increasing in a monotone fashion. In the measurement we measure the time duration required for the arrival rate to reach the level r_0. In the example which follows, we will assume that $r_0 = 10$ counts/s. The propagation properties of the unimpeded wavepacket are already known before the measurement. The peak of the wavefunction displaces with the group velocity imposed at launch. We can plot the wavepacket as a function of distance at the time determined for the detection of tunneling electrons. The two wavepackets are compared in Fig. 3.24. It can be seen that the magnitude of the wavepacket that represents the unimpeded electron is greater than that of the tunneling electron everywhere along the leading edge of the wavepacket. This follows from the result that the tunneling barrier attenuates the transmission coefficient when the incident kinetic energy is less than the barrier height. As a result, the reflection coefficient must be greater than zero in order to preserve the normalisation of the wavepacket to unity (that is, the total area under the square of the wavepacket must be 1). In fact the wavepacket is mostly reflected. Precisely what is attenuated is the arrival rate. The arrival rate of the unimpeded electron reaches the level of 10 counts/s in less time. The result of the above measurement is that passage through a barrier by tunneling retards the arrival of the electron at the detector.

In Fig. 3.24a we show a “snapshot” of two experiments taken at time $t = t_1$. Each experiment consists of measuring the arrival rate of electrons at the detector which is fixed in position. The electron source shown schematically in Fig. 3.20 emits single electron wave packets at the rate of 10^6 electrons/s, each with the same fixed velocity $v = v_0$. In the upper trace, the unimpeded wavepacket arrives at the detector. As time increases, the arrival rate increases until the expectation arrival time is reached, $t = \tau_{p2}$, and then the arrival rate decreases. After recording the passage of many identical wavepackets, we have a result similar to that shown in Fig. 3.22e, and we can determine the time when the arrival rate first reaches 10 counts/s. This time occurs at $t = t_1$, as shown in Fig. 3.24a. From Fig. 3.23 we know that this time corresponds to the arrival of the leading edge of the wavepacket.

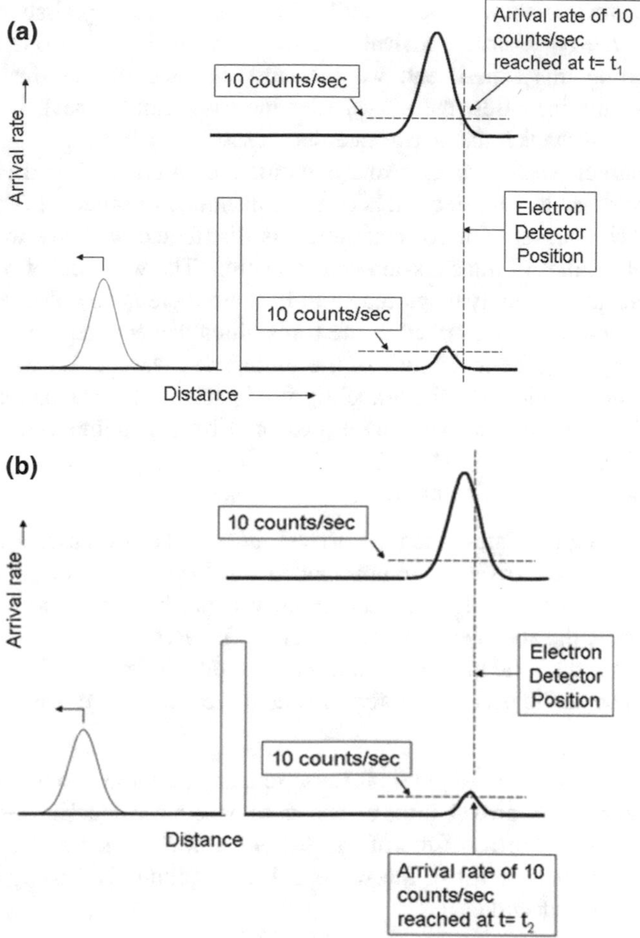

Fig. 3.24 Comparison of transmission of an electron wavepacket in the absence (shown in the *upper curve*) and in the presence of a tunneling barrier (shown in the *lower curve*). **a** At time $t = t_1$, the count rate of the upper, unimpeded wavepacket is sufficient to trigger the detector, while the tunneling wavepacket has not yet reached a sufficient rate. **b** At time $t = t_2$, the count rates of both wavepackets are sufficient to trigger the detector. Of course, the reduction in the transmission coefficient due to tunneling is exactly compensated by the magnitude of the reflected wavepacket moving to the left

Next, we repeat the same experiment, with a tunneling barrier in the path of the electron. Transmission through the tunneling barrier creates a small local peak in the wavepacket traveling toward the right, and another local peak representing the reflected part of the wavepacket that travels toward the left. Because the wavepacket magnitude is attenuated, the arrival rate of electrons at the detector will also be reduced. As a result, we can see in Fig. 3.24a, that at time $t = t_1$, the arrival rate for the tunneling electrons has not yet reached 10 counts/s.

In Fig. 3.24b we show a "snapshot" of the same two experiments, taken at a later time: $t = t_2$, when the arrival rate of the tunneling electrons reaches 10 counts/s. During this experiment, we measure the electrons as they reach the detector. As time increases, the arrival rate increases until a peak is reached at time $t = \tau_{p2}$, after which the arrival rate decreases.

When a particle such as an electron, is incident on a potential barrier, only two outcomes are possible. The electron is either transmitted or reflected. The quantum mechanical wavepacket of unity magnitude is distributed spatially to reflect the different probabilities of transmission and reflection. The wavepacket of the unimpeded particle also has unity magnitude, but because there is no reflected wave, the magnitude is entirely concentrated in the transmitted wavepacket. At a fixed position, and at any time, the magnitude of the unimpeded wavepacket will always be greater than the magnitude of the tunneling wavepacket. This means that the measurement of the arrival rate particles represented by the unimpeded wavepacket will also be larger.

The results of the experiment are.

1. By comparing the times when the arrival rates of the two cases are the same, we conclude that arrival of the unimpeded wavepacket is detected first at time $t = t_1$, while the arrival of the tunneling wavepacket arrives at a later time, $t = t_2$. From the above argument, $t_2 > t_1$ in all cases.
2. We note that the local maximum in the arrival rate of the tunneling wavepacket occurs *prior* to the peak in the arrival rate of the unimpeded wavepacket. That is: $\tau_{p2} < \tau_{p1}$.

However, it does not make physical sense to compare these two times. The time τ_{p1} is the expectation arrival time of the unimpeded wavepacket. On the other hand, the expectation arrival time of the tunneling wavepacket at the detector is infinitely long, because most of the wavepacket magnitude is moving toward the left as a result of reflection.

3. While it is straightforward to determine the relative times of arrival, such an experiment does not provide a satisfactory method for defining or determining the time of tunneling. To illustrate this point in the example above, suppose that we required an arrival rate of 100 counts/s as the criterion for detecting the arrival of the leading edge of a wavepacket. We could proceed as before for the measurement of the arrival time for the unimpeded wavepacket. However, for the case of the tunneling wavepacket, an arrival rate of 100 counts/s would probably never be reached, and one result of the experiment would be to conclude that no detectable tunneling occurred.

In the next section, we will consider tunneling transit time. We first look at tunneling through a single barrier to establish some basic ideas. Next in the case of double-barrier tunneling structures, three important effects occur: transmission, reflection and resonant storage of the wavepacket in the structure.

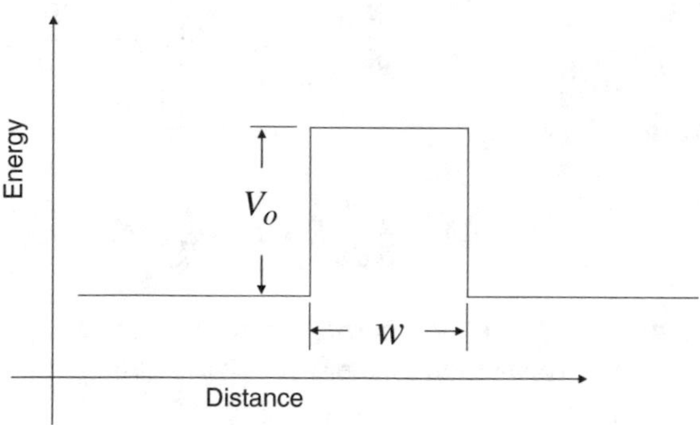

Fig. 3.25 One dimensional rectangular potential barrier of magnitude V_0 and width w

3.8.4 Analysis of the Wavepacket: Phase-Delay Tunneling Time

Analytic computation of tunneling time remains an unsolved problem in quantum mechanics. A excellent review of this important problem has been given by Hauge and Støvneng (1989). Those wishing to pursue this topic are referred to this review as a good foundation.

In Fig. 3.25 we show a simple rectangular one-dimensional potential barrier of height V_0 and width w.

$$\begin{aligned} E(k) &= \frac{\hbar^2 k^2}{2m} \\ V_0 &= \frac{\hbar^2 k_0^2}{2m} \\ \kappa^2 \equiv k_0^2 - k^2 &= \frac{2m(V_0 - E(k))}{\hbar^2} \end{aligned} \tag{3.48}$$

The particle wavefunction is represented by a wavepacket the peak of which can be characterized by a group velocity $\mathbf{v}_g$. The wavepacket is incident on the barrier and scatters into a superposition of transmitted and reflected components. We emphasize that the physical reality of the tunneling particle is represented by one wavefunction that incorporates these two correlated components. After some time, the wavepacket intensity in the barrier tends to zero, and the wavepacket of the particle is separated by the barrier into transmitted and reflected components. This is referred to as the asymptotic regime. The transmitted component can be expressed in the form;

$$\Psi_T(x,t) = \sqrt{T(k)}e^{ikx}e^{[i\phi x - i\omega t]} = \sqrt{T(k)}e^{\left[ikx(t) + i\phi(k) - i\frac{E(k)t}{\hbar}\right]} \tag{3.49}$$

In this situation we can apply the stationary phase approximation:

$$\frac{d}{dk}\left[ikx(t) + i\phi(k) - i\frac{E(k)t}{\hbar}\right] = 0 \tag{3.50}$$

which gives:

$$x(t) = \frac{1}{\hbar}\frac{d}{dk}E(k)t - \frac{d}{dk}\phi(k) = \left(\frac{\hbar k}{m}\right)t - \frac{d}{dk}\phi(k)$$

Setting $x(\tau) = w$, the barrier width, we can solve for a transit time τ associated with the propagation of the wavepacket through the barrier:

$$\tau = \frac{m}{\hbar k}\left(w + \frac{d\phi}{dk}\right) \tag{3.51}$$

The first term on the RHS is the thickness of the barrier divided by the group velocity of the wavepacket. The second term is the phase time and depends on how the wavepacket is "reshaped" by the barrier. The term $\frac{d\phi}{dk}$ has units of length and can be thought of as the extra "distance" the particle has to travel in crossing the barrier. It can be positive, corresponding to a time delay, or negative corresponding to an "acceleration" of the wavepacket by passage through the barrier.

The phase-time model can be applied to a monochromatic electron wave with a single k-vector at any point in space. In a real experiment involving electron passage through a resonant tunneling barrier, it is quite evident that the electron wavefunction has a finite width in momentum-space and in real-space, as required by the uncertainty principle. This approach confirms, however, that different k-components of the wavepacket will tunnel at different times resulting in a reshaping of the wavepacket. The large-k, high-energy components, at the leading edge of the wavepacket penetrate the barrier first.

In the discussion that follows, we will use a double barrier structure, with each barrier having a width of 0.5 nm and a potential of 2 eV. The separation between barriers is 1 nm. The effective mass of the electrons is unity. The phase method is used to calculate a transit time for each wavefunction in the wavepacket. We consider two important cases: tunneling of a wavepacket having a k-vector distribution centered on resonance, where the transmission probability is a maximum and tunneling of a wavepacket having a k-vector distribution centered in between two resonant levels with a much lower transmission probability.

The tunneling transmission coefficient is displayed in Fig. 3.26a and the "phase tunneling time" for a free electron corresponding to the phase model in Fig. 3.26b. There is a one-to-one relationship between the peaks in the transmission coefficient and the peaks in the phase tunneling time. The most tightly bound resonances have the longest tunneling times. In Fig. 3.26b we also show the time required for an unimpeded electron particle to traverse the same region of space. It can be seen that the time for an electron to transit the structure at resonance, when the transmission probability is high, is about 2 orders of magnitude longer than for the time required to transit the same region by the classical physics of electron

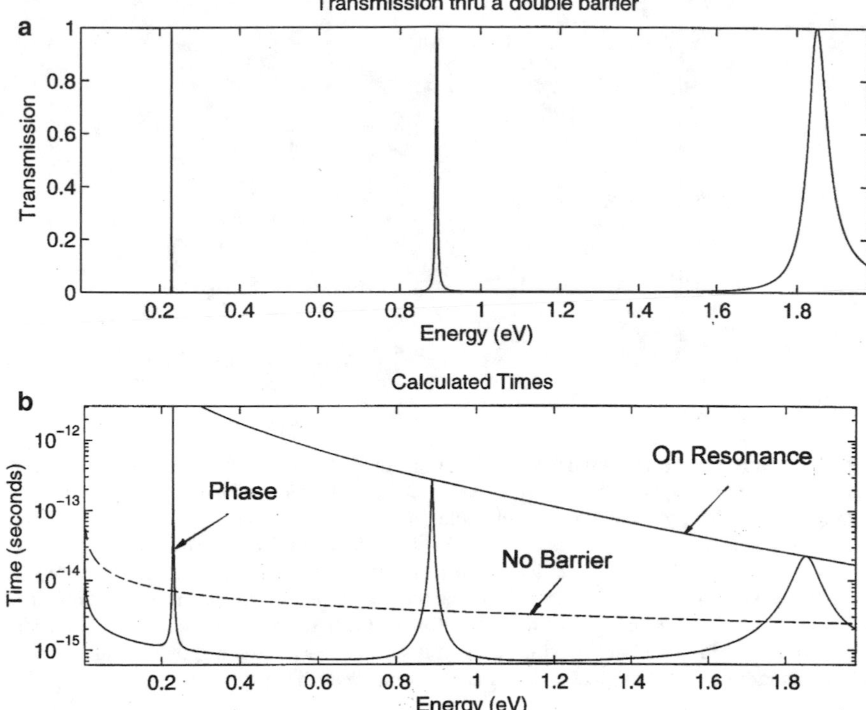

Fig. 3.26 **a** Transmission probability for an electron incident on a double potential barrier of height 2 eV and total width equal to 2 nm. **b** The phase tunneling time can be calculated as a function of kinetic energy from (3.51) presuming a perfectly monochromatic electron. This result shows that a non-resonant electron will tunnel through the barrier in a time that is about three orders of magnitude shorter than an electron that tunnels on resonance. In between we show the time for an electron to traverse the same region of space when no barrier is present

transport for an electron particle having the same kinetic energy. This result shows that there is storage of on-resonance wave-vectors in the potential well.

On the other hand, according to the phase time model, off-resonance tunneling results in a phase tunneling time that is shorter than that of classical transport, although the transmission probability is much lower. The electrons whose energy is resonant with the discrete eigenstates in the well will be filtered out of a wavepacket, and transmitted with a significant delay compared to electrons with off-resonance energies. As a result, significant reshaping of the wavepacket occurs.

This simple model shows some of the important features of resonant tunneling, but since the electron is modeled as a delocalized wave, the physical meaning of the phase tunneling time is not clear. Tunneling arrival time is fundamentally a statistical process, and analysis by continuum mathematics is certain to be inadequate.

An alternative to the analytic approach above is to discretize the Schrödinger equation for the wavepacket and follow the evolution of the wavepacket in space and in time as it interacts with the resonant tunneling barrier.

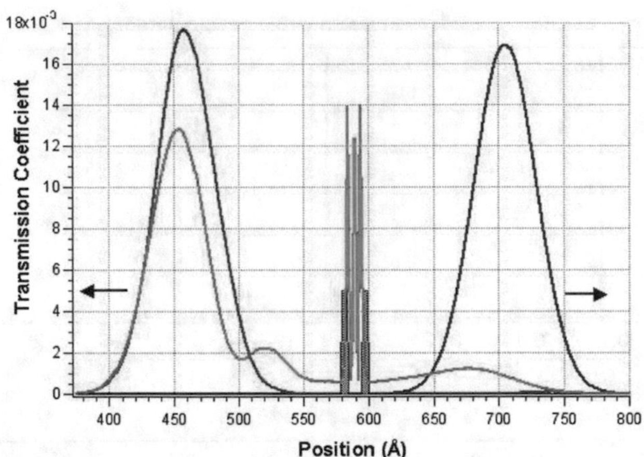

Fig. 3.27 Wavepacket magnitude as a function of distance. Each wavepacket has been launched toward the right with the same mean velocity. The *blue-colored* wavepacket is propagating toward the right without any obstacle in its path. The *red* and *green* wavepackets encounter a double tunneling barrier, located between 580 and 600 Å. The *green* wavepacket has its k-vector distribution centered on a resonant level of the double-barrier structure. The *red* wavepacket has its k-vector distribution centered between two resonant levels. Significant components of the resonant (*green*) and off-resonant (*red*) wavepackets are reflected on incidence with the barrier and are moving in the left direction. The figure shows a "snapshot" of the magnitudes of each wavepacket, taken at the same moment in time

$$i\hbar \frac{\Psi(x,t+\Delta t) - \Psi(x,t)}{\Delta t} = \frac{-\hbar^2}{2m} \left(\frac{\Psi(x,-\Delta x,t) - 2\Psi(x,t) + \Psi(x+\Delta x,t)}{(\Delta x)^2} \right) + V(x)\Psi(x,t) \tag{3.52}$$

The solution method is often abbreviated as FDTD for finite-difference time-domain. The details of the method are beyond the scope of this text. The result of this approach is a simulation of wavepacket propagation. The method itself is enabled by access to modern computer technology which was not available to the physicists who developed the phase time delay analysis. The accuracy of the solution is limited only by the available quantity of computer memory.

The results of one such simulation (Konsek and Pearsall 2003) are presented below as an illustration of the three interesting cases: (i) the propagation of an unimpeded wavepacket, (ii) a wavepacket with a k-vector distribution centered on a resonant state of the barrier, and (iii) A wavepacket with a k-vector distribution centered in between resonant states of the barrier (referred to as off-resonance).

The diagram in Fig. 3.27 shows the magnitude of the wavepacket as a function of distance. The two-barrier structure is located near the middle of the diagram between 570 and 600 Å. Three results are superimposed: (i) an unimpeded wavepacket (blue curve), a wavepacket with its k-vector distribution centered on a resonant tunneling level (green curve) and a wavepacket with its k-vector distribution centered off-resonance between resonant tunneling levels levels (red curve). The wavepackets are launched toward the double barrier with the same group velocity of the wavepacket peak magnitude. Each curve represents the spatial distribution of a wavepacket of a single electron, and so the total area under each wavepacket is unity. It is evident that the leading edge of the unimpeded wavepacket (blue) will be detected first before the arrival of the other two wavepackets.

Both the on-resonance and off-resonance wavepackets have transmitted and reflected components. As expected, there is a significant portion of the on-resonance wavepacket that is confined to the potential well in the interior of the two-barrier structure. The magnitude of this confined or stored portion depends, through the boundary conditions, on the magnitude of the wavefunction exterior to the double barrier. As the transmitted and reflected free-electron waves move away from the double barrier, the magnitude of the stored wavepacket will diminish near the barrier. To maintain the boundary conditions, the magnitude of the wavepacket inside the barrier must also decrease with time. Thus the storage of the wavepacket components is temporary, supporting the concept of a dwell-time introduced in the phase-delay analysis.

The off-resonance wavepacket is almost entirely reflected, but there is a transmitted part which can be seen near 740 Å. Storage of the off-resonance wavepacket inside the double barrier is negligible because the wavepacket contains relatively few on resonance k-vector components (Fig. 3.28).

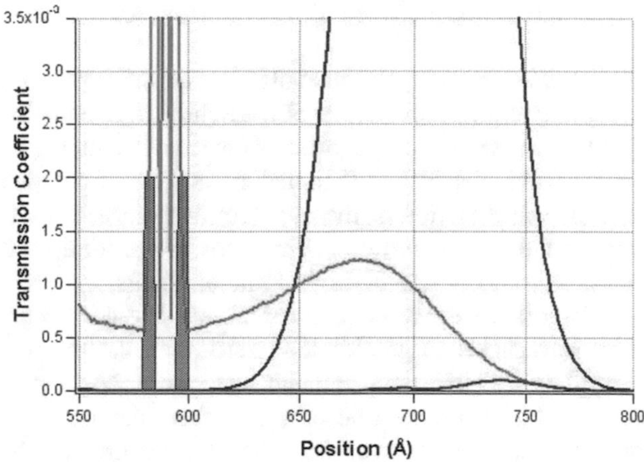

Fig. 3.28 Enlargement of the transmission region of Fig. 3.27. In this simulation, the unimpeded wavepacket (*blue*) will be detected first at any chosen arrival rate criterion. If the detection threshold can be set sufficiently low, (i.e. $y < 10^{-4}$ in this example) both the resonant and non-resonant wavepackets can be detected. Under this condition, it is not possible in this simulation to distinguish between the arrival times of the resonant and off-resonant wavepackets. If the detection criterion is set higher, for example, at $y = 0.5 \times 10^{-3}$, then the unimpeded (*blue*) and resonant (*green*) wavepackets can be detected. However, the non-resonant wavepacket (*red*) would not be detected. The result of such a measurement would be that the unimpeded wavepacket arrives well ahead of the resonant wavepacket, but that the off-resonant wavepacket is totally reflected

3.8.5 Tunneling Time: Some Conclusions

In our treatment the wavefunction is either transmitted through or reflected by the potential barrier. At all times, the sum of the reflection coefficient and transmission coefficient is equal to unity. If the probability density for reflection is non-zero, then the probability density from transmission is less than unity. Compared to the case of a quantum that encounters no barrier, the transmission coefficient of a tunneling particle will be less than that for the unscattered wavefunction. This implies that if we set up a detection station at any point in space to detect the arrival rate of quanta from an unscattered beam, and a beam that is scattered by a tunneling barrier, we will count a higher arrival rate of quanta from the unscattered beam. Otherwise stated, we will have to wait for a longer time to detect the same number of quanta from the beam of tunneling particles compared to the case for the beam of unscattered particles. In this simple model, the transit time of the particle is retarded by tunneling relative to the situation where the potential barrier is absent.

In our model we have represented the quantum by a wave packet as a summation of individual free-electron wavefunctions, each represented by its characteristic wavevector. The Schrödinger equation that we have used is a linear differential equation, and as such, it treats each wave vector component of the wavepacket independently of the others. The important result of this mathematical situation is that there is no interaction between different wavevector components of the wavepacket.

The FDTD simulation permits the visualization of the time-evolution of this wavepacket, according to the mathematics of the Schrödinger equation. This simulation shows that the wavepacket propagation is dispersive, with the larger wavevector components advancing faster than the smaller wavevector components. This is the case for both the wavefunction of the quantum that proceeds by tunneling and the wavefunction of the unscattered quantum. Thus, if we set up a detection station at any point in space to detect the arrival rate of quanta, the wavepacket components with the largest wavevectors will arrive first. If the reflection coefficient for the tunneling wavepacket is greater than zero, then the transmission coefficient will be less than that of an unscattered wavepacket. At any detector position, we will count a higher arrival rate of quanta from the unscattered beam. At equal detection threshold, the observer will always conclude that the unscattered wavepacket arrives before the tunneling wavepacket, in the few cases where the quantum tunnels through rather than being reflected by the barrier.

The result of using Schrödinger's equation for calculation of the transmission coefficient of quantum tunneling of a wavepacket through a potential barrier is similar to the action of a linear filter. Some wavevector components of the wavefunction are transmitted with a higher amplitude than others. All the wavevector components of the unscattered wave are transmitted with a higher amplitude than those of the tunneling wavepacket. Thus, regardless of the initial shape of the wavepacket or the spatial complexity of the barrier, a quantum that propagates by tunneling through a potential barrier will be retarded relative to a quantum that does not encounter such a barrier.

3.9 Summary

In this chapter we treat quanta having a non-zero mass, such as the electron, as a plane wave. The Schrödinger equation is a scalar differential equation that is used to calculate the stationary states of quanta, and we have explored a number of solutions of this equation in one-dimension.

Physical confinement of a quantum particle translates into potential energy. The dimensions of the confinement impose an upper limit on its wavelength and thus a minimum wavevector. The energy associated with a quantum with a non-zero mass is proportional to the square of its wavevector, and so confinement endows the electron with a minimum energy.

Tunneling of quantum particles through a potential barrier is an astonishing, and real feature of quantum behavior. Solutions of the Schrödinger equations show that the probability of finding a single quantum is non-zero on both sides of the barrier. Both reflection and transmission are possible. However a quantum is non divisible, and in a measurement, only one or the other result can be measured. Thus the wavefunction collapses to the eigenstate of the measurement, and it must collapse instantaneously in order to conserve mass, (the proof of which is the subject of Exercise 3.1).

Tunneling behavior can be described by calculation of the time-independent transmission coefficient. The transmission coefficient for two adjacent barriers is characterized by resonant energy levels where the transmission coefficient can reach unity. The transmission coefficient for three or more such barriers in tandem shows energy broadening of these resonant levels into energy bands.

Electron tunneling is used to make commercial photonic devices such as lasers and detectors which are capable of high-speed operation. The question of tunneling time has both fundamental and practical importance. Modeling the electron by a simple plane wave can be used successfully to calculate the stable energy states. However, since such a wavefunction is delocalized spatially, it cannot tell us about the time required to transit a tunneling structure. To approach this question, we have modeled the quantum as a wave packet, having a distribution of k-vectors, so that its spatial position can be determined. Numerical simulations using such wavepackets in the time-dependent Schrödinger equation show that quanta are retarded by passage through a tunneling barrier.

3.10 Exercises

3.1 The application of Schrödinger's equation to one-dimensional electron tunneling through a potential barrier shows that the electron wavefunction is non-zero on both sides of the potential barrier. However, experiment shows that the electron is either entirely on one side of the barrier or the other. The Copenhagen interpretation of quantum mechanics postulates that measurement causes the collapse of the wavefunction. Show that wavefunction collapse must occur instantaneously in space (and not at the speed of light), if matter is to be conserved.

3.2 See Fig. 3.29

Solve Schrödinger's equation and determine:

(a) quantization conditions
(b) energies of the stable states
(c) normalized wavefunctions

Compare your solution to that developed in Sect. 3.1.3. Which features are simpler? Which features are obscured? Which approach do you prefer?

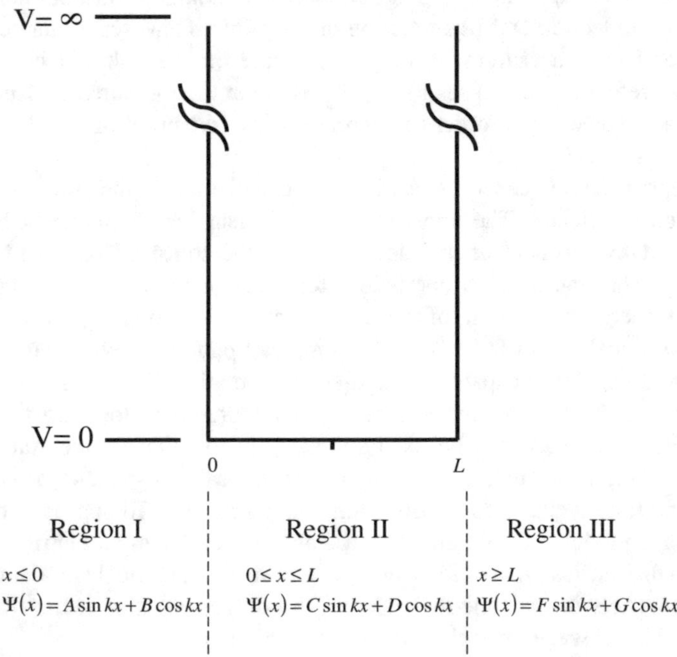

Fig. 3.29 Energy-distance diagram for an infinitely deep, 1 dimensional potential well of width = L

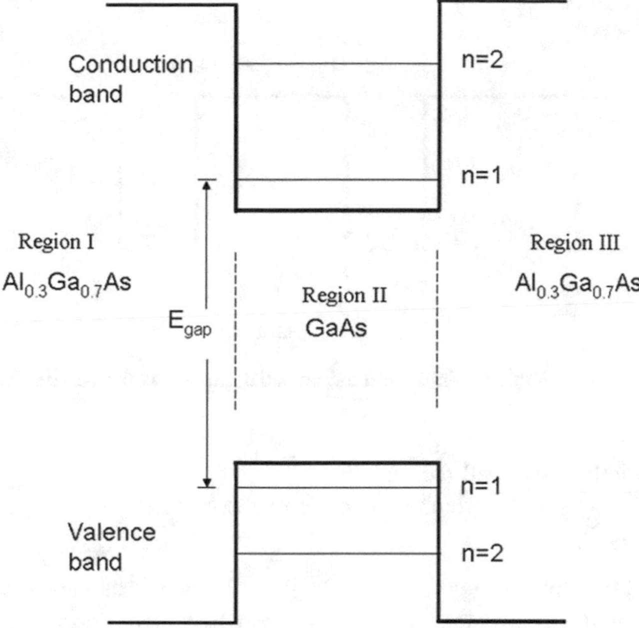

Fig. 3.30 A schematic diagram of a quantum well structure of AlxGa1-xAs/GaAs,prior to applying an electric field

3.3 Using a diagram of a wavefunction confined in an infinitely-deep potential well, demonstrate that the first-order Stark shift is zero, i.e. $\Delta E_1 = \langle \Psi j | q \mathcal{E} x | \Psi_j \rangle \equiv 0$ using a symmetry argument.

3.4 A double heterostructure of AlxGa1-xAs/GaAs forms two quantum wells, one in the conduction band and the other in the valence band, as shown in Fig. 3.30. Show that the Stark shift lowers E_{gap}, the minimum energy difference between the valence band and the conduction band states. This voltage-dependent shift in E_{gap} can be applied to make a high-speed absorption modulator for communications lasers.

3.5 See Fig. 3.31 The Krönig-Penney model treats the transport of electrons through a one dimensional infinite periodic potential. The result of the analysis gives relationship relating electron energy E to electron wavevector q:

$$\cos(qd) = \cos(\alpha(d-s))\cos(\beta s) - \frac{1}{2}\left(\frac{\alpha m_b^*}{\beta m_w^*} + \frac{\beta m_w^*}{\alpha m_b^*}\right)\sin(\alpha(d-s))\sin(\beta s) \tag{3.53}$$

(see Morrison et al. 1976, p. 465)

where:

d is the period

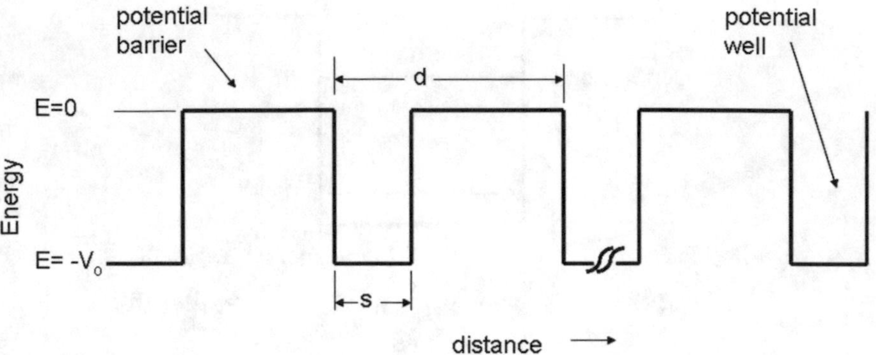

Fig. 3.31 A one-dimensional, infinite periodic potential, known as the Krönig-Penney model

s is the width of the well region
$\beta = \frac{1}{\hbar}\sqrt{2m_b^*(V_0 - E)}$, which can be a complex variable
$\alpha = \frac{1}{\hbar}\sqrt{2m_w^* E}$

(a) Use (3.53) to plot wavevector versus energy. Show that the allowed energy states form bands, separated by energy gaps where electron states do not exist.
(b) In Sect. 3.7, we showed that tunneling through a sequence of barriers and wells also gives rise to energy bands. By numerical simulation, compare the energy dependence of the transmission windows for sequential tunneling through a 4-barrier, 3-well structure, with the energy bands of the Kronig-Penney model having the same dimensions. Use GaAs and $Al_{0.3}Ga_{0.7}$ As for the materials:

$$V_0 = 0.35 \text{ eV}$$
$$m^*_{\text{GaAs}} = 0.066 m_e$$
$$m^*_{\text{AlGaAs}} = 0.092 m_e$$

3.6 The expression for the transmission coefficient for tunneling through a 1-dimensional potential barrier is $T \cong 16 e^{-2\kappa L}\left(\frac{k\kappa}{k^2+\kappa^2}\right)^2$ (see 3.39) show that the phase time delay is given by: $\tau_\phi = \frac{2mw}{\hbar\kappa}$.

This expression is known as the Larmor tunneling time.

Appendix: Source Code for Calculation of Sequential Tunneling Using MATLAB

```
%%%%%%%%% Sequential tunneling %%%%%%%%%%%
% In addition to this file there is a second file ''mat1.m''.
% In order to run, both files are to be in the directory in
% which you invoke Matlab.
```

```
% This program plots the transmission coefficients versus energy for a sequence of
% potential barriers of the same height vo, the same thickness, d2-d1=5 nm and
the same periodicity of 15nm
% These parameters can be changed to suit the example under consideration
% Initial conditions and definitions
% Initial energy in eV
Xa1 = 0.01;
%step in eV
st = 0.00003;
% final energy
Xa2 = 0.3;
pe = Xa1:st:Xa2;
y=[];
mt1 =[]; mt2=mt1;mt3=mt1; mt4=mt1;
h = 1/1.05e-34;
mo =9.11e-31;
% Barrier in eV
vo = 0.25;
vo = vo*1.6e-19
% Particle effective mass
m1 = 0.066;
m2 = 0.092;
m3 = m1;
m1 = m1*mo;
m3 = m3*mo;
m2 = m2*mo;

for k=1:length(pe),
E = pe(k)*1.6e-19;
k1 = h*sqrt(2*m1*E);
k2 = h*sqrt(2*m2*(E-vo));
k3 = k1;
re1 = (k2*m1)/(k1*m2);
re2 = (k3*m2)/(k2*m3);
d1=0;
d2=5e-9;
%initial calculation to set mt3
mt1 =mat1(vo,re1,k1,k2,d1);
mt2 =mat1(vo,re2,k2,k3,d2);
% cascade the barriers
mt1=mt1*mt2;
mt3=mt1;

% next barriers: nb = number of barriers
for nb=1:49;
d1=d1+15e-9;
d2=d2+15e-9;
mt1 =mat1(vo,re1,k1,k2,d1);
mt2 =mat1(vo,re2,k2,k3,d2);

% cascade the barriers
mt1=mt1*mt2;
mt3=mt3*mt1;
end;
y(k) = (abs(1/mt3(1,1)))^2;
end;
axis([0 0.3 log10(1e-8) 0]);
plot(pe,log10(y));

% This subroutine computes the transmission matrix from one medium to another
% It is needed for the sequential tunneling program

function[themt1]=mat1(Vo,R,k1,k2,x)
```

```
A1 = k1 + k2;
A2 = k2 - k1;

Row11 = (1 + R) * exp(j * A2 * x);
Row12 = (1 - R) * exp(-j * A1 * x);
Row21 = (1 - R) * exp(j * A1 * x);
Row22 = (1 + R) * exp(-j * A2 * x);

themt1 = 0.5 * [ Row11,Row12; Row21,Row22];
```

References

G. Bastard, Band structure in the envelope-function approximation. Phys. Rev. B **24**, 5693–5697 (1981). https://www.iop.vast.ac.vn/nhquang/MassMismatch/PRB81_24_5693Superlattice%20band%20structure%20in%20the%20envelope-function%20approximation.pdf

D.J. BenDaniel, C.B. Duke, Space charge cffects on electron tunneling, Phys. Rev. **152**, 683–692 (1966).[1] http://www.iop.vast.ac.vn/~nhquang/MassMismatch/PR66_152_683Space-Charge%20Effects%20on%20Electron%20Tunneling.pdf

L.L. Chang, L. Esaki, R. Tsu, Resonant tunneling in semiconductor double barriers, Appl. Phys. Lett. **24**, 593–595 (1974). https://www.expresssearch.com/sample-results/validity/npl/Chang%201974.pdf

C.B. Duke, Tunneling in solids, in *Solid State Physics, Supplement 10*, ed. by F. Seitz, D. Turnbull, H. Ehrenreich (Academic Press, New York, 1969). ISBN 0-12-607770-3

E.R. Fossum, Application of photon statistics to the quanta image sensor, in Proceedings of the International Image Sensor Workshop, Snowbird, Utah, June 2013. http://citeseerx.ist.psu.edu/viewdoc/download?doi=10.1.1.308.1840&rep=rep1&type=pdf

P. Harrison, *Quantum Wells, Wires and Dots*, 3rd edn. (Wiley Interscience, Chichester, 2006) ISBN 978-0470010808

E.H. Hauge, J.A. Støvneng, Tunneling times: a critical review. Rev. Mod. Phys. **61**, 917–936 (1989).[2] https://journals.aps.org/rmp/abstract/10.1103/RevModPhys.61.917

E.O. Kane, Zener tunneling in semiconductors, J. Phys. Chem. Solids **12**, 181–188 (1959).[3] https://www.sciencedirect.com/science/article/abs/pii/0022369760900354

[1]Note: Although the BenDaniel-Duke boundary conditions were developed about a half-century ago, this continues to be an area of current study. For example, the reader is also referred to G.T. Einvoll and L.J. Sham, Phys Rev B **49** 10533 (1994) for an excellent tutorial discussion on this topic. https://www.researchgate.net/profile/Gaute_Einevoll/publication/13279349Boundary_conditions_for_envelope_functions_at_interfaces_between_dissimilar_materials/links/0c960523757631db10000000.pdf.

[2]Note: This review is one of the best available treatments of the tunneling time problem in quantum mechanics.

[3]Note: The most popular model to calculate the generation is Kane's Model derived from the WKB method. Kane's model is derived for a direct gap semiconductor in a uniform electric field and is given by: $T(E) = A\frac{E^D}{\sqrt{E_g}}\exp\left(-\frac{BE_g^{1.5}}{E}\right)$ where E is the electric field and E_g the bandgap, while A and B are parameters depending on the effective mass of valence and conduction bands and D takes a default value of 2, but for the sake of generality it is left unspecified as an adjustable parameter.

S. L. Konsek, T. P. Pearsall, Dynamics of electron tunneling in semiconductor nanostructures, Phys. Rev. B **67**, 045306, 1–7 (2003). https://www.researchgate.net/profile/Thomas_Pearsall/publication/235523350_Dynamics_of_electron_tunneling_in_semiconductor_nanostructures/links/5574153508ae7521586a821f/Dynamics-of-electron-tunneling-in-semiconductor-nanostructures.pdf

M.A. Morrison, T.L. Estle, N.F. Lane, *Quantum States of Atoms Molecules and Solids* (Prentice-Hall, Englewood Cliffs, 1976). ISBN 0-471-87474-4

Resonant tunneling Diode, Wikipedia. http://en.wikipedia.org/wiki/Resonant-tunneling_diode

R. Tsu, L. Esaki, Tunneling in a finite superlattice. Appl. Phys. Lett. **22**, 562–564 (1973). https://aip.scitation.org/doi/abs/10.1063/1.1654509

M. Wartak, *Computational Photonics: an Introduction with MATLAB* (Cambridge University Press, Cambridge, 2013). ISBN -13 978-1107005525. This text treats topics related to fiber optic transmission, principally optical wave propagation, but includes analysis of lasers and photodiodes

Chapter 4
Electronic Energy Levels in Crystalline Semiconductors

Abstract Electronic bandstructure is the energy-momentum relationship for an electron. The behavior of a totally-free electron in a crystal lattice shows the constraints that symmetry alone places on the energy-momentum relationship. Electrons in semiconductors are not free. Their wavefunctions take the form of Bloch functions containing a part that has the periodicity of the crystal structure and a free-electron term. The electronic bandstructure of many semiconductors of technological importance can be calculated using the pseudopotential method. The power of the pseudopotential method resides in its ability to substitute a fictitious potential that accounts for the behavior of the electrons in the weakly-bound valence band states. A pseudopotential can be calculated from first principles if the ground-state configuration of the material can be determined. In a spirit similar to that of the pseudopotential method, Pierre Hohenberg and Walter Kohn proved that the ground state configuration of an ensemble of atoms, such as a crystal or a molecule, is a functional of the correct electron charge density: that is, a function of only three spatial dimensions. Density functional theory has enabled the determination of the ground-state configuration not only of semiconductors whose structure is relatively simple, but more importantly of complex molecules and proteins with relevance to biochemistry.

4.1 Introduction

The electronic and photonic properties of semiconductors are a direct function of their electronic energy levels. The wavefunction of the electron in a crystalline material can be represented by the product of a free-electron term and a term that has the periodicity of the crystal lattice. The solution to Schrödinger's equation using this wavefunction gives the energy levels, and the energy-momentum relationship of electronic states; that is, the electronic bandstructure, to an excellent level of accuracy.

In this chapter we investigate the electronic bandstructure of crystalline semiconductors. The behavior of a totally free electron in a crystal lattice shows the

© Springer Nature Switzerland AG 2020

T. P. Pearsall, *Quantum Photonics*, Graduate Texts in Physics,
https://doi.org/10.1007/978-3-030-47325-9_4

constraints that symmetry alone places on the energy momentum relationship. Electrons in semiconductors are not free. Their wavefunctions take the form of Bloch functions containing a part that has the periodicity of the crystal structure and a free-electron term. A realistic and accurate energy-momentum relationship can be calculated for an electron by taking account of the periodic potential of the particular semiconductor under study. The electronic bandstructure of many semiconductors of technological importance can be calculated using the pseudopotential method. The power of the pseudopotential method resides in its ability to substitute a fictitious potential that accounts for the behavior of the electrons in the weakly-bound valence band states.

A pseudopotential can be calculated from first principles if the ground-state configuration of the material can be determined. In a spirit similar to that of the pseudopotential method, Pierre Hohenberg and Walter Kohn proved that the ground state configuration of an ensemble of atoms, such as a crystal or a molecule, is a functional of the correct electron charge density: that is, a function of three spatial dimensions. Density functional theory has enabled the determination of the ground-state configuration not only of semiconductors whose structure is relatively simple, but more importantly of complex molecules and proteins with relevance to biochemistry.

4.2 Periodicity

A crystal is a periodic arrangement of atoms called a lattice. Each point in the lattice can be represented by a vector: $\mathbf{R}_n = n_1\mathbf{a}_1 + n_2\mathbf{a}_2 + n_3\mathbf{a}_3$, where the n_i are integers. The $\mathbf{a}_i$ are the set of non co-planar vectors that form a unit cell of the lattice. The unit cell contains all the symmetry elements of the crystal structure. The choice of $\mathbf{a}_i$ is not unique. The sole requirement is that starting from a lattice point any other point in the lattice can be reached by a linear combination of these vectors. In practice, one chooses the set of the shortest vectors that meet this requirement. In Fig. 4.1 we show an example to illustrate this concept for a simple two-dimensional face-centered lattice.

The Wigner-Seitz cell is a useful concept to describe the area of "unique space" in the lattice. This cell is bounded on all sides by the perpendicular bisectors of all the lines that connect the nearest neighbors of a single lattice point. This procedure is illustrated in Fig. 4.2 for 2-dimensional face-centred square lattice.

The most commonly-encountered crystal lattices in 3-d solid-state materials are face-centered cubic (fcc), hexagonal (hex) and body-centered cubic (bcc). Silicon, diamond, germanium, and many III-V semiconductors, such as GaAs, crystallize in the fcc lattice. Some III-V materials, such as GaN, ZnS and ZnSe crystallize principally in the hex lattice, but can also be made to crystallize in the fcc structure. Illustrations of these structures are shown in Fig. 4.3. Each crystal structure is accompanied by its Wigner-Seitz cell.

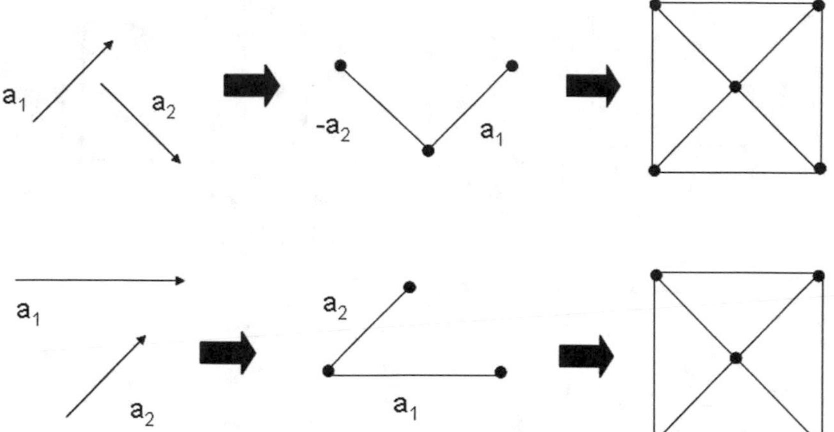

Fig. 4.1 Some lattice translation vectors for a 2-dimensional face-centred lattice. Note that the choice of vectors is not unique

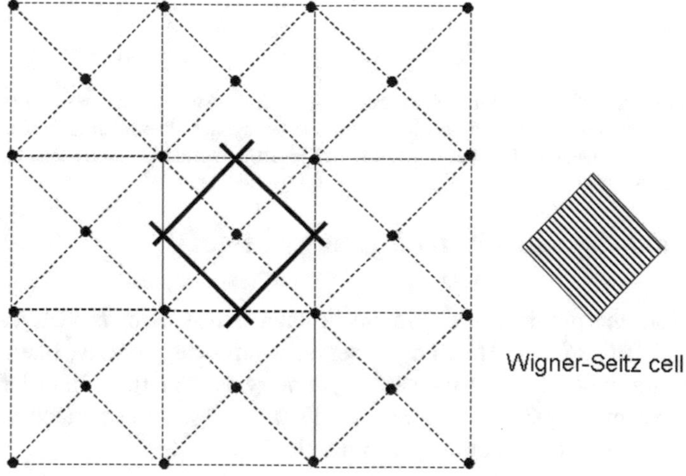

Fig. 4.2 Construction of the Wigner-Seitz cell for a 2-dimensional face-centred lattice. The cell is bounded by the perpendiculars to the lines joining nearest-neighbor lattice points

The Wigner-Seitz cell in three dimensions is constructed in an analogous manner. It is the volume bounded by the perpendicular planes that bisect the lines that connect nearest neighbors in the crystal lattice.

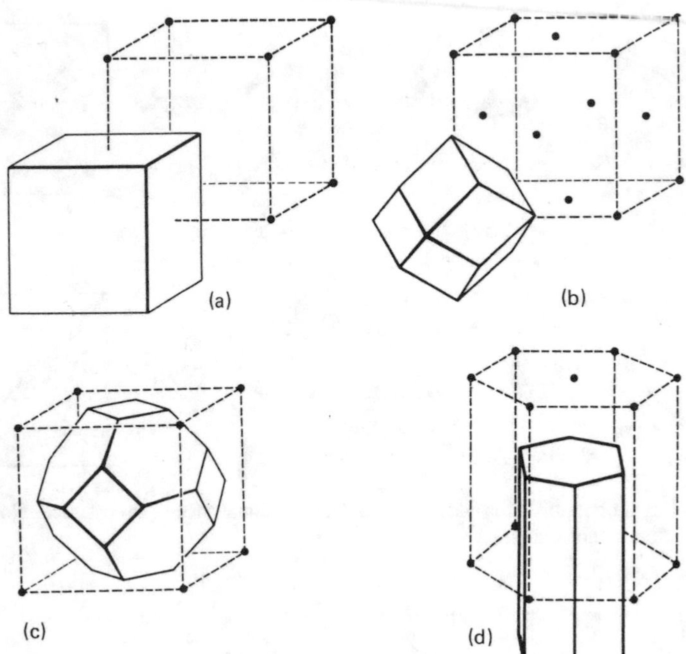

Fig. 4.3 Some examples of Wigner-Seitz cells for 3-dimensional crystal structures. **a** simple cubic, **b** face-centred cubic, **c** body-centred cubic, **d** hexagonal. The boundary of each cell is determined by the planes that bisect the lines that join the atomic positions that make up the crystal lattice structure

4.3 Periodicity and Electron Energy States

By definition, the physical environment at each lattice point is identical to that of any other lattice point. Thus any observable physical property, like energy or charge density must also have the same symmetry as the lattice. In order to make this statement more quantitative, we introduce a translation operator $\mathcal{T}_{R_i}$ which translates a function by the primitive translation vector $\mathbf{R}_i$.

$$\mathcal{T}_{R_i} f(\mathbf{r}) = f(\mathbf{r} + \mathbf{R}_i) \tag{4.1}$$

Starting with the Schrödinger equation:

$$\begin{aligned} \mathcal{H}\Psi_n &= E_n \Psi_n, \\ \mathcal{T}_{R_i}(\mathcal{H}\Psi_n) &= \mathcal{T}_{R_i}(E_n \Psi_n) = E_n \mathcal{T}_{R_i}(\Psi_n) \end{aligned} \tag{4.2}$$

That is, the energy eigenvalues E_n of the Schrödinger equation are the same (invariant) when the wavefunction is translated by the periodicity of the lattice. The charge density is given by the squared amplitude of the wavefunction. This implies that the wavefunctions of the initial and the translated state are the same within a phase factor of unity magnitude.

$$\text{i.e } \mathcal{T}_{R_i}(\Psi_n) = \Psi(\mathbf{r} + \mathbf{R}_i) = e^{i\varphi}\Psi_n = e^{i\mathbf{k}\cdot\mathbf{R}_i}\Psi_n \tag{4.3}$$

Equation 4.3 is also known as Bloch's theorem.

The electronic properties of semiconductors are determined by the wavelike properties of delocalised electrons and holes. We can separate the wavefunction into two parts one with the periodic behavior of the crystal lattice, and the other with wavelike behavior.

$$\Psi_n(\mathbf{k}, \mathbf{r}) = e^{i\mathbf{k}\cdot\mathbf{r}} u_n(\mathbf{k}, \mathbf{r}) \tag{4.4}$$

Substituting this form into (4.3) gives

$$\begin{aligned}\Psi(\mathbf{k}, \mathbf{r} + \mathbf{R}_i) &= \mathbf{e}^{i\mathbf{k}\cdot(\mathbf{r}+\mathbf{R}_i)} u_n(\mathbf{k}, \mathbf{r} + \mathbf{R}_i) = \mathbf{e}^{i\mathbf{k}\cdot\mathbf{R}_i} \mathbf{e}^{i\mathbf{k}\cdot\mathbf{r}} u_n(\mathbf{k}, \mathbf{r} + \mathbf{R}_i) \\ &= \mathbf{e}^{i\mathbf{k}\cdot\mathbf{R}_i} \mathbf{e}^{i\mathbf{k}\cdot\mathbf{r}} u_n(\mathbf{k}, \mathbf{r}) \\ &= \mathbf{e}^{i\mathbf{k}\cdot\mathbf{R}_i} \Psi(\mathbf{k}, \mathbf{r})\end{aligned} \tag{4.5}$$

where we have used the periodic property of $u_n(\mathbf{k}, \mathbf{r})$.

Wavefunctions of the form of (4.4) are called Bloch functions.

Electronic states in crystals are defined by energy and momentum ($= \hbar\mathbf{k}$). We can now show that there are a set of basis vectors in $\mathbf{k}$-space; $\mathbf{G}_m$ which are also periodic:

$$\mathcal{T}_{R_i}\Psi(\mathbf{k} + \mathbf{G}_m, \mathbf{r}) = \mathbf{e}^{i(\mathbf{k}+\mathbf{G}_m)\cdot\mathbf{R}_i}\Psi(\mathbf{k} + \mathbf{G}_m, \mathbf{r}) = \mathbf{e}^{i\mathbf{G}_m\cdot\mathbf{R}_i}\mathbf{e}^{i\mathbf{k}\cdot\mathbf{R}_i}\Psi(\mathbf{k} + \mathbf{G}_m, \mathbf{r}) \tag{4.6}$$

Bloch's theorem holds:

$$\mathcal{T}_{R_i}\Psi(\mathbf{k} + \mathbf{G}_m, \mathbf{r}) = \mathbf{e}^{i\mathbf{k}\cdot\mathbf{R}_i}\Psi(\mathbf{k} + \mathbf{G}_m, \mathbf{r}), \quad \text{provided that } \mathbf{G}_m \cdot \mathbf{R}_i = 2n\pi$$

We can construct the vectors of $\mathbf{k}$-space using the condition of (4.6):

$$\mathbf{G}_m = m_1\mathbf{b}_1 + m_2\mathbf{b}_2 + m_3\mathbf{b}_3$$

where $\mathbf{b}_1$, $\mathbf{b}_2$ and $\mathbf{b}_3$ are by analogy the primitive vectors of the lattice in $\mathbf{k}$-space.

The periodicity condition requires that the vector $\mathbf{b}_1$ be perpendicular to both $\mathbf{a}_2$ and $\mathbf{a}_3$, and similarly for $\mathbf{b}_2$ and $\mathbf{b}_3$.

$$\mathbf{b}_1 = 2\pi\frac{(\mathbf{a}_2 \times \mathbf{a}_3)}{\mathbf{a}_1 \cdot (\mathbf{a}_2 \times \mathbf{a}_3)} = \frac{2\pi}{V}(\mathbf{a}_2 \times \mathbf{a}_3) \tag{4.7a}$$

$$\mathbf{b}_2 = \frac{2\pi}{V}(\mathbf{a}_3 \times \mathbf{a}_1) \tag{4.7b}$$

$$\mathbf{b}_3 = \frac{2\pi}{V}(\mathbf{a}_1 \times \mathbf{a}_2) \tag{4.7c}$$

where V is the volume of the Wigner-Seitz unit cell.

Example: The reciprocal lattice of an fcc crystal has a body-centered structure

As an important example, we will consider the reciprocal lattice for Si, Ge and other semiconductor crystals. Si, Ge and most III-V semiconductors crystallize in the face-centered cubic (fcc) lattice structure. A set of real-space primitive lattice vectors could be:

$$\begin{aligned} \mathbf{a}_1 &= \frac{a_0}{2}(1 \;\; 1 \;\; 0) \\ \mathbf{a}_2 &= \frac{a_0}{2}(0 \;\; 1 \;\; 1) \\ \mathbf{a}_3 &= \frac{a_0}{2}(1 \;\; 0 \;\; 1) \end{aligned} \tag{4.8}$$

where a_0 is the lattice parameter of the fcc structure.

To find a corresponding primitive vector of the reciprocal lattice, we apply (4.7a)

$$\mathbf{b}_1 = \frac{2\pi}{\left(\frac{a_0^3}{4}\right)} \det \begin{vmatrix} i & j & k \\ 0 & \frac{a_0}{2} & \frac{a_0}{2} \\ \frac{a_0}{2} & 0 & \frac{a_0}{2} \end{vmatrix} = \frac{2\pi\left(\frac{a_0^2}{4}\right)}{\frac{a_0^3}{4}}(1 \;\; 1 \;\; -1) = \frac{2\pi}{a_0}(1 \;\; 1 \;\; -1) \tag{4.9}$$

Similar operations are used to find $\mathbf{b}_2$ and $\mathbf{b}_3$. The result shows that the reciprocal lattice has primitive translation vectors that correspond to a body-centred cubic structure.

In units of $\frac{2\pi}{a_0}$, some of the points of the body-centered reciprocal lattice are in order of length:

$$\begin{aligned} \mathbf{G}_0 &= (0 \;\; 0 \;\; 0) \\ \mathbf{G}_3 &= (1 \;\; 1 \;\; 1) \\ \mathbf{G}_4 &= (2 \;\; 0 \;\; 0) \\ \mathbf{G}_8 &= (2 \;\; 2 \;\; 0) \\ \mathbf{G}_{11} &= (3 \;\; 1 \;\; 1) \\ \mathbf{G}_{12} &= (2 \;\; 2 \;\; 2) \end{aligned} \tag{4.10}$$

where the subscript on **G** refers to the square of the vector length. In general there are multiple vectors for a given length. For example, there are 12 distinct **G**-vectors that have length 8.

$$\begin{aligned} &\mathbf{G}_0 : 1 \\ &\mathbf{G}_3 : 8 \\ &\mathbf{G}_4 : 6 \\ &\mathbf{G}_8 : 12 \\ &\mathbf{G}_{11} : 24 \\ &\mathbf{G}_{12} : 8 \end{aligned} \tag{4.11}$$

Note that some lattice points such as (1 0 0) or (1 1 0), which exist in the real space lattice, are not present in the reciprocal lattice.

4.4 Brillouin Zones

The lattice in reciprocal or **k**-space (**k** has units of $\frac{\pi}{a}$) is periodic, and so we can identify a unit cell, called the Brillouin zone that is analogous to the Wigner-Seitz cell in real space. Léon Brillouin was a classmate of Louis de Broglie and authored many key texts in solid-state physics, including the landmark study of wave propagation in periodic structures (Brillouin 1946). The importance of the Brillouin zone concept can be appreciated by considering the relationship between momentum ($p = \hbar\mathbf{k}$) and energy ($E = \frac{\hbar^2 k^2}{2m}$) for a free electron in a periodic 1-dimensional environment, as diagrammed in Fig. 4.4. Starting from the point $\mathbf{k} = 0$, we have plotted energy as a function of momentum in Fig. 4.4a. Because the structure is periodic in k-space, all the lattice points having the same $\mathbf{G}_m$ are equivalent, and we can plot out the same energy-momentum relationship starting from any one of them. This is illustrated in Fig. 4.4b for the case of a simple one-dimensional lattice. Finally, it is clear that all the unique information about energy and momentum is contained in a region indicated by the boundaries shown in Fig. 4.4c. This region is the Brillouin zone. One could identify many such zones in the structure shown in Fig. 4.4. Because of periodicity, each one contains all the essential information (Jones 1975).

One could say that the energy-momentum relationship in a single Brillouin zone is the result of folding the diagram of Fig. 4.4b so that the adjacent portions overlap each other with the periodicity of the reciprocal lattice. However, such zone-folding is more than a geometric artifice. It is clear from Fig. 4.4c that periodicity imposes that for any given value of electron momentum $\hbar\mathbf{k}_i$, there is a multiple of possible energy values:

$$\begin{aligned} E_0(\mathbf{k}_i) &= \frac{\hbar^2 \mathbf{k}_i^2}{2m} \\ E_1(\mathbf{k}_i) &= \frac{\hbar^2 (\mathbf{k}_i \pm \mathbf{G}_1)^2}{2m}, \\ E_2(\mathbf{k}_i) &= \frac{\hbar^2 (\mathbf{k}_i \pm \mathbf{G}_2)^2}{2m}, \text{ etc.} \end{aligned} \tag{4.12}$$

These multiple values are indicated in Fig. 4.4d. The calculation of the electron energy levels in real crystals follows this route, as will be demonstrated in Sect. 4.5.

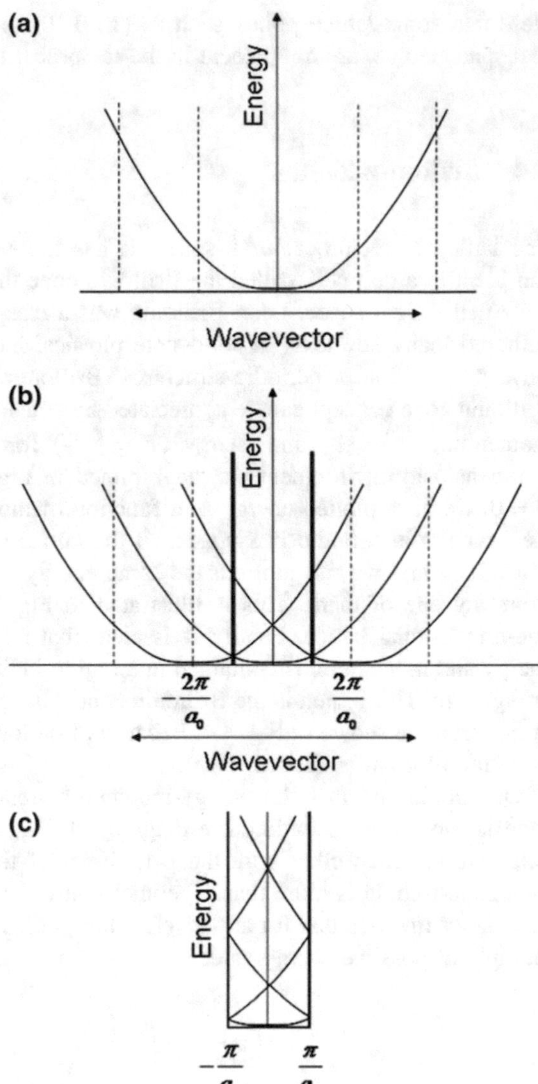

Fig. 4.4 A simple 1-dimensional periodic structure in **k**-space. **a** The energy momentum relationship for a particle is parabolic. The *vertical lines* indicate wavevectors that are displaced from the origin by a reciprocal lattice vector $\mathbf{G}_n = \frac{2n\pi}{a}$. **b** In the environment of a crystal lattice, this parabolic relationship repeats with the period of the reciprocal lattice: $\frac{2\pi}{a_0}$. **c** Note that the Brillouin Zone can be identified readily as a region of width $\frac{2\pi}{a_0}$ delimited by the bold-face vertical boundaries in Fig. 4.4b

Example: The role of symmetry in 3 dimensions, the free electron solution using parabolic equations

The fcc crystal structure is of primary importance for solid-state photonics because it is the crystal structure for a large number of photonic semiconductors such as Si, Ge, GaAs, InAs CdTe, ZnS to name a few. In (4.7) we have shown that the reciprocal lattice of the fcc structure is a body-centred cubic lattice. The Brillouin zone for the bcc structure is identical in shape to the Wigner-Seitz cell for a body-centered cubic, and it has already been shown in Fig. 4.3d. All that is needed is to specify the boundaries of the Brillouin zone in terms of the lattice parameter a_0 of the real-space fcc structure. Recalling that the boundaries of the Brillouin zone are determined by the point of bisection between adjacent points of the reciprocal lattice, we can determine the points along high-symmetry axes using simple geometry.

The centre of the Brillouin zone at $\frac{2\pi}{a_0}(0 \quad 0 \quad 0)$ is called the Γ-point. In the $\langle 100\rangle$ direction, the nearest neighbor in **k**-space occurs at $\frac{2\pi}{a_0}(2 \quad 0 \quad 0)$. This high-symmetry direction is called the Δ-axis. Therefore the boundary of the Brillouin zone occurs at $\frac{2\pi}{a_0}(1 \quad 0 \quad 0)$, which is called the X-point. The distance in **k**-space from Γ to X is $\frac{2\pi}{a_0}$. Along the $\langle 111\rangle$ direction, called the Λ-axis, the boundary of the Brillouin zone occurs at $\frac{2\pi}{a_0}\left(\frac{1}{2} \quad \frac{1}{2} \quad \frac{1}{2}\right)$, called the L-point. The distance in **k**-space from Γ to L is $\frac{\pi}{a_0}\sqrt{3}$. The Brillouin zone, which has the appearance of a truncated octahedron, is shown in Fig. 4.5, where we have identified the high-symmetry axes and boundary points (Jones 1975).

With this information we can calculate the energy-momentum relationship for a free electron, subject only to the condition of motion in a lattice with the periodicity of an fcc crystal structure, but having a uniform potential throughout.

The energy-momentum relationship has been given in (4.12): $E_n(\mathbf{k}_i) = \frac{\hbar^2(\mathbf{k}_i \pm \mathbf{G}_n)^2}{2m}$.

For $\mathbf{G}_0 = \frac{\pi}{a_0}(0 \quad 0 \quad 0)$, and **k** along the Δ-axis,

$$E_0(\mathbf{k}_\Delta) = \frac{\hbar^2(\mathbf{k}_\Delta)^2}{2m}, \text{ from } \mathbf{k}_\Delta = \frac{2\pi}{a_0}(0 \quad 0 \quad 0), \text{ to } \mathbf{k}_\Delta = \frac{2\pi}{a_0}(1 \quad 0 \quad 0)$$

$$\text{When } \mathbf{k}_\Delta = \frac{\pi}{a_0}(0 \quad 0 \quad 0),\ E_0(\mathbf{k}_\Delta) = 0$$

$$\text{For } \mathbf{k}_\Delta = \frac{2\pi}{a_0}(1 \quad 0 \quad 0), \text{ at the } X\text{-point } E_0(\mathbf{k}_\Delta) = \frac{2\hbar^2\pi^2}{ma_0^2}$$

For $\mathbf{G}_3 = \frac{2\pi}{a_0}(\pm 1 \quad \pm 1 \quad \pm 1)$, and **k** along the Δ-axis, there are 8 calculations, one corresponding to each $\mathbf{G}_3$-vector. As before:

$$E_3(\mathbf{k}_\Delta) = \frac{\hbar^2(\mathbf{k}_\Delta)^2}{2m}, \text{ from } \mathbf{k}_\Delta = \frac{2\pi}{a_0}(0 \quad 0 \quad 0), \text{ to } \mathbf{k}_\Delta = \frac{2\pi}{a_0}(1 \quad 0 \quad 0)$$

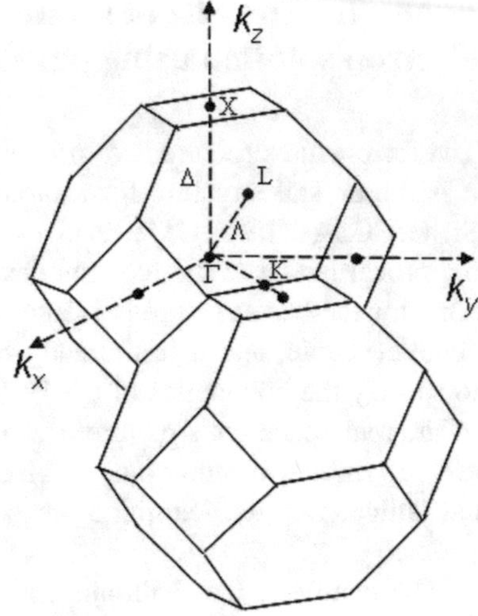

Fig. 4.5 The Brillouin zone for the fcc real-space lattice is an octahedron, truncated along the ⟨100⟩ directions. The Γ-point is located at the center of the zone and has the highest symmetry (Madelung 1981)

For $\mathbf{k}_\Delta = \frac{\pi}{a_0}(0 \quad 0 \quad 0)$, we are at the center of the Brillouin zone, and

$$E_3(k_\Delta) = \frac{\hbar^2}{2m}\left(\frac{4\pi^2}{a_0^2}\right)\left((0 \pm 1)^2 + (0 \pm 1)^2 + (0 \pm 1)^2\right) = \frac{6\hbar^2\pi^2}{ma_0^2}, \text{ (8-fold degenerate)}$$

$$\mathbf{k}_X = \frac{2\pi}{a_0}(1 \quad 0 \quad 0):$$

$$E_3(\mathbf{k}_X) = \frac{\hbar^2}{2m}\left(\frac{4\pi^2}{a_0^2}\right)\left((1+1)^2 + (0 \pm 1)^2 + (0 \pm 1)^2\right) = \frac{12\hbar^2\pi^2}{ma_0^2}, \text{ (4-fold degenerate)}$$

$$\text{for } \mathbf{G}_3 = \frac{2\pi}{a_0}(+1 \quad \pm 1 \quad \pm 1)$$

and,

$$E_3(\mathbf{k}_X) = \frac{4\hbar^2\pi^2}{ma_0^2}, \text{ (4-fold degenerate) for } \mathbf{G}_3 = \frac{2\pi}{a_0}(-1 \quad \pm 1 \quad \pm 1)$$

The cubic symmetry of the Brillouin zone splits the band into two parts. At the zone center the energy level is 8-fold degenerate. At the X-point, there are two bands, each of which is 4-fold degenerate.

Next we would use the set of $\mathbf{G}_4$ vectors.

$$\mathbf{G}_4 = \frac{2\pi}{a_0}(\pm 2 \quad 0 \quad 0), \quad \mathbf{G}_4 = \frac{2\pi}{a_0}(0 \quad \pm 2 \quad 0) \text{ and } \mathbf{G}_4 = \frac{2\pi}{a_0}(0 \quad 0 \quad \pm 2).$$

and calculate the bandstructure along the X-direction as shown above.

We can repeat this procedure along the Λ-direction.

For $\mathbf{G}_0 = \frac{2\pi}{a_0}(0 \quad 0 \quad 0)$, and $\mathbf{k}$ along the Λ-axis,

$$E_0(\mathbf{k}_\Lambda) = \frac{\hbar^2(\mathbf{k}_\Lambda)^2}{2m}, \text{ from } \mathbf{k}_\Lambda = \frac{\pi}{a_0}(0 \quad 0 \quad 0), \text{ to } \mathbf{k}_\Lambda = \frac{2\pi}{a_0}\left(\tfrac{1}{2} \quad \tfrac{1}{2} \quad \tfrac{1}{2}\right)$$

$$\text{For } \mathbf{k}_\Lambda = \frac{\pi}{a_0}(0 \quad 0 \quad 0),\ E_0(\mathbf{k}_\Lambda) = 0$$

$$\text{For } \mathbf{k}_L = \frac{2\pi}{a_0}\left(\tfrac{1}{2} \quad \tfrac{1}{2} \quad \tfrac{1}{2}\right),\ E_0(\mathbf{k}_L) = \frac{3\hbar^2\pi^2}{2ma_0^2}$$

For $\mathbf{G}_3 = \frac{2\pi}{a_0}(\pm 1 \quad \pm 1 \quad \pm 1)$, and $\mathbf{k}$ along the Λ-axis, as before, there are 8 calculations, one corresponding to each $\mathbf{G}_3$-vector. However, the symmetry along the Λ-axis is different from that along the Δ-axis, and the 8 energy levels are split into 4 bands.

$$E_3(\mathbf{k}_\Lambda) = \frac{\hbar^2(k_\Lambda)^2}{2m}, \text{ from } \mathbf{k}_\Lambda = \frac{\pi}{a_0}(0 \quad 0 \quad 0), \text{ to } \mathbf{k}_L\frac{2\pi}{a_0}\left(\tfrac{1}{2} \quad \tfrac{1}{2} \quad \tfrac{1}{2}\right)$$

$$\text{For } \mathbf{k}_\Lambda = \frac{2\pi}{a_0}(0 \quad 0 \quad 0),\ E_3(\mathbf{k}_\Lambda) = \frac{\hbar^2}{2m}\left(\frac{4\pi^2}{a_0^2}\right)(1 \quad 1 \quad 1)^2 = \frac{6\hbar^2\pi^2}{ma_0^2}$$

$$E_3(\mathbf{k}_L) = \frac{2\hbar^2\pi^2}{ma_0^2}\left(\left(\frac{3}{2}\right)^2 + \left(\frac{3}{2}\right)^2 + \left(\frac{3}{2}\right)^2\right) = \frac{27\hbar^2\pi^2}{2ma_0^2}, \text{ non degenerate, for}$$

$$\mathbf{G}_3 = \frac{2\pi}{a_0}(+1 \quad +1 \quad +1)$$

and,

$$E_3(\mathbf{k}_L) = \frac{3\hbar^2\pi^2}{2ma_0^2}, \text{ non degenerate, for } \mathbf{G}_3 = \frac{2\pi}{a_0}(-1 \quad -1 \quad -1)$$

$$E_3(\mathbf{k}_L) = \frac{19\hbar^2\pi^2}{2ma_0^2}, \text{ 3-fold degenerate for } \mathbf{G}_3 = \frac{2\pi}{a_0}(-1 \quad +1 \quad +1) \text{ and permutations}$$

and

$$E_3(\mathbf{k}_L) = \frac{11\hbar^2\pi^2}{2ma_0^2}, \text{ 3-fold degenerate, for } \mathbf{G}_3 = \frac{2\pi}{a_0}(-1 \quad -1 \quad +1) \text{ and permutations.}$$

There are 8 distinct $\mathbf{G}_3$ vectors at the Γ-point, all having the same energy, resulting in 8-fold degeneracy. Moving away from the Γ-point in any direction lowers the symmetry, and lifts the degeneracy to some degree. For example: along the Λ-direction toward the L-point, the lower symmetry environment results in 4 energy-

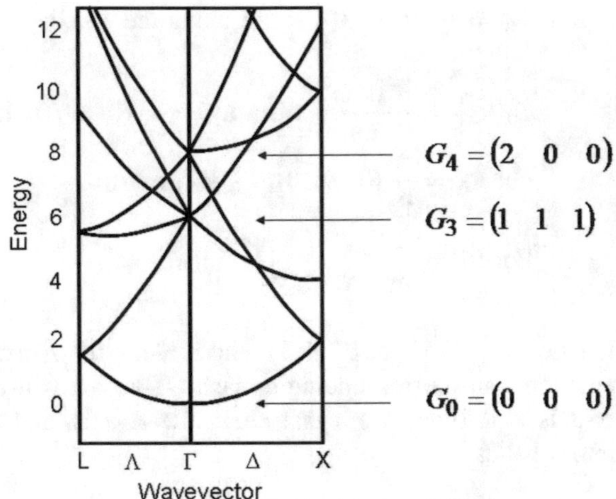

Fig. 4.6 The energy-momentum relationship for a free electron in a 3-dimensional fcc lattice. This diagram applies for any material and shows only the constraints imposed by symmetry

momentum bands. The upper and lower energy bands are non-degenerate, while the two intermediate energy bands are each 3-fold degenerate. On the other hand, along the Δ-direction toward X, more symmetries are preserved, and the 8-fold degenerate state at Γ is split into 2 four-fold degenerate bands.

A diagram of these energy free-electron energy bands for the fcc crystal structure is shown in Fig. 4.6 (Jones 1975).

The bandstructure diagram of Fig. 4.6 shows the behavior of the energy-momentum ($\mathbf{p} = \hbar\mathbf{k}$) relationship for a free electron in a three-dimensional fcc symmetry environment. It is not material-specific. The crystal lattice is "empty" because we have not yet integrated the effect of atomic potential. At the Γ-point at the center of the Brillouin zone, the electronic wavefunctions see the maximum symmetry of the fcc structure: 48 operations (rotation, reflection, inversion) that leave the structure invariant. As the wavevector increases away from the zone center in any direction, the symmetry is decreased, the degeneracy of the Γ-point states is removed and the energy states split up. In Fig. 4.6 these states are plotted for two high-symmetry directions. Some splitting of the states occurs, but the degeneracies are not completely removed. Incorporation of the atomic potential will modify the energy of the electron states. The atomic potential will further reduce symmetries, and in particular those at the zone center, splitting the state into different energy levels. For the semiconductor materials of interest, the splitting creates an energy gap between the eight $\mathbf{G}_3$ states at the Γ-point. This energy gap is the fundamental feature of the photonic properties of these materials.

4.5 Electronic Bandstructure of Specific Materials: Integrating the Crystal Potential

4.5.1 Introduction

Our objective is to determine the relationship between energy and momentum for stable states of electrons in a crystalline semiconductor such as silicon, gallium arsenide and the related photonic materials. These states are solutions of the Schrödinger equation for electrons moving in the periodic potential of the constituent atoms. However, we are not interested to know the solutions in this case, but rather *the relationship between the electron wave-vector and energy* that is required by the solutions. The photonic behavior of the material as a whole can be understood by knowing the energy of the stable states and the probability that these states are occupied by an electron (or hole).

In the case of a silicon atom, there are 14 electrons that surround the nucleus. The atomic configuration is: 1s 2s 2p, accounting for 10 core-state electrons, and 4 electrons in the 3s and 3p states. The core-state electrons are tightly bound to the nucleus. Their major role in our case is to shield the nuclear potential. When silicon atoms bond to form a crystal, the valence-band electrons form s-p hybrid orbitals. They are responsible for determining the tetrahedral symmetry of the crystal structure. Four of the orbitals are completely filled by electrons and constitute the valence band (or bonding state) of the solid. One orbital is empty and constitutes the conduction band (or anti-bonding state). A perfect silicon crystal at absolute zero temperature would appear to be an insulator with its valence states fully occupied and its conduction states completely unoccupied and separated from the valence states by an energy gap equal to the energy required to break a crystal bond.

4.5.2 Schrödinger's Equation in a Periodic Potential

In principle, one could solve the Schrödinger equation for all the electronic states in the crystal. This would prove to be an immense task, and a large number of the results would concern core electron states, whereas we are interested only in the properties of electrons in the valence-band and conduction band states. A more interesting and efficient approach is to describe these electrons with Bloch functions in the presence of a periodic potential having the symmetry of the crystal lattice.

Schrödinger's equation says that the stable states of an electron are those for which total energy is conserved:

$$-\frac{\hbar^2}{2m}\nabla^2\Psi(\mathbf{r}) + U(\mathbf{r})\Psi(\mathbf{r}) = E(\mathbf{r})\Psi(\mathbf{r}) \tag{4.13}$$

We can use the periodic behavior of the potential to advantage by using Fourier analysis to decompose the potential and Schrodinger's equation into Fourier components (Kittel 1986).

The periodic potential can be rewritten:

$$U(\mathbf{r}) = \sum_{\mathbf{G}} U_{\mathbf{G}} e^{i\mathbf{G}\cdot\mathbf{r}} \tag{4.14}$$

where the $\mathbf{G}$ are the vectors of the reciprocal lattice.

The wave-function states can be expressed,

$$\Psi(\mathbf{r}) = u_{\mathbf{k}}(\mathbf{r})e^{i\mathbf{k}\cdot\mathbf{r}} = \sum_{\mathbf{k}} C(\mathbf{k})e^{i\mathbf{k}\cdot\mathbf{r}} \tag{4.15}$$

The periodic nature of $u_{\mathbf{k}}(\mathbf{x})$ is contained in the Fourier components: $C(\mathbf{k})$. Substituting these results gives:

$$\sum_{\mathbf{k}} \frac{\hbar^2}{2m} k^2 C(\mathbf{k}) e^{i\mathbf{k}\cdot\mathbf{r}} + \sum_{\mathbf{G}} U_{\mathbf{G}} \sum_{\mathbf{k}} C(\mathbf{k}) e^{i(\mathbf{k}+\mathbf{G})\cdot\mathbf{r}} = E_{\mathbf{k}} \sum_{\mathbf{k}} C(\mathbf{k}) e^{i\mathbf{k}\cdot\mathbf{r}} \tag{4.16}$$

This equation is true for each Fourier component $C(\mathbf{k})$, which gives:

$$\begin{gathered} \left(\frac{\hbar^2}{2m}k^2 - E_{\mathbf{k}}\right) C(\mathbf{k})e^{i\mathbf{k}\cdot\mathbf{r}} + \sum_{\mathbf{G}} U_{\mathbf{G}} C(\mathbf{k}-\mathbf{G}) e^{i(\mathbf{k}-\mathbf{G}+\mathbf{G})\cdot\mathbf{r}} = 0 \\ \left(\frac{\hbar^2}{2m}k^2 - E_{\mathbf{k}}\right) C(\mathbf{k}) + \sum_{\mathbf{G}} U_{\mathbf{G}} C(\mathbf{k}-\mathbf{G}) = 0 \end{gathered} \tag{4.17}$$

Thus, by Fourier analysis we have transformed a single differential equation into a series of linear homogenous equations, one for each reciprocal lattice vector. There is a different series for each value of $\mathbf{k}$ in the Brillouin zone. This shows the tremendous advantage of working in reciprocal (Fourier) space.

The solution to these equations will give each of the coefficients $C(\mathbf{k})$. However, we are not interested here in knowing this solution. Organising the equations in matrix form, we can solve for the eigenvalues of the matrix. This procedure gives all the energies that are associated with each value of wave vector $\mathbf{k}$. In so doing we can determine the relationship between $\hbar\mathbf{k}$, the electron momentum and $E_{\mathbf{k}}$, the energy of the state with this momentum (Hamaguchi 2010).

$$\begin{bmatrix} \frac{\hbar^2 k_1^2}{2m} - E_1 & & \\ & 51\ x\ 51 & \\ & & \frac{\hbar^2 k_1^2}{2m} - E_{51} \end{bmatrix} \begin{pmatrix} C_1 \\ C_2 \\ \cdots \\ C_{50} \\ C_{51} \end{pmatrix} = \begin{pmatrix} 0 \\ 0 \\ \cdots \\ 0 \\ 0 \end{pmatrix} \tag{4.18}$$

4.5.3 The Pseudopotential Method

The calculation proceeds as follows. First, all the reciprocal lattice vectors to be used in the calculation are determined. In the example that follows, we will use all reciprocal lattice vectors from (0 0 0) through (3 1 1). There are 51 in all, which is determined by adding the number of reciprocal lattice vectors in (4.11) for $\mathbf{G}_0$ through $\mathbf{G}_{11}$. Next we write the 51 equations in matrix form. Then we will pick a value for **k**, the wavevector. The eigenvalues of the matrix can be calculated using efficient existing computational routines, determining the values of energy of the stable states of the system for that particular wavevector. There are 51 energies, but some values will be identical (degeneracies). Next, **k** is incremented and the calculation is repeated. This procedure, while it appears tedious, can be completed in a few seconds on a personal computer for **k** vectors lying along the major symmetry axes of the Brillouin zone. The values of $\mathbf{U}_\mathbf{G}$ are the Fourier components of the potential that is calculated to represent the environment that an outer shell electron would experience. It is called a pseudopotential because it replaces the actual atomic potential and the screening action of the core electrons around the nucleus. As a result, the energies determined by taking the determinant of coefficients apply to the states of the electrons in the conduction and valence bands.

The pseudopotential method can be used when the core electron states shield conduction and valence band electrons from the nuclear potential (Cohen and Heine 1970). This would be a good approximation for Si, Ge or GaAs, but not so obvious for diamond or graphene because the n = 2 shell is not complete (see Exercise 4.3).

The pseudopotential method was developed through the efforts of many condensed matter theorists. An excellent review of this work is given in Volume 24 of the Solid-State Physics series (Heine 1970), and in the monograph by M. L. Cohen and J. R. Chelikowsky (1989) cited in the references to this chapter. This is highly recommended reading in order to better understand the physics behind this important work.

For a bandstructure calculation of silicon using the first 5 groups of reciprocal lattice vectors $\mathbf{G}_0$ through $\mathbf{G}_{11}$, there are three non-zero Fourier components of the potential.

The pseudopotential form factors energies are referenced to U_0 which is set to zero for convenience. U_4 is zero by symmetry for all fcc crystals. Thus, there are only three parameters needed to determine the pseudopotential needed to perform the calculation. These parameters can be calculated from first principles, but are usually adjusted by using experimental data. The values in Table 4.1 give the bandstructure energies of silicon within a few percent over the entire Brillouin zone, using only the first five wave vectors of the lattice. Greater accuracy can be obtained by extending the calculation to include more reciprocal lattice vectors at the price of longer computing time.

The actual determinant has too many terms to write explicitly here. To give an example, we will write the matrix of coefficients using only the first two groups

Table 4.1 Pseudopotential form factors in eV for the first 5 reciprocal lattice vectors of silicon

	Length in units of $\frac{2\pi}{a_0}$	Number of equivalent wavevectors	U_G, Silicon pseudopotential component (eV)
$\mathbf{G}_0$	0	1	0
$\mathbf{G}_3$	3	8	−2.856
$\mathbf{G}_4$	4	6	0
$\mathbf{G}_8$	8	12	0.544
$\mathbf{G}_{11}$	11	24	1.088

of lattice vectors, $\mathbf{G}_0 : \frac{2\pi}{a_0}(0 \quad 0 \quad 0)$ and $\mathbf{G}_3 : \frac{2\pi}{a_0}(1 \quad 1 \quad 1)$. In this example we pick $\mathbf{k} = \frac{2\pi}{a_0}(0.1 \quad 0 \quad 0)$.

There are two Fourier components of potential. $U_{\mathbf{G_0}}$ and $U_{\mathbf{G_3}}$. We can choose $U_{\mathbf{G_0}} = 0$. In the case of silicon $U_{\mathbf{G_3}} = -2.856\,\text{eV}$, and the matrix size is 7×7. It is still too large to write explicitly on a single page. To illustrate, we write out the first few terms below:

$$\begin{pmatrix} \left(\frac{\hbar^2\left(\frac{2\pi}{a_0}\right)^2(0.1^2)}{2m} - E_k\right) & U_{\mathbf{G_3}} & U_{\mathbf{G_3}} & U_{\mathbf{G_3}}\ U_{\mathbf{G_3}} \cdots \\ U_{\mathbf{G_0}} & \left(\frac{\hbar^2\left(\frac{2\pi}{a_0}\right)^2(0.9^2+2)}{2m} - E_k\right) & U_{\mathbf{G_3}} & U_{\mathbf{G_3}}\ U_{\mathbf{G_3}} \cdots \\ U_{\mathbf{G_3}} & U_{\mathbf{G_0}} & \left(\frac{\hbar^2\left(\frac{2\pi}{a_0}\right)^2(1.1^2+2)}{2m} - E_k\right) & U_{\mathbf{G_3}}\ U_{\mathbf{G_3}} \cdots \\ \cdots & \cdots & \cdots & \cdots \ \cdots \ \cdots \end{pmatrix} \tag{4.19}$$

where $\mathbf{G}_0 = (0 \quad 0 \quad 0)$ in line 1, $\mathbf{G}_3 = (-1 \quad -1 \quad -1)$ in line 2, and $\mathbf{G}_3 = (1 \quad -1 \quad -1)$ in line 3.

The energies, E_k are the eigenvalues of the matrix. These can be determined from the EIG(M) command in MATLAB in a single step. In the appendix to this chapter, we give a MATLAB code for the solution of the bandstructure for fcc semiconductors, using the first 5 lattice vectors. This code has been used to produce the bandstructure diagrams in the following figures.

4.6 Energy Bandstructure of Silicon

To solve for the eigenvalues of the pseudopotential equations we need to know the lattice constant of silicon and the Fourier components of the pseudopotential.

The lattice constant of silicon at 25 °C is 5.43 Å.

The pseudopotential components are: (Cohen and Bergstresser 1966)

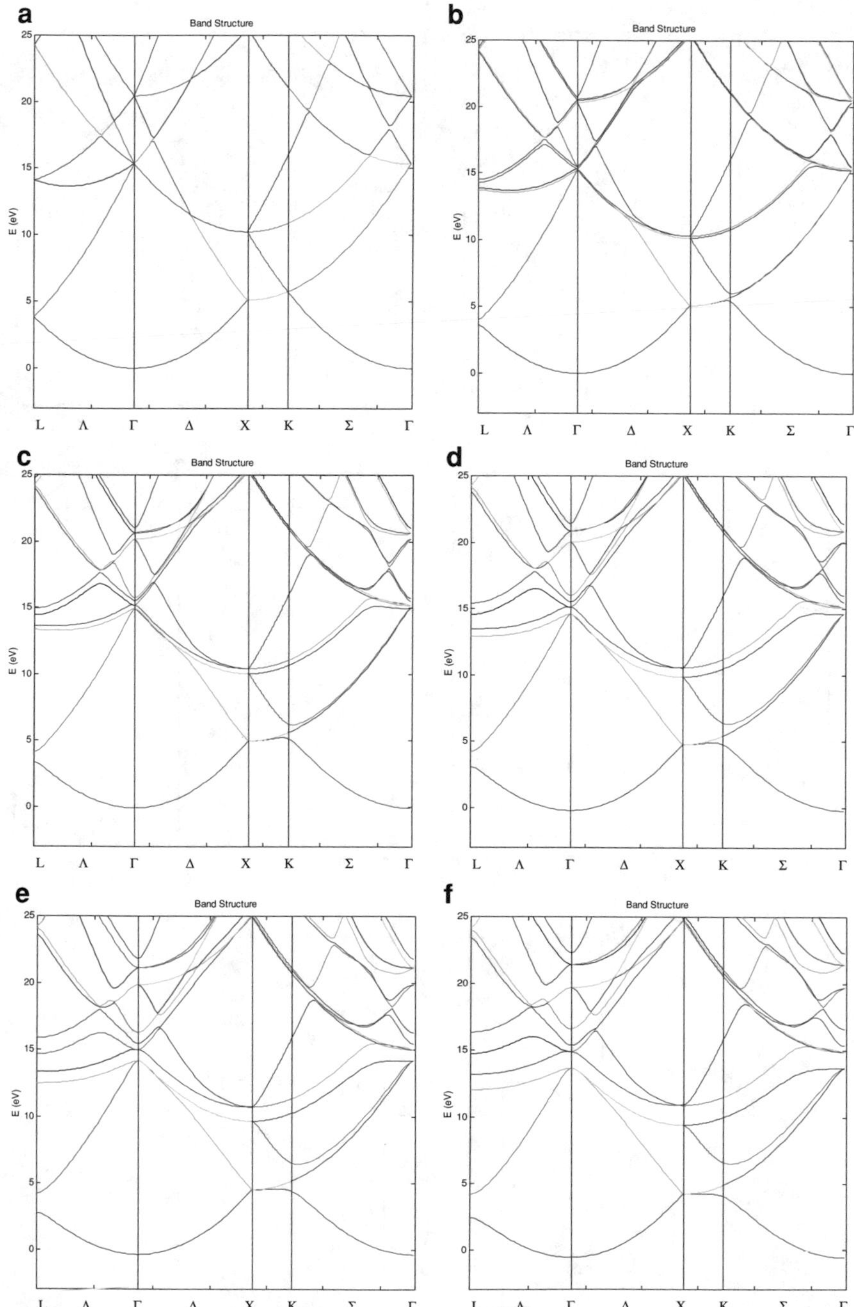

Fig. 4.7 a–k Bandstructure of silicon calculated using the pseudopotential method. In this series of diagrams, the pseudopotential is augmented in 10 steps, showing how the atomic potential modifies the free-electron bandstructure, shown in (**a**)

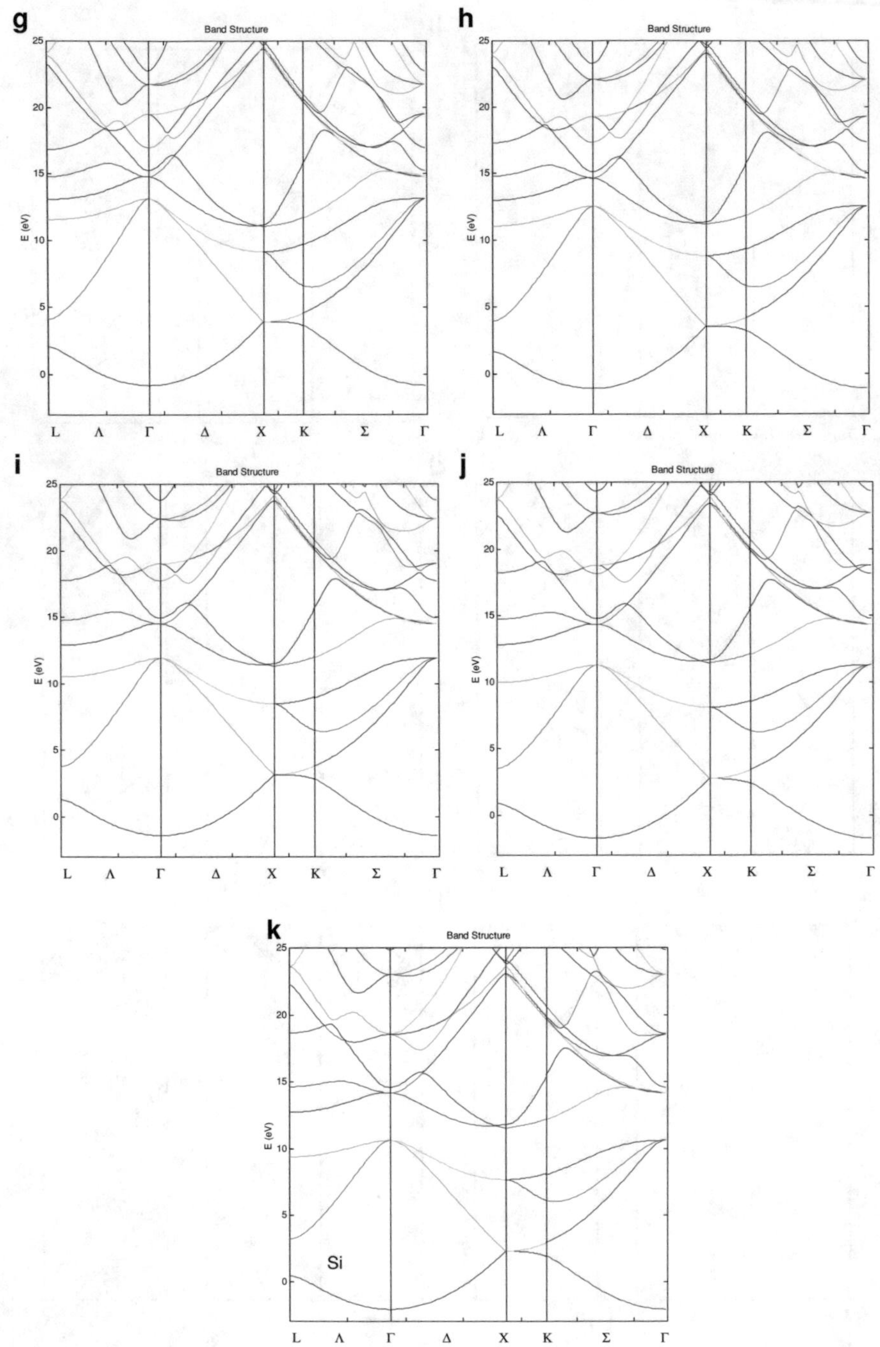

Fig. 4.7 (continued)

$$
\begin{aligned}
U_0 &= 0 \\
U_3 &= -2.856\,\text{eV} \\
U_4 &= 0\,\text{eV} \\
U_8 &= 0.544\,\text{eV} \\
U_{11} &= 1.088\,\text{eV}
\end{aligned}
\tag{4.20}
$$

In Fig. 4.7, we show results for the series of 51 eigenvalue calculations along 4 different directions in k-space. In this series we have added the atomic potential in increments of 10% of the full atomic potential to illustrate how this potential modifies the free-electron bandstructure which is determined only by crystal symmetry. The actual electronic bandstructure of silicon is the last figure in the series.

The fundamental bandgap develops from splitting of the 8-fold degenerate G_3 manifold (Fig. 4.6). Four eigenvalue solutions form the conduction band, and the remaining 4 solutions form the valence band The pseudopotential does not completely lift all the degeneracy. The energy bandgap of silicon (E = 1.1 eV) is the difference between the maximum energy of the valence band, (E = 10.8 eV) which occurs at $\mathbf{k} = \Gamma$, and the minimum of the conduction band, (E = 11.9 eV) which occurs near $\mathbf{k} = X$. The maximum of the valence band energies and the minimum of the conduction band energies do not occur at the same point in k-space, and thus silicon is called an indirect bandgap semiconductor. This indirect bandgap, which occurs at 1.18 eV at room temperature dominates the optical absorption properties of silicon. Silicon does have a direct bandgap of 3.41 eV which occurs between the maximum energy of the valence band and the minimum energy of the conduction band at the Γ-point. Although the optical absorption spectrum is determined by transitions across the indirect bandgap, the optical properties of silicon, such as the refractive index and dielectric loss are determined by optical transitions across the direct bandgap.

4.7 Energy Bandstructure of Germanium

To solve for the eigenvalues of the pseudopotential equations we need to know the lattice constant of germanium and the Fourier components of the pseudopotential.

The lattice constant of germanium at 25 °C is 5.66 Å.

The components of the pseudopotential form factors are: (Cohen and Bergstresser 1966)

$$
\begin{aligned}
U_0 &= 0 \\
U_3 &= -3.128\,\text{eV} \\
U_4 &= 0\,\text{eV} \\
U_8 &= 0.136\,\text{eV} \\
U_{11} &= 0.816\,\text{eV}
\end{aligned}
\tag{4.21}
$$

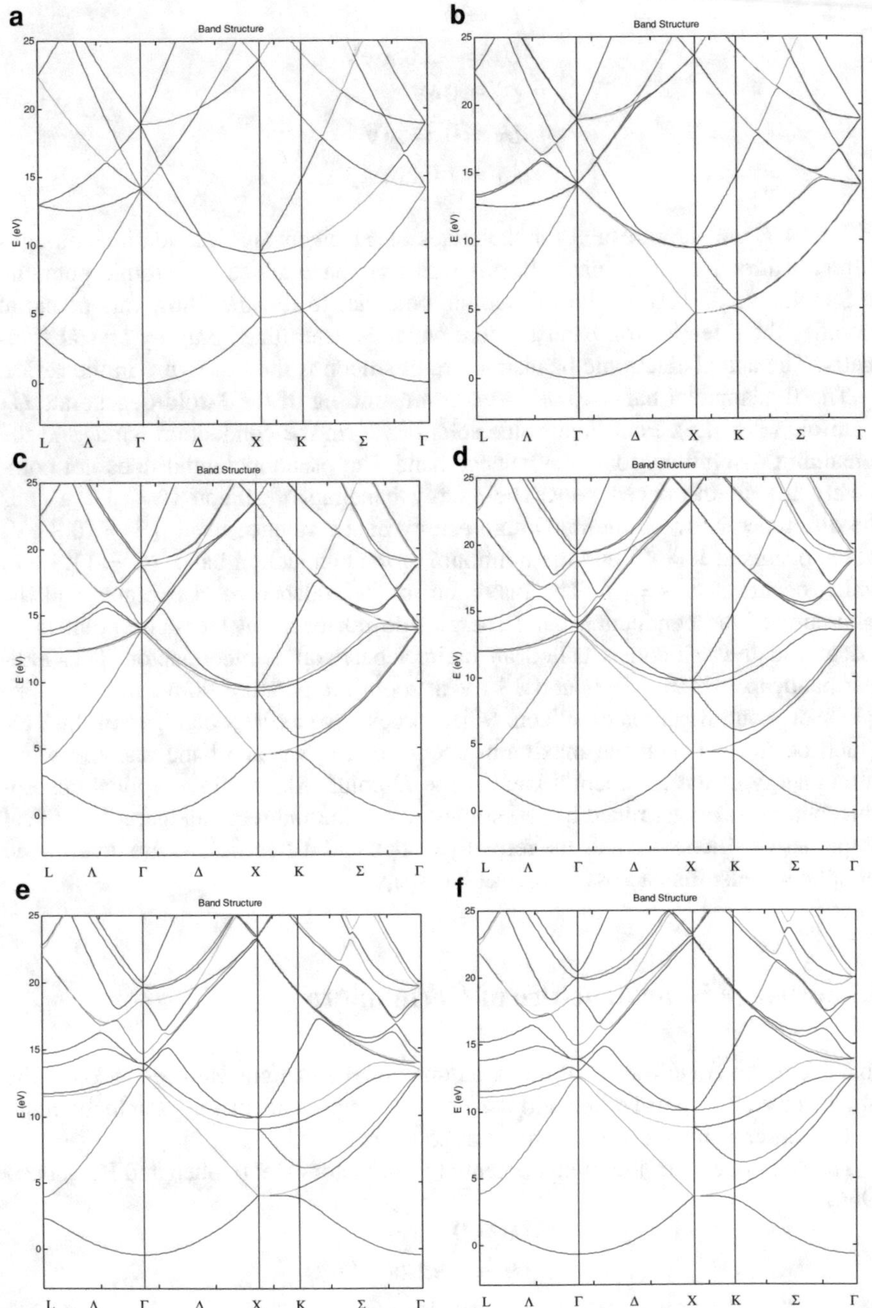

Fig. 4.8 a–k Bandstructure of germanium calculated using the pseudopotential method. In this series of diagrams, the pseudo potential is augmented in 10 steps, showing how the atomic potential modifies the free-electron bandstructure, shown in (**a**)

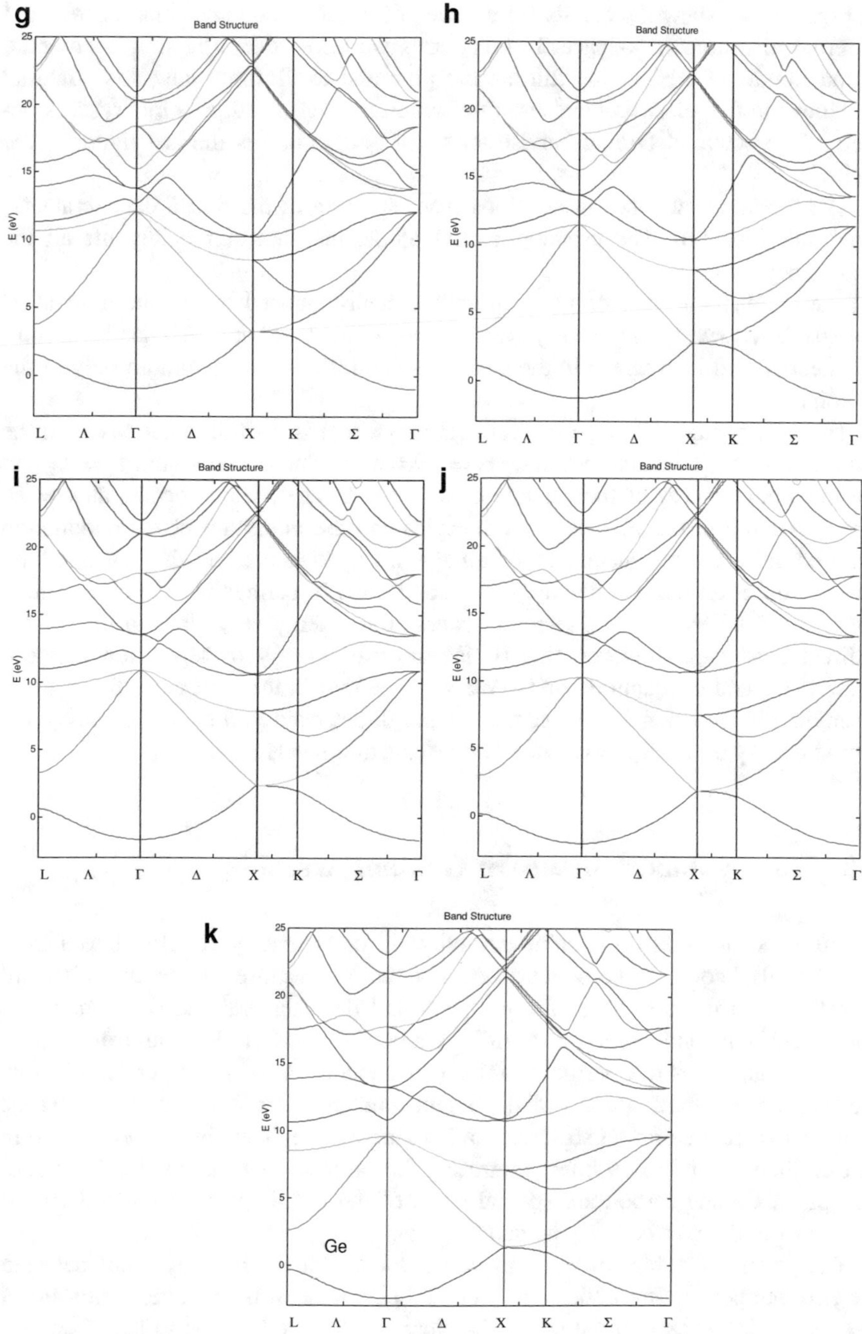

Fig. 4.8 (continued)

In Fig. 4.8, we show the results for a series of 51 eigenvalue calculations along 4 different directions in **k**-space. In this series we have added the atomic potential in increments of 10% of the full atomic potential to illustrate how this potential modifies the free-electron bandstructure which is determined only by crystal symmetry. The actual electronic bandstructure of germanium is the last figure in the series.

The fundamental bandgap develops from splitting of the 8-fold degenerate $\mathbf{G}_3$ manifold (Fig. 4.6). The pseudopotential alone does not completely lift all the degeneracy.

The conduction band of germanium has a pronounced local minimum at Γ ($\mathbf{k} = 0$). The lowest lying energy band in the conduction band also has local minima near X and L, whereas in the case of silicon, there is a minimum only at the X-point

The energy bandgap of germanium (E = 0.8 eV) is the difference between the minimum of the conduction band, (E = 10.6 eV) which occurs near $\mathbf{k} = L$, and the maximum energy of the valence band, (E = 9.8 eV) which occurs at $\mathbf{k} = \Gamma$. The maximum of the valence band energies and the minimum of the conduction band energies do not occur at the same point in **k**-space, and thus germanium, like silicon, is an indirect bandgap semiconductor. In germanium there is a direct bandgap of 0.9 eV at $\mathbf{k} = \Gamma$ which occurs at an energy very close to that of the indirect band gap. However, this is not the least energy of separation between the valence and conduction bands. We will see that in the next case, for gallium arsenide, this situation is reversed, with the direct band gap at $\mathbf{k} = \Gamma$ being the lowest energy between the valence and conduction bands.

4.8 Energy Bandstructure of Gallium Arsenide

Gallium arsenide (GaAs) is cubic crystal with fcc symmetry. It belongs to a class of materials known as compound semiconductors because of the two different atoms that form the basis of the unit cell and the chemical composition. Some compound semiconductors are formed by a combination of elements from group-III and group V of the periodic table. Other examples of III-V semiconductors are InAs and GaSb. Examples of II-VI semiconductors are ZnSe and CdTe. These materials share a common structure called zinc-blende. The zinc-blende structure can be thought of as two interpenetrating fcc lattices. The diamond cubic structure of Si, Ge and carbon are special cases of the zinc-blende structure when all the atoms in the unit cell are the same element.

The geometric center of the unit cell is located at the halfway point between the two interpenetrating lattices at $\frac{a_0}{8}\langle 1 \quad 1 \quad 1\rangle$, where a_0 is the lattice constant of the crystal. In the case of Ge or Si, the atom on either side is the same, of course. In the case of zinc-blende materials, the atoms on either side are different, and the crystal as a result lacks inversion symmetry. The absence of inversion symmetry is a necessary requirement for nonlinear optical behavior. In fact, zinc-blende materials are excellent non-linear materials of significant technological importance. In contrast, non-linear optical behavior is forbidden to the first order by symmetry in Si, Ge and diamond.

The calculation of the energy bandstructure of zinc-blende materials must take into account the reduced symmetry of the structure compared to that of Si or Ge. Basically two similar calculations are needed, one for each sublattice. This can be taken into account in an efficient manner by writing the pseudopotential in two parts, symmetric and anti-symmetric. Denoting the two atomic species by B and C, this gives for the form factor of the potential:

$$V_S(\mathbf{G}) = \frac{1}{2}(V_B(\mathbf{G}) + V_C(\mathbf{G}))$$

and (4.22)

$$V_A(\mathbf{G}) = \frac{1}{2}(V_B(\mathbf{G}) - V_C(\mathbf{G}))$$

Here it is easily seen that the antisymmetric component vanishes in the case of Si, Ge and diamond where atoms B and C are the same.

The pseudopotential can then be written:

$$V(\mathbf{G}) = V_S(\mathbf{G})\cos(\mathbf{G}\cdot\mathbf{r}_i) + i\sin(\mathbf{G}\cdot\mathbf{r}_i) \tag{4.23}$$

where $\mathbf{r}_i$ is the real-space coordinate of an atom in the lattice.

This approach increases the number of elements in the matrix of equations, but the number of unknowns the $C(\mathbf{k})$ remains the same. The MATLAB routine takes a few additional milliseconds to extract the eigenvalues.

To solve for the eigenvalues of the pseudopotential equations we need to know the lattice constant of GaAs and the Fourier components of the pseudopotential.

The lattice constant of GaAs at 25 °C is 5.66 Å, nearly identical to that of germanium.

The pseudopotential form factors are: (Cohen and Bergstresser 1966)

Symmetric

$$\begin{aligned} U_0 &= 0 \\ U_3 &= -3.128\,\mathrm{eV} \\ U_4 &= 0\,\mathrm{eV} \\ U_8 &= 0.136\,\mathrm{eV} \\ U_{11} &= 0.816\,\mathrm{eV} \end{aligned} \tag{4.24}$$

Antisymmetric

$$\begin{aligned} U_0 &= 0 \\ U_3 &= 0.952\,\mathrm{eV} \\ U_4 &= 0.680\,\mathrm{eV} \\ U_8 &= 0\,\mathrm{eV} \\ U_{11} &= 0.136\,\mathrm{eV} \end{aligned} \tag{4.25}$$

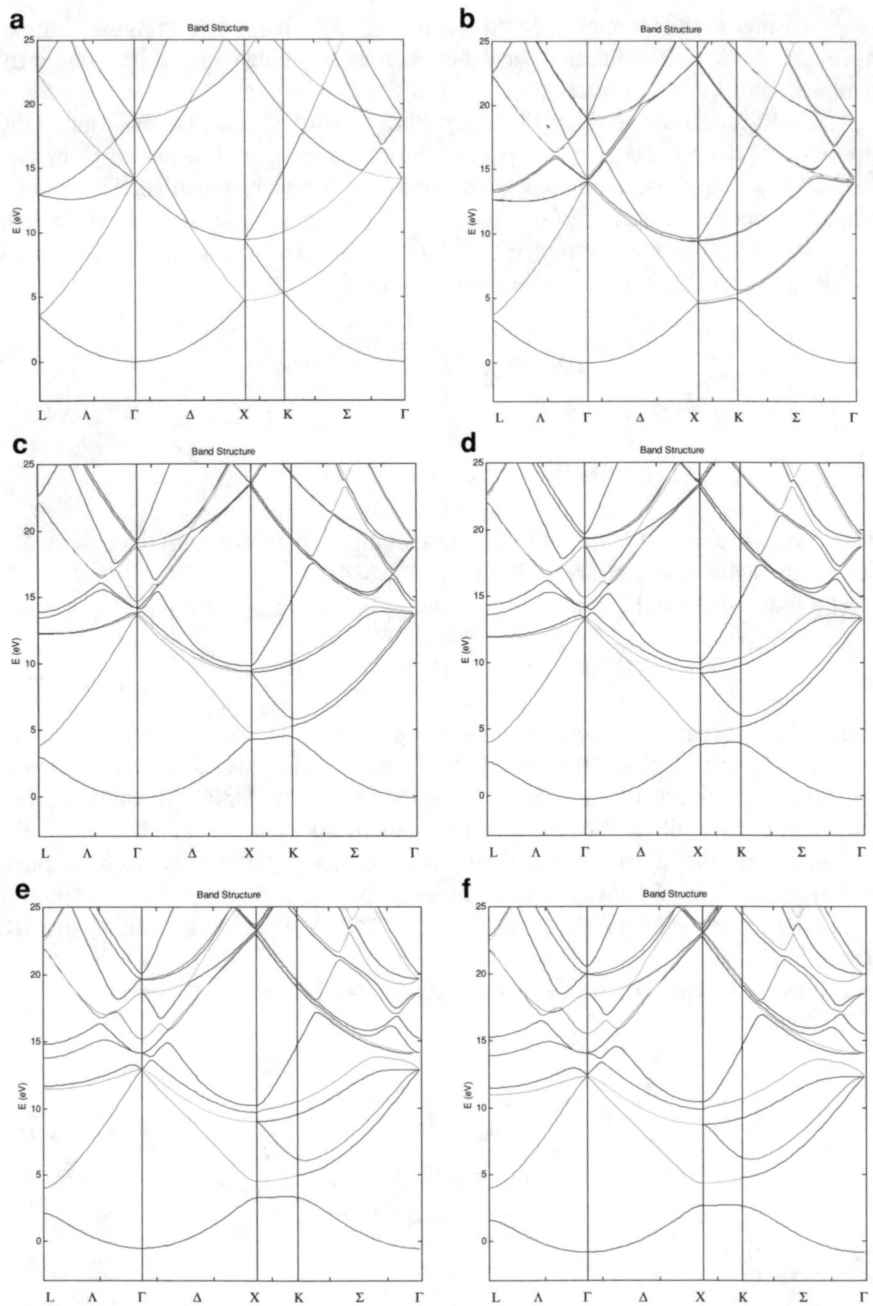

Fig. 4.9 a–k Bandstructure of gallium arsenide calculated using the pseudopotential method. In this series of diagrams, the pseudo potential is augmented in 10 steps, showing how the atomic potential modifies the free-electron bandstructure, shown in (**a**)

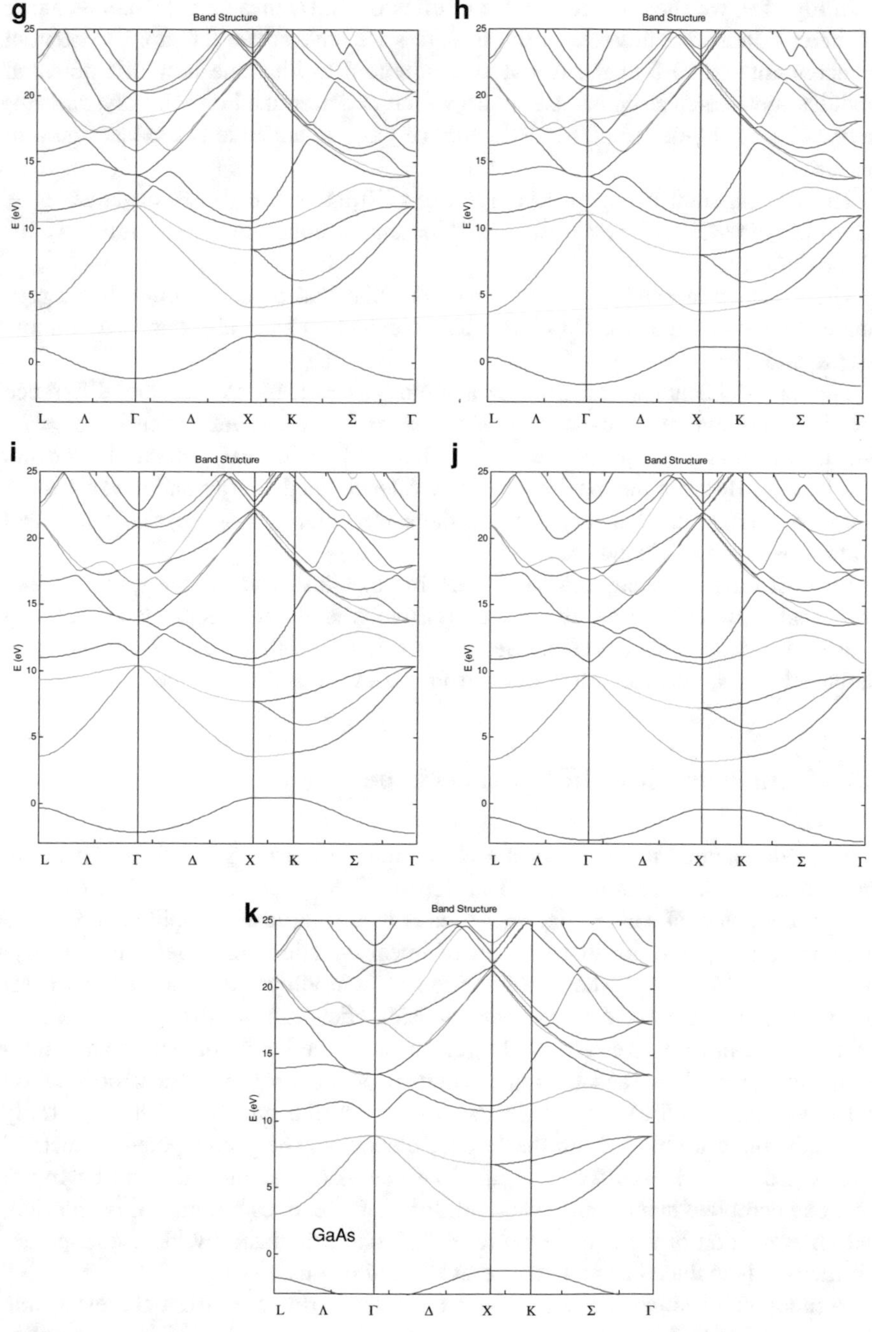

Fig. 4.9 (continued)

In Fig. 4.9, we show the results for a series of 51 eigenvalue calculations along 4 different directions in k-space. In this series we have added the atomic potential in increments of 10% of the full atomic potential to illustrate how this potential modifies the free-electron bandstructure which is determined only by crystal symmetry. The actual electronic bandstructure of gallium arsenide is the last figure in the series.

The fundamental bandgap develops from lifting of the 8-fold degenerate G_3 manifold (Fig. 4.6). However, the pseudopotential alone does not completely lift all the degeneracy.

The conduction band of gallium arsenide, like that of germanium, has a pronounced local minimum at Γ ($\mathbf{k} = 0$). The conduction band also has local minima near X and L.

The energy bandgap of gallium arsenide (E = 1.42 eV) is the difference between minimum of the conduction band, (E = 10.3 eV) which occurs at $\mathbf{k} = \Gamma$, and the maximum energy of the valence band, (E = 8.9 eV) which also occurs near $\mathbf{k} = \Gamma$. Because the maximum of the valence band energy and the minimum of the conduction band energy occur at the same point in $\mathbf{k}$-space, GaAs is called a direct bandgap semiconductor.

In most other respects, the bandstructures of GaAs and Ge are quite similar. This small difference of a few hundred millivolts is all that keeps Ge from being a direct bandgap semiconductor. Because of this fact, there are no conventional Ge-based lasers, whereas lasers made from GaAs are commonplace.

4.9 Materials with Multiple Electrons

The pseudopotential method shows how the many-electron Schrödinger equation, for a material like GaAs or Si, can be replaced by an equation for one electron in a pseudopotential This replacement leads to a significant simplification in the mathematics required, leading to a set of linear equations that yields the energy-momentum relationship. The simplification of Schrödinger equation has its basis in observation that the electrons under consideration can be divided into valence electrons and inner core electrons. The electrons in the inner core states are tightly bound to the nucleus, and do not participate in bonding, electrical conduction and low-energy (<6 eV) photonic transitions. On the other hand they partially screen the nuclear charge from the valence electrons. The pseudopotential method is expected to work well for materials like metals and semiconductors having at least one complete inner shell. Diamond, for example, does not meet this criterion, and an evaluation of the bandstructure of diamond calculated by the pseudopotential method is included as an exercise at the end of this chapter.

Pseudopotentials can be determined from first principles from the electronic charge distribution which defines the structure of the material. Finding the equilibrium structure means calculating the ground state energy of the ensemble of atoms under study. This calculation involves accounting for the multi-body interactions between all the electrons and nuclei.

Hohenberg and Kohn (1964) proved that there is a unique charge distribution that gives the correct ground state energy. Stated simply: In a system having N electrons, the ground state energy does not depend on 3N coordinates of the electrons, but rather on the charge distribution, which has only 3 spatial coordinates. The ground state energy is a functional of the charge density, giving rise to the name density functional theory. This simplification has revolutionized the determination of the structure of complex molecules. Next, Hohenberg and Kohn proved that the correct ground state electron density minimizes this energy functional.

Almost all materials, including most of the elements on one hand, and complex organic molecules on the other, are multi-body systems. Computations on such systems represent a considerable challenge to both mathematics and physics. Yet, common proteins, active in the human body, know how to solve this problem by folding into a minimum energy configuration in a fraction of a second. To appreciate the challenge, we shall examine the Hamiltonian for a multi-electron system of materials. Such a system is made of M nuclei and N electrons. For simplicity, we will assume that the nuclei are frozen in place, and consider only the contributions that electrons make to the overall energy.

First of all there is a component of the total energy coming from the kinetic energy of the electrons:

$$\mathcal{H}_{\text{kinetic}} = \sum_{i=1}^{N} -\frac{\hbar^2}{2m_e}\nabla_i^2 \tag{4.26}$$

There is a term coming from the potential of the nuclei:

$$\mathcal{H}_{\text{potential}} = \sum_{i=1}^{N} V(\mathbf{r}_i) = -\sum_{i=1}^{N}\sum_{k=1}^{M} \frac{Z_k q^2}{|\mathbf{R}_k - \mathbf{r}_i|} \tag{4.27}$$

There is a term coming from the interaction between the electrons themselves:

$$\mathcal{H}_{e-e} = \sum_{i=1}^{N}\sum_{j=i+1}^{N} \frac{q^2}{|\mathbf{r}_i - \mathbf{r}_j|} \tag{4.28}$$

The Schrödinger equation is:

$$(\mathcal{H}_{\text{kinetic}} + \mathcal{H}_{\text{potential}} + \mathcal{H}_{e-e})\Psi = E\Psi, \tag{4.29}$$

where $\Psi = \Psi(\mathbf{r}_1, \mathbf{r}_2 \ldots \mathbf{r}_N)$ is the electron wave function that depends on each of the coordinates on the N electrons and E is the ground-state energy of the system.

However, solving (4.29) is impeded by the fact that this differential equation is not separable, because the electron-electron interaction depends on $\frac{1}{|\mathbf{r}_i - \mathbf{r}_j|}$.

To proceed further, we could propose, as an approximation to substitute an average potential $= v_{AV}(\mathbf{r}_i)$ for each electron in the place of this term. On this basis:

$$\begin{aligned}\mathcal{H} &= \sum_{i=1}^{N} -\frac{\hbar^2}{2m_e}\nabla_i^2 + \sum_{i=1}^{N} V(\mathbf{r}_i) + \sum_{i=1}^{N} v_{AV}(\mathbf{r}_i) \\ &= \sum_{i=1}^{N}\left\{-\frac{\hbar^2}{2m}\nabla^2 + V_{\text{effective}}(\mathbf{r}_i)\right\} = \sum_{i=1}^{N} \boldsymbol{h}_i\end{aligned} \tag{4.30}$$

That is, the Hamiltonian can be reduced to the sum of individual Hamiltonians for each electron. In this approximation, the wavefunction Ψ is expressed as the product of the individual one-electron solutions

$$\boldsymbol{h}_i \psi_i(\mathbf{r}_i) = \varepsilon_i \psi_i(\mathbf{r}_i)$$

$$\begin{aligned}\mathcal{H}\Psi &= \mathcal{H}\{\psi_1(\mathbf{r}_1)\psi_2(\mathbf{r}_2)\ldots\psi_N(\mathbf{r}_N)\} \\ &= (\varepsilon_1 + \varepsilon_2 + \cdots + \varepsilon_N)\{\psi_1(\mathbf{r}_1)\psi_2(\mathbf{r}_2)\ldots\psi_N(\mathbf{r}_N)\}\end{aligned} \tag{4.31}$$

The proposed wavefunction, $\Psi = \{\psi_1(\mathbf{r}_1)\psi_2(\mathbf{r}_2)\ldots\psi_N(\mathbf{r}_N)\}$, called a Hartree product, is not an acceptable choice because its structure does not comply with the Pauli exclusion principle, namely that it is not antisymmetric with the exchange of electrons. However, another combination of one electron wavefunctions could serve as a solution to (4.31) and satisfy the Pauli exclusion principle:

$$\Psi = \frac{1}{\sqrt{N!}}\begin{vmatrix}\psi_1(\mathbf{r}_1) & \psi_2(\mathbf{r}_1) & \ldots & \psi_N(\mathbf{r}_1) \\ \psi_1(\mathbf{r}_2) & \psi_2(\mathbf{r}_2) & \ldots & \psi_N(\mathbf{r}_2) \\ \ldots & \ldots & \ldots & \ldots \\ \psi_1(\mathbf{r}_N) & \psi_2(\mathbf{r}_N) & \ldots & \psi_N(\mathbf{r}_N)\end{vmatrix} \tag{4.32}$$

Such a wavefunction is called a Slater determinant. A wavefunction of this form could be used as a solution for (4.31). Note that it contains $N!$ terms, each term depends on 3 spatial variables as well as spin. If we were to try to solve the Schrödinger equation for a cluster of only 10 atoms, we would have to imagine dealing with hundreds of electrons, depending on the atomic number of the elements in the cluster. If we take the example of silicon, with N = 14, this means treating 140 electrons, and more than 560 dimensions including spin. The number of terms in the Slater determinant would exceed 10^{241}. It would require a great deal of patience to achieve a result for systems having more than a handful of electrons.

This brute-force analytical approach leads to dramatically increasing complexity, and it would be better to look for an alternate route to the principal objective which is to determine the ground-state energy of the system. In order to derive the energy eigenstates of the Schrödinger equation, it is necessary to have a set of wavefunctions. In order to determine the correct ground-state energy, it is necessary to use the correct wavefunctions. Finding these wavefunctions is an iterative process. In many cases, we are not even interested in the wavefunctions of the electrons. These are a means to an end. As in the case of the pseudopotential method, we should develop a different approach by searching for a variable that reflects the behavior of electrons in the ground-state.

In the face of this situation, Hohenberg and Kohn proposed in 1964 such an alternative analysis. They proposed and proved that

- There exists a variable which is a function of the electron charge density distribution.
- The energy of the multi-electron system depends on this variable.
- The minimum of this energy is the true ground state energy of the system.

Their proof, which we reproduce below is exemplary for both its elegance and clarity.

4.9.1 The Hohenberg-Kohn Theorem and Density Functional Theory

There is a universal functional $F|\rho(\mathbf{r})|$ of the electron charge density distribution $\rho(\mathbf{r})$ that defines the total energy of the electron distribution by

$$E = \int v(\mathbf{r})\rho(\mathbf{r})d\mathbf{r} + F|\rho(\mathbf{r})| \tag{4.33}$$

The total energy E has a minimum when the charge density $\rho(\mathbf{r})$ coincides with the true charge density distribution in the potential $v(\mathbf{r})$.

The electronic charge density distribution $\rho(\mathbf{r})$ is related to the true N-body wavefunction $\Psi(\mathbf{r}_1, \mathbf{r}_2 \ldots \mathbf{r}_N)$ via

$$\rho(\mathbf{r}) = \int \Psi^*(\mathbf{r}_1, \mathbf{r}_2 \ldots \mathbf{r}_N)\Psi(\mathbf{r}_1, \mathbf{r}_2 \ldots \mathbf{r}_N)d\mathbf{r}_1 d\mathbf{r}_2 \ldots d\mathbf{r}_N \tag{4.34}$$

Then the expectation value for the one-body potential $v(\mathbf{r})$ is

$$\langle\Psi|V|\Psi\rangle = \int \rho(\mathbf{r})v(\mathbf{r})d\mathbf{r} = \sum_{i=1}^{N}\langle\Psi|v(\mathbf{r}_i)|\Psi\rangle \tag{4.35}$$

The true wavefunction is a unique functional of the external potential

$$\Psi(\mathbf{r}_1, \mathbf{r}_2 \ldots \mathbf{r}_N) = F'|v(\mathbf{r})| \tag{4.36}$$

The charge density is a unique functional of the external potential

$$\rho(\mathbf{r}) = F''|v(\mathbf{r})| \tag{4.37}$$

On the other hand, is it true that the potential $v(\mathbf{r})$ is a unique functional of the charge density distribution?

To answer this question let us suppose that a different potential $v'(\mathbf{r})$ gives the same ground-state charge density distribution $\rho(\mathbf{r})$.

Let the Hamilitonians whose potentials are given by $v(\mathbf{r})$ and $v'(\mathbf{r})$ be given by $\mathcal{H}$ and $\mathcal{H}'$ respectively. Similarly, let us write their eigenvalues and eigenfunctions as E and E' and Ψ and Ψ'.

We have

$$\mathcal{H}\Psi = E\Psi$$

and

$$\mathcal{H}'\Psi' = E'\Psi' \tag{4.38}$$

From the variational principle of the ground state, we expect

$$\begin{aligned} E' &= \langle\Psi'|\mathcal{H}'|\Psi'\rangle < \langle\Psi|\mathcal{H}'|\Psi\rangle \\ \langle\Psi|\mathcal{H}'|\Psi\rangle &= \langle\Psi|\mathcal{H} + V' - V|\Psi\rangle \end{aligned} \tag{4.39}$$

$$E' < \langle\Psi|\mathcal{H}|\Psi\rangle + \int [v'(\mathbf{r}) - v(\mathbf{r})]\rho(\mathbf{r})d\mathbf{r} = E + \int [v'(\mathbf{r}) - v(\mathbf{r})]\rho(\mathbf{r})d\mathbf{r} \tag{4.40}$$

And simultaneously,

$$\begin{aligned} E &< \langle\Psi|\mathcal{H}|\Psi\rangle < \langle\Psi|'\mathcal{H}|\Psi'\rangle \\ E &< \langle\Psi'|\mathcal{H}'|\Psi'\rangle + \textstyle\int[v(\mathbf{r}) - v'(\mathbf{r})]\rho(\mathbf{r})d\mathbf{r} = E' - \int[v'(\mathbf{r}) - v(\mathbf{r})]\rho(\mathbf{r})d\mathbf{r} \end{aligned} \tag{4.41}$$

From these 2 equations we must draw the self-contradictory result:

$$E' + E < E + E' \tag{4.42}$$

So we must abandon the original presumption. Thus $v(\mathbf{r})$ is a unique functional of $\rho(\mathbf{r})$. Since $v(\mathbf{r})$ determines the form of the Hamiltonian $\mathcal{H}$ uniquely, then the wavefunction Ψ in the ground state is a unique functional of $\rho(\mathbf{r})$.

It follows that we can write the ground-state energy as a unique functional of the charge distribution, which will depend on the wavefunction and the Hamiltonian $\mathcal{H} = T + U$, where **T** represents kinetic energy and U represents the interaction energy between electrons.

$$F|\rho(\mathbf{r})| = F|\Psi, (T + U)\Psi| \tag{4.43}$$

The energy of the system of electrons is expressed as a functional of the charge density

$$E_v|\rho(\mathbf{r})| = \int v(\mathbf{r})\rho(\mathbf{r})d\mathbf{r} + F|\rho(\mathbf{r})| \tag{4.44}$$

This functional can be minimized by an appropriate iteration algorithm, yielding the ground state energy. In a few more lines, Hohenberg and Kohn show that the electron density $\rho(\mathbf{r})$ that minimizes (4.44) is the true electron density.

This result is an intellectual breakthrough that has dramatically changed the course of chemistry and physics. While the brute-force approach, which determines a wavefunction for each electron and accounts for the interactions between all the electrons, would produce a correct result for the energies of the systems of particles, this method cannot be applied to real systems because it is too time-consuming.

Hohenberg and Kohn showed that there is another route to the answer. The charge density distribution depends on 3 spatial variables (4, if spin is taken into account), *no matter how many electrons are involved*. The ground state energy can be determined by minimizing a functional that depends only on this charge density. We remember that the actual particle cluster or molecule under consideration shows us yet another way to determine the ground-state energy in less than a picosecond without recourse to the brute-force procedure either.

4.9.2 The Kohn-Sham Equations

The paper by Hohenberg and Kohn showed that there is an alternate route to determination of the ground-state energy and the importance of the charge density, but it did not show how to derive the charge density distribution that is necessary to make the approach practical. A procedure was given in the following year by Kohn and Sham (1965). These two seminal papers form the basis of density functional theory.

Kohn and Sham confirmed the importance of decoupling the interactions between electrons that we evoked in (4.29). However, instead of substituting an average interaction potential, they introduced the charge density function $\rho(\mathbf{r})$.

$$\mathcal{H}_{e-e} = \sum_{i=1}^{N}\sum_{j=i+1}^{N} \frac{q^2}{|\mathbf{r}_i - \mathbf{r}_j|} \rightarrow \int \frac{q^2\rho(\mathbf{r}')}{|\mathbf{r} - \mathbf{r}'|}d\mathbf{r}' \tag{4.45}$$

That is, the electron-electron interaction is replaced by the interaction between an electron and the charge density function. With this substitution, the Schrödinger equation can be separated into a system of independent and linear one-electron equations. These electrons are not "real" electrons, but rather Kohn-Sham non-interacting electrons. A simple Hamiltonian for each electron can be written down as:

$$\mathcal{H}_{K-S} = T(\mathbf{r}) + V(\mathbf{r}) + \mathcal{H}_{e-e}(\mathbf{r}) + V_{\text{exch}}(\mathbf{r}) \tag{4.46}$$

where

$T(\mathbf{r})$ is the kinetic energy
$V(\mathbf{r})$ is the Coulomb potential of the nuclei

and $V_{\text{exch}}(\mathbf{r})$ is an electron exchange-energy term that corrects for the self-interaction of an electron included in the expression of (4.45), as well as correlation effects that need to be taken into account because they are ignored by the non-interacting assumption. This term is the subject of numerous studies and is often represented by an empirical model appropriate to the system under study.

The Kohn-Sham equations, one for each electron can be written down:

$$\left[-\frac{\hbar^2}{2m}\nabla^2 + V(\mathbf{r}) + \mathcal{H}_{e-e}(\mathbf{r}) + V_{\text{exch}}(\mathbf{r})\right]\varphi_i(\mathbf{r}) = \varepsilon_i\varphi_i(\mathbf{r}) \tag{4.47}$$

This set of linear uncoupled equations can be set up in matrix form and solved for the eigenvalues, and the set of fictitious non-interacting wavefunctions can be determined. The charge density is determined from the summation:

$$\rho(r) = 2\sum_{i=1}^{N}\varphi_i^*\varphi_i$$

Of course, the fictitious wavefunctions do not represent an approximation to the real electron wavefunctions, and the eigenvalues determined by this process do not correspond to the real energy states either. What Kohn and Sham proved is that the electron density determined in this process, by using a recursive calculation can be made to converge to the real electron density (Lesar 2013). This is the key result that is needed to exploit the Hohenberg-Kohn theorem. An iteration procedure is in Fig. 4.10.

4.9.3 Using Density Functional Theory to Analyze Real Systems

Realistic modeling and simulation of real atomic and molecular systems involves significant computational resources, even with the benefit of density functional theory. There are a number of centers which have developed computational infrastuctures for such studies, some of which can be used by interested scientists. Scholl and Steckel (2009) have written an excellent introduction to density functional calculations for the prospective user that make use of available computational platforms. These platforms provide an entry to density-functional analysis without having to reinvent the software routines needed for implementation, thus saving years of effort to develop and de-bug the calculational routines.

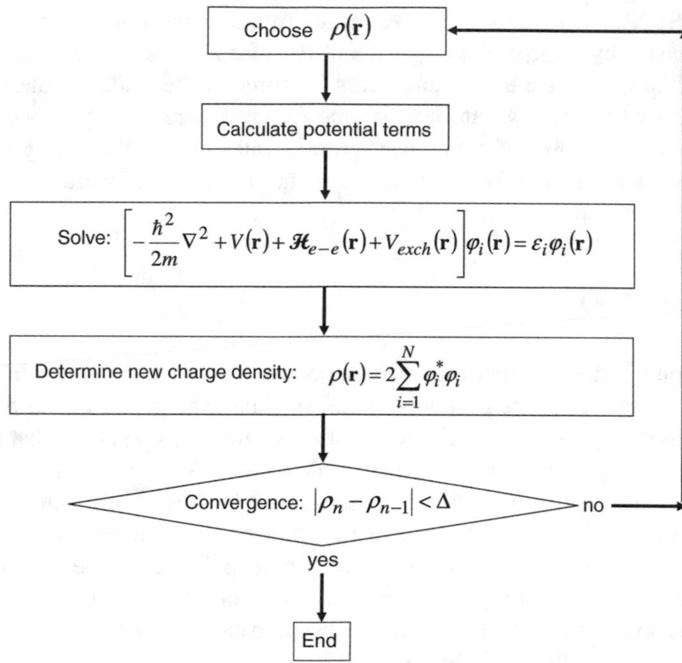

Fig. 4.10 Flow diagram for a recursive solution of the Kohn-Sham equations

Two such initiatives are mentioned here: VASP and Quantum Espresso. Additional software routines can be found at SourceForge.net. https://sourceforge.net/directory/os:windows/?q=density%20functional%20theory.

The Vienna Ab initio Simulation Package (VASP) is a computer program for atomic scale materials modelling, e.g. electronic structure calculations and quantum-mechanical molecular dynamics, from first principles. https://en.wikipedia.org/wiki/Vienna_Ab_initio_Simulation_Package.

VASP computes an approximate solution to the many-body Schrödinger equation, either within density functional theory (DFT), solving the Kohn-Sham equations, or within the Hartree-Fock (HF) approximation, solving the Roothaan equations. Hybrid functionals that mix the Hartree-Fock approach with density functional theory are implemented as well. Access to VASP tools and capabilities is by paid subscription and comes with significant support from VASP.

Quantum ESPRESSO is an integrated suite of computer codes for electronic-structure calculations and materials modeling, based on density-functional theory, plane waves, and pseudopotentials (norm-conserving, ultrasoft, and projector-augmented wave). http://www.quantum-espresso.org/.

Quantum ESPRESSO stands for “opEn Source Package for Research in Electronic Structure, Simulation, and optimization”. It is available at no cost to researchers around the world under the terms of the GNU General Public License.

Quantum ESPRESSO builds upon electronic-structure codes that have been developed and tested by some of the original authors of novel electronic-structure algorithms and applied in the last twenty years by some of the leading materials modeling groups worldwide. As an open-source project, there is no paid support service. There are a number of public user-groups. Interaction with these groups may provide a level of support for understanding and using the software.

4.10 Summary

The electronic bandstructure of many semiconductors of technological importance can be calculated using the pseudopotential method. The power of the pseudopotential method resides in its ability to substitute a fictitious potential that accounts for the behavior of the electrons in the weakly-bound valence band states. In the case of germanium for example, the interactions between the electron of interest and the nuclear potential as well as the interactions between the 32 electrons that make up the atom are summarized in a potential that can be expanded in a Fourier series based on the periodicity of the crystal lattice. Only a few terms of this expansion are needed to give an electronic bandstructure that is in excellent agreement with experimental measurement.

A pseudopotential can be calculated from first principles if the ground-state configuration of the material can be determined. Hohenberg and Kohn proved that the ground state configuration of an ensemble of atoms, such as a crystal or a molecule, is a functional of the correct electron charge density: that is, a function of three spatial dimensions. Density functional theory has enabled the determination of the ground-state configuration not only of semiconductors whose structure is relatively simple, but more importantly of complex molecules and proteins with relevance to biochemistry. In 1998 Walter Kohn received the Nobel Prize in Chemistry for the development of density functional theory.

4.11 Exercises

4.1 Prove that the energy eigenvalues of $\Psi(\mathbf{k}, \mathbf{r})$ are identical to those of $\Psi(\mathbf{k} + \mathbf{G}_m, \mathbf{r})$.

4.2 Graphene is a material of great technological promise. It was studied as a model system for many years before and André Geim and Konstantin Novoselov showed how to make real samples. Their discovery was recognised by the 2010 Nobel Prize in physics.
A diagram of the graphene structure is shown in Fig. 4.11. Note that the unit cell contains two carbon atoms.

4.2a Using the primitive vectors $\mathbf{a}_1, \mathbf{a}_2$ and $\mathbf{a}_3$, find the three basis vectors that span the reciprocal lattice. Show that the resulting structure is independent of

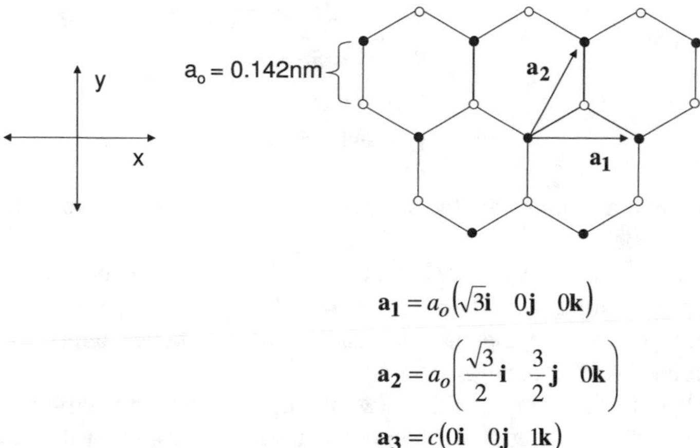

Fig. 4.11 Crystal structure of graphene in the x-y plane. Graphene is a 2-dimensional structure in a 3-dimensional world. Thus, the 3-dimensional structure is completed by supposing parallel sheets of graphene separated in the 3rd dimension by $z = c$

the spacing between parallel sheets, and becomes 2-dimensional in the limit of very large c. In the 2-dimensional limit, make a diagram of the points in reciprocal space obtained by adding and subtracting the two basis vectors.

4.2b Make a graph of the Brillouin zone for graphene, and label the high-symmetry corners and faces. Include the coordinates of these points.

4.3 Using the results in Exercise 4.2, make a calculation of the free-electron bandstructure for graphene. Compare your calculation with the results of Fischetti et al. (2013).

4.4 Calculate the energy band structures for Ge, using MATLAB, and incrementing the periodic potential from 0 to 1 (full potential), as shown in Sect. 4.7. Integrate your results into a video animation using the moviemaker function of MATLAB.

4.5 The pseudopotential method is based on the assumption of the separation of electrons into two categories: valence states and core states. The core states are contained in a complete and closed shell. The assumption is valid for silicon. In the case of diamond the core states are the 1s and 2s electrons. The valence states are the 2p electrons. In this case there is less complete screening of the nuclear potential.

Give an evaluation of the limitations of the pseudopotential method using diamond as an example.

4.5a Perform a calculation of the bandstructure of diamond using the pseudopotential form factors proposed by Saslow et al. (1966)

$$V_{111} = -11.0296$$
$$V_{220} = 4.58$$
$$V_{311} = 1.7952$$
$$V_{222} = 0.5576.$$

In order to accommodate the V_{222} form factor, you will have to modify the Matlab code.

4.5b Compare your result to more recent calculations of the diamond bandstructure, for example Calzaferri and Rytz (1996). What adjustments could be made to the pseudopotential form factors to give better agreement with the direct and indirect bandgaps?

4.5c The V_{222} form factor should be identically zero by symmetry. Criticize the rationale of Saslow et al. for proposing a non-zero value for this form factor. Does this feature improve the overall result in a significant way?

Appendix: Source Code for Bandstructure Calculation Using MATLAB

```
% This program calculates the band structure of a F.C.C material using
% pseudopotentials to model the potential due to atoms at lattice sites in the
% material. The program plots E vs k for four k-directions with calculations
% through the (311) family of reciprocal lattice vectors.
% This is a Matlab program. In order to run it you must save the file with
% a .m extension. After loading Matlab, type the filename without .m and it
% will run.

% Example:
%        >> filename
%user-defined variables.

clear;
Gmax =11;               %Maximum of desired |G|

% Symmetric and antisymmetric Fourier coefficients of pseudopotentials
% (Ge as example)
blank = ' ';
intro1 = 'Please enter the symmetric and anti-symmetric potential';
intro2 = 'coefficients corresponding to the proper values of |G|^2';
intro3 = 'in eV. The prompts will have the form "V-sym(3)"';
disp('                Pseudopotential Coefficients in eV' );
disp('Material   V-sym(3)  V-sym(8)  V-sym(11)   V-asym(3)  V-asym(4)   V-asym(11)');
disp('-------- -------- -------- ---------- ---------- ---------- ------------');
disp(' Si         -2.856      0.544      1.088         0          0           0');
disp(' Ge         -3.128      0.136      0.816         0          0           0');
disp(' Sn         -2.720        0        0.544         0          0           0');
disp(' ');
disp(' GaP        -2.992    0.408      0.952       1.632       0.952       0.272');
disp(' GaAs       -3.128    0.136      0.816       0.952       0.680       0.136');
disp(' AlSb       -2.856    0.272      0.816       0.816       0.544       0.272');
disp(' InP        -3.128    0.136      0.816       0.952       0.680       0.136');
disp(' GaSb       -2.992     0.0       0.680       0.816       0.680       0.136');
disp(' InAs       -2.992     0.0       0.680       1.088       0.680       0.408');
disp(' InSb       -2.720     0.0       0.544       0.816       0.680       0.136');
```

```
disp(' ');
disp(' ZnS      -2.992   0.408    0.952      3.264     1.904      0.544');
disp(' ZnSe     -3.128   0.136    0.816      2.448     1.632      0.408');
disp(' ZnTe     -2.992    0.0     0.680      1.768     1.360      0.136');
disp(' CdTe     -2.720    0.0     0.544      2.040     1.224      0.544');
disp(' ');
disp('==================================================================');

disp(blank);
disp(blank);
disp(intro1);
disp(intro2);
disp(intro3);
v1 = input('V-sym(3) = ');
v2 = input('V-sym(4) = ');
v3 = input('V-sym(8) = ');
v4 = input('V-sym(11) = ');
v5 = input('V-antisym(3) = ');
v6 = input('V-antisym(4) = ');
v7 = input('V-antisym(8) = ');
v8 = input('V-antisym(11) = ');

C= input('Enter the potential scaling factor (0=no potential, 1=full potential): ');
% Scaling factor for potentials.
% C=0 is free electron, C=1 is full potential.
S_pot_3 =C*(v1);       %G = (111);
A_pot_3 =C*( v5);
A_pot_4 =C*( v6);      %G = (200);
S_pot_4 =C*( v2);
S_pot_8 =C*( v3);      %G = (220);
S_pot_11=C*( v4);      %G = (311);
A_pot_8 =C*( v7);
A_pot_11=C*( v8);

lattice= input('lattice constant in angstroms');
lattice=lattice*1e-10;          %Lattice constant in m.
sample=40;                     %Sampling points per leg.
leg_num=5;                     %Plotting leg number.
Emax=17;                       %Plotting energy maximum.
Emin=4;                        %Plotting energy minimum.

BZ=[1/2 -1/2 1/2; 0 0 0; 1 0 0; 1 1/4 -1/4;0 3/4 3/4; 0 0 0]; %Define zone boundary.
T=(1.054e-34*2*pi/lattice)^2/(2*9.1e-31*1.6e-19);
i=sqrt(-1);

%Start calculating
       %Find desired G's
count=1;
for n1 =-3:1:3;
       for n2=-3:1:3
              for n3=-3:1:3;
                     x=n1-n2+n3;
                     y=n1+n2-n3;
                     z=-n1+n2+n3;
                     if ((x)^2+y^2+z^2)<=Gmax;
                            G(count,1)=x;
                            G(count,2)=y;
                            G(count,3)=z;
                            count=count+1;
                     end;
              end;
       end;
end;
vec_num=count-1;               %Total desired G number.
M(vec_num,vec_num)=0;   %Reset matrix
```

```
%Set up off diagonal elements.
for column=1:1:vec_num;
        for row=column+1:1:vec_num;
                x=G(column,1)-G(row,1);
                y=G(column,2)-G(row,2);
                z=G(column,3)-G(row,3);
                potential=0;
                potential2=0;
                leng2=x^2+y^2+z^2;
                L=pi/4*(x+y+z);
                if leng2==3,
                        potential=S_pot_3*cos(L)+i*A_pot_3*sin(L);
                        potential2=S_pot_3*cos(-L)+i*A_pot_3*sin(-L); end;
                if leng2==4,
                        potential=S_pot_4*cos(L)+i*A_pot_4*sin(L);
                        potential2=S_pot_4*cos(-L)+i*A_pot_4*sin(-L); end;
                if leng2==8,
                        potential=S_pot_8*cos(L)+i*A_pot_8*sin(L);
                        potential2=S_pot_8*cos(-L)+i*A_pot_8*sin(-L); end;
                if leng2==11,
                        potential=S_pot_11*cos(L)+i*A_pot_11*sin(L);
                        potential2=S_pot_11*cos(-L)+i*A_pot_11*sin(-L); end;
                M(column,row)=potential;
                M(row,column)=potential2;
        end;
end;

%Eigenvalue calculation routine.
pointa=0;
ks = sqrt(3/4);                 %Scaling factor for k values.
zb = 0;
interval = 0;
for leg=1:1:leg_num;
         if leg~=4,
         x_interval=(BZ(leg+1,1)-BZ(leg,1))/sample;
         y_interval=(BZ(leg+1,2)-BZ(leg,2))/sample;
         z_interval=(BZ(leg+1,3)-BZ(leg,3))/sample;
         if leg==2, zb = k(leg-1,pointa+1); interval = 0; ks = 1.0; end;
         if leg==3, ks = sqrt(1/8); zb = k(leg-1,pointa+1); interval = 0; end;
         if leg==5, ks = sqrt(9/8); zb=k(leg-2,pointa+1); interval=0; end;
         pointa = 0;
         for n=0:1:sample;
                x=BZ(leg,1)+n*x_interval;
                y=BZ(leg,2)+n*y_interval;
                z=BZ(leg,3)+n*z_interval;
                for m=1:1:vec_num;              %Set up diagonal elements
                       M(m,m)=T*((G(m,1)+x)^2+(G(m,2)+y)^2+(G(m,3)+z)^2);
                end;
                Egl=eig(M);
                Eg=sort(Egl);
                for m=1:1:vec_num;r
                        if leg==1, p1(pointa+1,m)=Eg(m); end;
                        if leg==2, p2(pointa+1,m)=Eg(m); end;
                        if leg==3, p3(pointa+1,m)=Eg(m); end;
                        if leg==5, p4(pointa+1,m)=Eg(m); end;
                end;
                k(leg,pointa+1)=zb+(ks*interval/sample);
                pointa=pointa+1;
                interval = interval+1;
        end;
        pointa=pointa-1;
        end;
end;

%plotting routine.
plot(k(1,:),p1,'-',k(2,:),p2,'-',k(3,:),p3,'-',k(5,:),p4,'-');
```

```
axis([0, 3.3, -3, 30]);

set (gca,'FontName','Symbol');
xlabel('L                    G                    C
K                    G');

set (gca,'FontName','Helvetica');
ylabel('E (eV)');
title('Band Structure ');
                 if leg==1, p1(pointa+1,m)=Eg(m); end;
```

References

L. Brillouin, *Wave Propagation in Periodic Structures* (McGraw-Hill, New York, 1946). Dover reprint of 2nd ed. (Dover, Mineola, 2003) ISBN 0486495566

G. Calzaferri, R. Rytz, The band structure of diamond. J. Phys. Chem. **100**, 11122–11124 (1996). https://www.researchgate.net/publication/231657003_The_Band_Structure_of_Diamond

M.L. Cohen, T.K. Bergstresser, Bandstructures and pseudopotential form factors for fourteen semiconductors of the diamond and zinc-blende structures. Phys. Rev. **141**, 789–796 (1966). http://www.fisica.uniud.it/~giannozz/Corsi/MQ/LectureNotes/cohenbergstresser.pdf

M.L. Cohen and V. Heine, The Fitting of Pseudopotentials to Experimental Data and their Subsequent Application in *Solid State Physics*, vol. 24, ed. by Henry Ehrenreich, Frederick Seitz and David Turnbull (New York, Academic Press, 1970), pp 37–248. ISBN-13: 978-0126077247. https://www.sciencedirect.com/science/article/pii/S0081194708600703

M.L. Cohen, J.R. Chelikowsky, *Electronic Structure and Optical Properties of Semiconductors*, 2nd edn. (Springer, Berlin, 1989). ISBN 3-540-51391-4

M.V. Fischetti, J. Kim, S. Narayanan, Z.-Y. Ong, C. Sachs, D.K. Ferry, S.J. Aboud, Pseudopotential-based studies of electron transport in graphene and graphene nanoribbons. J. Phys. Condens. Matter **25**, 473202, 1–37 (2013). https://3.amazonaws.com/academia.edu.documents/ 43565390/Pseudopotential-based.pdf?response-content-disposition=inline%3B%20filename% 3DPseudopotential-based_studies_of_electro.pdf&X-Amz-Algorithm=AWS4-HMAC-SHA256 &X-Amz-Credential=AKIAIWOWYYGZ2Y53UL3A%2F20200209%2Fus-east-1%2Fs3%2 Faws4_request&X-Amz-Date=20200209T105748Z&X-Amz-Expires=3600&X-Amz- Signed-Headers=host&X-Amz-Signature=3be8da87be8b9bac430c6d9062344819e77a836c8a57fb2ef 57db75ba07faace

J. Hafner, *Foundations of Density Functional Theory*. Institut für Materialphysik and Center for Computational Material Science, University of Vienna. Vienna. https://www.vasp.at/vasp-workshop/slides/dft_introd.pdf

C. Hamaguchi, *Basic Semiconductor Physics*, 2nd edn. (Springer, Heidelberg, 2010). ISBN 978-3-642-03302-5

V. Heine, The Pseudopotential Concept, in *Solid State Physics*, vol. 24, ed. by Henry Ehrenreich, Frederick Seitz and David Turnbull (New York, Academic Press, 1970), pp 37–248. ISBN-13: 978-01 26077247

P. Hohenberg, W. Kohn, Inhomogeneous electron gas. Phys. Rev. **136**, B864–B867 (1964). http://yclept.ucdavis.edu/course/240C/Notes/DFT/HohenbergKohn.pdf

H. Jones, *The Theory of Brillouin Zones and Electronic States in Crystals* (North Holland Publishing Co., Amsterdam, 1975). ISBN 0-444-10639-1

C. Kittel, *Introduction to Solid-State Physics*, 6th edn. (Wiley, New York, 1986). ISBN 0-87474-4

W. Kohn, L. Sham, Self-consistent equations including exchange and correlation effects. Phys. Rev. 140 A1133–1138 (1965). http://lptms.u-psud.fr/nicolas_pavloff/files/2010/03/PhysRev.140.A11331.pdf

R. LeSar, *Introduction to Computational Materials Science* (Cambridge University Press, 2013). ISBN -13 978-0521845878

O. Madelung, *Introduction to Solid-State Theory* (Springer, Berlin, 1981). ISBN 3-540-08516-5

W. Saslow, T.K. Bergstresser, M.L. Cohen, Bandstructure and optical properties of diamond. Phy. Rev. Lett. **16**, 354–356 (1966); erratum: Phys. Rev. Lett. **21**, 715 (1968) https://journals.aps.org/prl/abstract/10.1103/PhysRevLett.16.354

D.S. Scholl, J.A. Steckel, *Density Functional Theory: a Practical Introduction* (Wiley, Hoboken, 2009). ISBN 978-0-470-37317-0

Chapter 5
The Harmonic Oscillator and Quantization of Electromagnetic Fields

Abstract Maxwell's equations give a description of electricity and magnetism in terms of the electric and magnetic fields, but say very little about the quanta: electrons and photons, that constitute the fields. This field description treats electric and magnetic phenomena like continuum fluids. By quantization, we can treat electromagnetic phenomena from the perspective of the quanta. This is the basis of quantum photonics. We show that photon behavior can be described mathematically by the formalism of the harmonic oscillator. This important principle enables the description of single photon; that is, quantum photon behavior and the quantization of the electromagnetic field. Creation or annihilation of photons is the way that the electromagnetic field exchanges energy with matter. The creation and annihilation operators can be used to describe the emission and absorption of photons in a direct and transparent way, without having to derive the quantum wavefunction in explicit form. We use these operators to analyze stimulated and spontaneous emission (and absorption) of photons, and thus the performance of Lasers, LEDs and photodiodes.

5.1 Introduction

The time-independent Schrödinger equation for a particle in the presence of a one-dimensional potential is written:

$$\frac{p^2}{2m}\Psi + V(x)\Psi = -\frac{\hbar^2}{2m}\frac{d^2}{dx^2}\Psi(x) + V(x)\Psi(x) = E\Psi(x). \tag{5.1}$$

The total energy E is shared between kinetic $(= \frac{p^2}{2m})$ and potential $(V(x))$ energies. In Chap. 3, we examined the properties of a particle confined by a constant potential (particle in a well). In this environment, the allowed quantized total energy scales according to the square of the quantum number:

$$E_n = \frac{n^2\pi^2\hbar^2}{2L^2m}, \tag{5.2}$$

© Springer Nature Switzerland AG 2020

T. P. Pearsall, *Quantum Photonics*, Graduate Texts in Physics,
https://doi.org/10.1007/978-3-030-47325-9_5

where L is the width of the potential well. The dependence of the allowed energies on the quantum number is a function of the geometry of the potential.

In this chapter we will study the behavior of a particle confined by a potential well, the width of which varies as a function of distance.

A simple example of a harmonic oscillator is the movement of a mass connected to a spring. When displaced from equilibrium the spring exerts a restoring force that is proportional to the displacement: $F(x) = -Kx$. The corresponding potential energy stored in the spring is: $V(x) = \frac{1}{2}Kx^2$. When released, the mass begins to move, and energy oscillates between reservoirs of kinetic and potential contributions with a frequency: $\omega = \sqrt{\frac{K}{m}}$.

The vibration of atoms in a crystalline solid is an important example in physics which can be treated accurately by harmonic oscillator analysis.

Maxwell's equations describe the propagation of a photon as the periodic exchange of energy stored in its electric and magnetic field. We will argue in Sect. 5.4 that photon behavior can be described mathematically by the formalism of the harmonic oscillator. This important principle enables the description of single photon; that is, quantum photon behavior and the quantization of the electromagnetic field. This is the subject of this chapter.

5.2 The Harmonic Oscillator

The Hamiltonian operator for the total energy of a simple one-dimensional harmonic oscillator is:

$$\mathcal{H} = \frac{p^2}{2m} + \frac{1}{2}Kx^2 \tag{5.3}$$

The time-independent Schrödinger equation is

$$-\frac{\hbar^2}{2m}\frac{d^2}{dx^2}\Psi(x) + V(x)\Psi(x) = E\Psi(x) \tag{5.4}$$

For the case of a quantum harmonic oscillator, (such as an electron confined to a quantum well having a parabolic potential), we can write:

$$-\frac{d^2}{dx^2}\Psi(x) + \frac{mK}{\hbar^2}x^2\Psi(x) = \frac{2mE}{\hbar^2}\Psi(x) \tag{5.5}$$

$$-\frac{d^2}{dx^2}\Psi(x) + \alpha^4 x^2\Psi(x) = \frac{2mE}{\hbar^2}\Psi(x) \tag{5.6}$$

$$\text{where } \alpha^4 = \left(\frac{mK}{\hbar^2}\right) = \left(\frac{m\omega}{\hbar}\right)^2.$$

The solution to this equation is straightforward if somewhat labor-intensive. For the details, we refer the reader to any standard text on quantum mechanics, several of which are given in the references to this chapter (Dirac 1958; Landau and Lifshitz 1958; Merzbacher 1960; Yariv 1975).

The elements are as follows: The eigen functions are composed of the product of a Gaussian function that declines to zero at $x = \pm\infty$, permitting normalization of the wavefunction, and an oscillating polynomial part inside the potential well where the particle is confined.

$$\Psi(x) = Ce^{-\frac{(\alpha x)^2}{2}} P(x), \text{ where } C \text{ is the constant of normalization.} \tag{5.7}$$

$P(x)$ is a polynomial of order n for the nth energy level.

These polynomials are called Hermite polynomials, $H_n(\alpha x)$. The first few Hermite polynomials are:

$$\begin{aligned} H_0(\alpha x) &= 1 \\ H_1(\alpha x) &= 2\alpha x \\ H_2(\alpha x) &= 4(\alpha x)^2 - 2 \end{aligned} \tag{5.8}$$

However, as is often the case, we are not as interested in the wavefunction as we are to know the quantized energies of the particle. Using Schrödinger's equation, we can solve for the energies:

$$\begin{aligned} E_0 &= \hbar\omega\left(\tfrac{1}{2}\right) \\ E_1 &= \hbar\omega\left(1 + \tfrac{1}{2}\right) \\ E_2 &= \hbar\omega\left(2 + \tfrac{1}{2}\right) \\ \ldots \\ E_n &= \hbar\omega\left(n + \tfrac{1}{2}\right) \end{aligned} \tag{5.9}$$

The stable energies or modes of the quantum harmonic oscillator are equally-spaced in energy $\Delta(E_n - E_{n-1}) = \hbar\omega$. This is a consequence of the quadratic dependence of the potential on distance. The energy level quantum number n can be any positive integer, including zero. In the case of lattice vibrations (phonons) and photons, the quantum number n also gives the number of quanta present at a particular frequency ω. Note that there is an energy of $\frac{1}{2}\hbar\omega$ associated with the vacuum state where no quantum is present. The proof of this result is the subject of Exercise 5.1. This is a manifestation of the Heisenberg uncertainty principle which states that the product of the minimum uncertainties is

$$\Delta p \Delta x = \frac{\hbar}{2}. \tag{5.10}$$

The Heisenberg uncertainty principle also teaches that the result of the measurement of position, followed by measurement of momentum is not the same as that of measuring first the momentum and then the position. That is, the position and momentum operations do not commute. The commutator of p and x is the mathematical expression of this result:

$$[p,x]\,\Psi \equiv p(x\Psi) - x(p\Psi) = -i\hbar\frac{d}{dx}(x\Psi) - x\left(-ih\frac{d}{dx}\Psi\right)$$
$$= -i\hbar\left(\Psi + x\frac{d}{dx}\Psi\right) + ihx\frac{d\Psi}{dx} = -i\hbar\Psi$$

Therefore we write:

$$[p,x] = -i\hbar \tag{5.11}$$

5.3 Raising and Lowering Operators

We introduce two operators: a^+ and a.

$$a^+ = \frac{\alpha}{\sqrt{2}}x - \frac{1}{\alpha\sqrt{2}}\frac{\partial}{\partial x} = \frac{\alpha}{\sqrt{2}}x - \frac{i}{\alpha\hbar\sqrt{2}}p \tag{5.12}$$

where $p = -i\hbar\frac{\partial}{\partial x}$ is the quantum mechanical momentum operator

$$a = \frac{\alpha}{\sqrt{2}}x + \frac{1}{\alpha\sqrt{2}}\frac{\partial}{\partial x} = \frac{\alpha}{\sqrt{2}}x + \frac{i}{\alpha\hbar\sqrt{2}}p \tag{5.13}$$

Note that a^+ and a do not commute. The commutator:

$$[a,a^+] = 1 \tag{5.14}$$

The proof of which is the subject of Exercise 5.2.

The particular operators a^+ and a have been developed in particular for study of the harmonic oscillator (Dirac 1958). They enable us to extract many properties of the eigenfunctions without having to write down the eigenfunctions explicitly. Thus, we can study the energy eigenvalues without having to solve for the eigenfunctions.

We will state the principal properties of these two operators. A full derivation can be found in the references on quantum mechanics that are cited in this chapter.

1. If ϕ_n is an eigenfunction of the harmonic oscillator, then

$$a^+\phi_n = (\sqrt{n+1})\phi_{n+1} \tag{5.15}$$

That is, the operator a^+ transforms a given eigenstate to the next higher eigenstate multiplied by the square root of its occupation number.

2. If ϕ_n is an eigenfunction of the harmonic oscillator, then

$$a\phi_n = (\sqrt{n})\phi_{n-1} \tag{5.16}$$

   That is, the operator a transforms a given eigenstate to the next lower eigenstate multiplied by the square root of the occupation number of the original state.

The operators a^+ and a are referred to as raising and lowering operators for obvious reasons.

Next, we solve (5.12) and (5.13) for p and x,

$$p = \frac{i\hbar\alpha}{\sqrt{2}}(a^+ - a) \tag{5.17}$$

$$x = \frac{1}{\alpha\sqrt{2}}(a^+ + a) \tag{5.18}$$

Substituting these expressions for p and x into (5.3) we obtain;

$$\mathcal{H} = \frac{\hbar\omega}{2}(aa^+ + a^+a) \tag{5.19}$$

Using the expression for the commutator of (5.12)

$$\mathcal{H} = \hbar\omega\left(a^+a + \frac{1}{2}\right) \tag{5.20}$$

Finally, we use (5.15) and (5.16) to show (see Exercise 5.4)

$$a^+(a\phi_n) = n\phi_n \tag{5.21}$$

So that we find directly the results of (5.7)

$$\mathcal{H} = \hbar\omega\left(n + \frac{1}{2}\right) \tag{5.22}$$

The operation $a^+(a\phi_n) = n\phi_n$ gives the number of quanta in the eigenstate ϕ_n. Since the total energy is directly related to the number of quanta, we have two important results that characterise the harmonic oscillator without having to work explicitly with the eigenstate wavefunction itself. This is an immense simplification that enables a clearer view of the physics. Next, we will apply this methodology to the characterisation of photons as the quanta that constitute the electromagnetic field.

5.4 Quantization of the Electromagnetic Fields

Maxwell's equations give a description of electricity and magnetism in terms of the electric and magnetic fields, but say very little about the quanta: electrons and photons, that constitute the fields. This field description treats electric and magnetic phenomena like continuum fluids. By quantization, we can treat electromagnetic phenomena from the perspective of the quanta. This is the basis of quantum photonics.

A laser (**l**ight **a**mplification by **s**timulated **e**mission of **r**adiation) is a device whose operation is based on the behavior of an electromagnetic field inside an optical resonator cavity. Laser emission is based on the presence of both spontaneous and stimulated emission. Stimulated emission does not occur on its own. It requires the presence of a photon. These photons are generated by spontaneous emission. A laser cannot function without the occurrence of spontaneous emission.

The macroscopic description given by Maxwell's equations is inadequate to describe spontaneous emission, whereas, spontaneous emission is a natural consequence of the quantization of the electromagnetic fields. We will show here that the quantum analysis of the electromagnetic field inside a laser cavity is formally equivalent to the treatment of the harmonic oscillator. This results directly from the fact that the electromagnetic energy oscillates between electric and magnetic fields in a way that is mathematically identical to the oscillation of energy between kinetic and potential components in a harmonic oscillator.

The Hamiltonian operator for the total energy is:

$$\mathcal{H} = \frac{1}{2}\int_V (\mu \mathbf{H}\cdot\mathbf{H} + \varepsilon \mathbf{E}\cdot\mathbf{E})dV \tag{5.23}$$

We seek conjugate variables P and Q that will let us write the Hamiltonian operator in the form:

$$\mathcal{H} = \sum_a P_a^2 + c^2 Q_a^2 \tag{5.24}$$

With the desired commutator properties:

$$[P_a, Q_b] = -i\hbar\delta_{a,b} \tag{5.25}$$

and

$$[Q_a, Q_b] = [P_a, P_b] = 0 \tag{5.26}$$

Of course Q_a and P_a no longer correspond to position and momentum. We wish to describe elementary excitations of the electromagnetic field. These are photons.

Without any loss of generality, we can suppose that the electric and magnetic fields exist inside a volume of arbitrary dimensions. This volume forms a cavity. To develop the formalism, the electric and magnetic fields can be expressed as an expansion of the cavity modes: $\mathbf{E}_a(\mathbf{r})$ and $\mathbf{H}_a(\mathbf{r})$, which are orthogonal by design. We can propose

$$\mathbf{H}(\mathbf{r}, t) = \sum_a \frac{1}{\sqrt{\mu}} \omega_a Q_a(t) \mathbf{H}_a(\mathbf{r}) \tag{5.27}$$

and

$$\mathbf{E}(\mathbf{r}, t) = -\sum_a \frac{1}{\sqrt{\varepsilon}} P_a(t) \mathbf{E}_a(\mathbf{r}) \tag{5.28}$$

where as usual $\omega = \frac{k_a}{\sqrt{\mu\varepsilon}}$.

Substituting (5.27) and (5.28) in (5.23), and using the orthogonality of cavity modes, we obtain:

$$\mathcal{H} = \sum_a \frac{1}{2}(P_a^2(t) + \omega_a^2 Q_a^2(t)) \tag{5.29}$$

It can be seen by comparison with (5.3) that (5.29) is expressed in the desired form for a harmonic oscillator. By considering P_a and Q_a as canonically conjugate variables, we can use all the results (5.10 through 5.22) to quantize the electromagnetic fields, based on the individual quantum: that is, the photon (Grynberg et al. 2010).

Following (5.11), the basic commutator is:

$$[P_a, Q_b] = -i\hbar\delta_{a,b} \tag{5.30}$$

We define raising and lowering operators a_l^+ and a_l. These operators are time-dependent, because $P_a(t)$ and $Q_a(t)$ are time-dependent.

$$a_l^+(t) = \sqrt{\frac{1}{2\hbar\omega_l}}[\omega_n Q_l(t) - iP_l(t)] \tag{5.31}$$

and

$$a_l(t) = \sqrt{\frac{1}{2\hbar\omega_l}}[\omega_n Q_l(t) + iP_l(t)] \tag{5.32}$$

As is the case with (5.14), the commutator $[a_l(t), a_l^+(t)] = 1$.

Solving for P and Q,

$$P_l(t) = i\sqrt{\frac{\hbar\omega_l}{2}}[a_l^+(t) - a_l(t)] \tag{5.33}$$

and

$$Q_l(t) = \sqrt{\frac{\hbar}{2\omega_l}}[a_l^+(t) + a_l(t)] \tag{5.34}$$

Substitution of these expressions into (5.24) gives straight-away:

$$\mathcal{H} = \sum_l \hbar\omega_l \left(a_l^+ a_l + \frac{1}{2} \right) \tag{5.35}$$

The eigenfunctions of this Hamiltonian are the stationary states of the electromagnetic field, for which total energy is conserved as it oscillates between electrical and magnetic fields. (The time dependence is of the form $e^{i\omega t}$, and can be treated separately).

If ϕ_n is such a stationary state, with characteristic frequency $= \omega_l$, then:

$$a_l^+ \phi_n = \sqrt{n+1}\phi_{n+1} \tag{5.36}$$

$$a_l \phi_n = \sqrt{n}\phi_{n-1} \tag{5.37}$$

and

$$a_l^+(a\phi_n) = n\phi_n \tag{5.38}$$

These results give a complete description of the electro-magnetic field from the quantum; that is, from the photon point of view. A given stable state of the field $=$ ϕ_n consists of n quanta or photons. Each photon has an energy $= \hbar\omega_l$. The energy of the state ϕ_n is the sum of the energy of the photons, plus $\frac{1}{2}\hbar\omega_l$. Photons are excitations of the electromagnetic field. Excitations can be created or destroyed. On the other hand there is no mechanism for interaction of these excitations.

The operator a_l^+ changes the stable state ϕ_n to the stable state ϕ_{n+1}. This action creates an additional photon of frequency ω_l, because the state ϕ_{n+1} must have $n+1$ quanta. For this reason, we can call the a_l^+ the creation operator. In a similar fashion, the operator a_l changes the stable state ϕ_n to the state ϕ_{n-1}. This removes a photon from the field, and a_l is known as an annihilation operator. The actions of creation or annihilation correspond experimentally to the emission or absorption of a photon. Note from (5.31), that the application of the creation operator on the vacuum state ($n = 0$) results in the creation of a photon in the $n = 1$ state. On the other hand, the application of the annihilation operator on the vacuum state returns 0 photons.

We can use these results to describe entangled photons, lasers and photodetectors.

5.5 Spontaneous Parametric Down-Conversion

Creation or annihilation of photons is the way that the electromagnetic field exchanges energy with matter. For example, a photon can be destroyed, promoting an electron from one energy state to another, provided that the energy difference between the initial and final states is equal to the energy of the photon. An important example for quantum photonics is spontaneous parametric down-conversion (SPDC) which is a useful method for creating entangled photons. Such photons are the workhorse for the examination of Bell's Theorem, which is the subject of Chap. 6.

Spontaneous parametric down-conversion is a special case of parametric optical amplification, which occurs in non-linear optical materials. In Chap. 9 we will study some important non-linear optical effects using mostly a classic analysis. However, the case of spontaneous parametric down-conversion cannot be described by Maxwell's equations because it involves interactions with the vacuum state, where classically speaking, the electric field is zero. In this section we can show in a straightforward way, using the quantized treatment of photons, some fundamental aspects of parametric down-conversion.

With the quantized representation of the electromagnetic field it is natural to think of the photon as a particle with energy $E = \hbar\omega$ and momentum $\mathbf{p} = \hbar\mathbf{k}$. We will now consider the situation where a photon of energy $E = \hbar\omega_3$ is annihilated, creating in its place two photons with energies $E_1 = \hbar\omega_1$ and $E_2 = \hbar\omega_2$, where $\hbar\omega_3 = \hbar\omega_1 + \hbar\omega_2$ conserving energy and where $\hbar\mathbf{k}_3 = \hbar\mathbf{k}_1 + \hbar\mathbf{k}_2$ conserving momentum. These two photons must be created within a wavelength of each other in order to conserve momentum. This means that the two lower energy photons are created in pairs simultaneously, (that is within a time interval $\Delta\tau < \frac{1}{\omega_3}$) with the annihilation of the higher energy photon in order to conserve energy. The Hamiltonian for this interaction can be written down straight away as:

$$\mathcal{H} = \sum_l \hbar\omega_l \left(a_l^+ a_l + \frac{1}{2} \right) + \hbar\gamma (a_1^+ a_2^+ a_3 + a_1 a_2 a_3^+) = \mathcal{H}_o + \mathcal{H}_1 \tag{5.39}$$

The additional term contains a coupling constant γ and corresponds to the annihilation of a photon in the mode 3 and the creation of two photons from the vacuum in modes 1 and 2 where no photons were present initially. It contains a second term corresponding to the annihilation of two photons in the vacuum state, but this returns 0. Since the interaction is simultaneous, the order of the events is not important.

The initial state is written:

$$\phi_i(t = 0) = |0, 0, N_3\rangle \tag{5.40}$$

The effect of the Hamiltonian $\mathcal{H}_1$ acting on this state is:

$$\mathcal{H}_1 |\phi_i(t)\rangle = \hbar\gamma\sqrt{N_3}|1, 1, N_3 - 1\rangle + 0 \tag{5.41}$$

The intensity of the down-converted beam is proportional to the square modulus of this state.

$$I \propto |\mathcal{H}_1 |\phi_i(t)\rangle^2| \propto N_3 \tag{5.42}$$

and is linearly proportional to the intensity of the incident beam. As a result, the conversion efficiency is not dependent on pump beam intensity (Fig. 5.1).

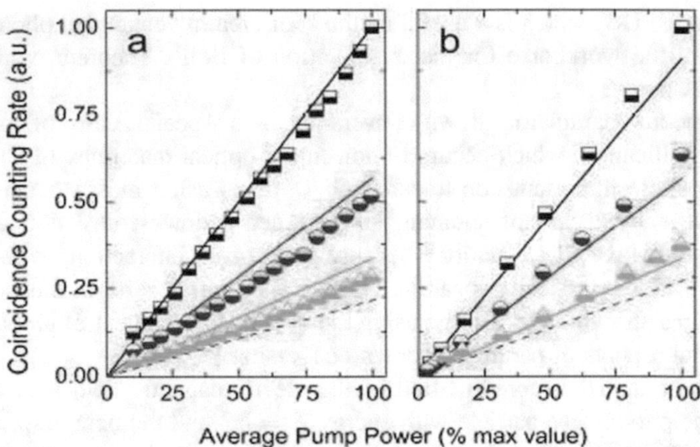

Fig. 5.1 The detection rate of down-converted photons is linearly proportional to the pump power, in agreement with (5.42). **a**, **b** correspond to two different sample thicknesses. In each case, the three curves correspond to different conditions of pump beam focus. Reproduced from Ö. Süzer and T.G. Goodson III, Optics Express, **16**, pp. 20, 166–20, 175 (2008) by permission of the Optical Society of America, 2016

In practice, the down-conversion effect is very weak, with an efficiency of about 1 part in 10^{12}. That is, γ (the coupling constant) is small. This is in part due to the 3-particle interaction which requires all three particles to be in the same place (within a wavelength) and the same time (within a reciprocal frequency). Thus it can be treated as a perturbation. Such a low rate assures that any down-converted photons exit the conversion crystal before the generation of a subsequent pair. For measurements on distinguishable photon pairs, such a low efficiency results in an arrival rate of several thousand per second. This is an advantage in experiment because it allows substantial "dead time" between the arrival of photon pairs at the detectors, and this allows the clean identification of correlated photon pairs.

Birefringent crystalline materials provide a medium in which both energy conservation and momentum conservation can be achieved. Beta Barium Borate ($\beta - BaB_2O_4$) or BBO is a birefringent non-linear crystal widely used for this application. Its wavelength range of transparency is $300 < \lambda < 2000$ nm. Silicon avalanche photodiodes are a good choice to detect single photons. Their sensitivity is optimal at wavelengths shorter than 900 nm. For example, a He-Cd laser emits at $\lambda = 325$ nm in vacuum ($\hbar\omega = 3.82\,\text{eV}$). It can generate two photons having half the energy ($\hbar\omega = 1.91\,\text{eV}, \lambda = 650\,\text{nm}$ in vacuum). Using type-I phase matching, the polarization of the two photons is the same. When type-II phase-matching is used, one photon will be polarized at 90° with respect to the other. This is the desired situation for generating an entangled photon pair.

Momentum conservation is achieved by considering the wavevectors of the three photons. The total momentum, i.e. the wavevector of the incident beam, is determined by the experimental choice of its propagation direction in the

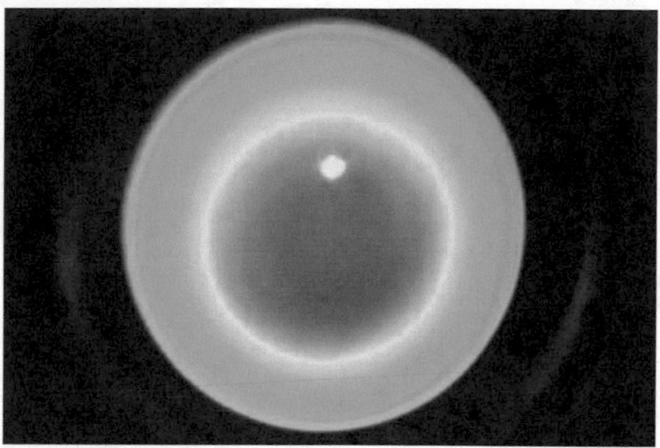

Fig. 5.2 Photograph of parametric down-conversion in BBO, using type-I phase—matching. The laser pump beam (energy = 3.82 eV) interacts with BBO through its 2nd order nonlinear optic coefficient to annihilate a pump photon and create a pair of photons from the vacuum state. Many possibilities exist for the wavelength and emission directions of the down-converted photons provided that total energy and momentum are conserved. (Photograph courtesy of Prof. Alexander Ling, Center for Quantum Technologies, National University of Singapore, reproduced by the kind permission of the author)

birefringent crystal relative to its optical axis. With the energy of the incident photon invariant and the wavevector determined by the propagation direction, a number of solutions for momentum conservation are possible.

The down-converted photons span a range of allowed photon energies and emission angles. When type-I phase matching is used, the two down-converted photons are emitted on the surface of a cone. A presentation of this effect by Alan Migdall can be viewed on Youtube (Migdall 2010). An image of this fluorescence is shown in Fig. 5.2 (Ling 2008).

To illustrate, we suppose the case of degenerate down-conversion, where the energies of the two down-converted photons are the same.

$$\begin{aligned} &\text{Conservation of energy: } \omega_3 = 2\omega_1 = 2\omega_2 \\ &\text{Conservation of momentum } \mathbf{k}_3 = \mathbf{k}_1 + \mathbf{k}_2 \end{aligned} \tag{5.43}$$

where

$|\mathbf{k}_3| = \frac{\omega_3 n_3}{c}$: n_3 is the index of refraction of the pump beam at the higher frequency and in the direction of propagation in the crystal. One photon of the down-converted pair is emitted along the extraordinary axis.

$|\mathbf{k}_1| = \frac{\omega_1 n_e}{c}$: n_e is the extraordinary index of refraction at $\omega_1 = \omega_2 = \frac{\omega_3}{2}$. It is polarized parallel to the plane defined by the optic axis of the crystal and the pump beam.

The other photon of the pair is emitted along the ordinary axis:

$|\mathbf{k}_2| = \frac{\omega_1 n_o}{c}$: n_o is the ordinary index of refraction at $\omega_2 = \frac{\omega_3}{2}$. This photon is polarized perpendicular to the plane defined by the optic axis and the pump beam.

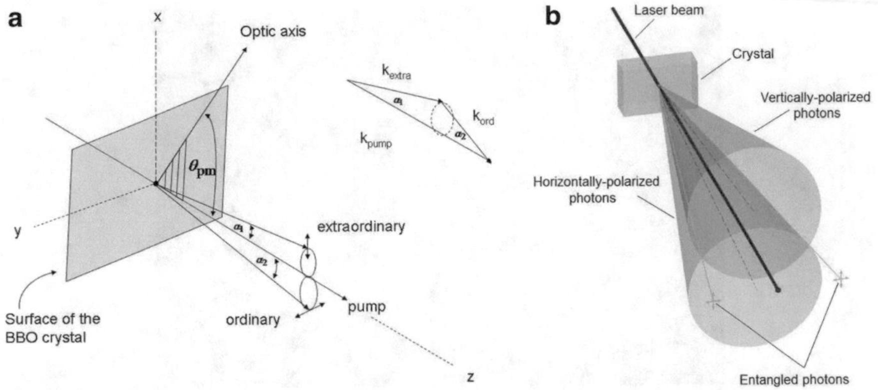

Fig. 5.3 A schematic diagram of spontaneous parametric down-conversion in a non-linear crystal like BBO, using type-II phase-matching. Down-converted photons are emitted along the surface of a cone only when momentum and energy are simultaneously conserved. (Fig. 5.3b copyright J-Wiki at English Wikipedia, 2011, https://commons.wikimedia.org/wiki/File:SPDC_figure.png)

In the case of type-II conservation of momentum, $|\mathbf{k}_1| + |\mathbf{k}_2| > |\mathbf{k}_3|$. In this case, both $\mathbf{k}_1$ and $\mathbf{k}_2$ must be oriented at angles with respect to $\mathbf{k}_3$. This means that the down-converted photons are emitted in different directions, away from each other and from the pump beam. This is a convenient way to separate the photons (Kwiat et al. 1995). Momentum conservation is achieved when:

$$|\mathbf{k}_3| = |\mathbf{k}_1| \cos(\alpha_1) + |\mathbf{k}_2| \cos(\alpha_2) \tag{5.44}$$

The photon pairs that satisfy both energy and momentum conservation are emitted in directions that lie along the surface of a cone, one for each photon of the down-converted pair. The angle of each cone (α_1 or α_2) depends on the wavevectors of all three photons.

When the pump laser beam at $\lambda = 325\,\text{nm}$ (in vacuum) passes through a BBO crystal, some of the photons undergo spontaneous down-conversion with trajectories that are constrained by conservation of momentum to lie along the surface of two cones, as shown in Figs. 5.3 and 5.4. The polarization of the photon on one cone is correlated and perpendicular to the polarization of the photon on the second cone. Because of conservation of momentum, each photon of a pair travels on a trajectory that is symmetric to that of the second photon with respect to the axis defined by the trajectory of the pump beam.

Changing the angle of the pump beam relative to the optical axis also changes its wavevector. While the energy conservation condition remains unchanged, this modifies the condition for momentum conservation, resulting in a change in the angle between the two emission cones. An example is diagrammed in Fig. 5.4b where the two emission cones intersect. A video of this experiment has been prepared by the group of Prof. Alexander Ling at the Center for Quantum

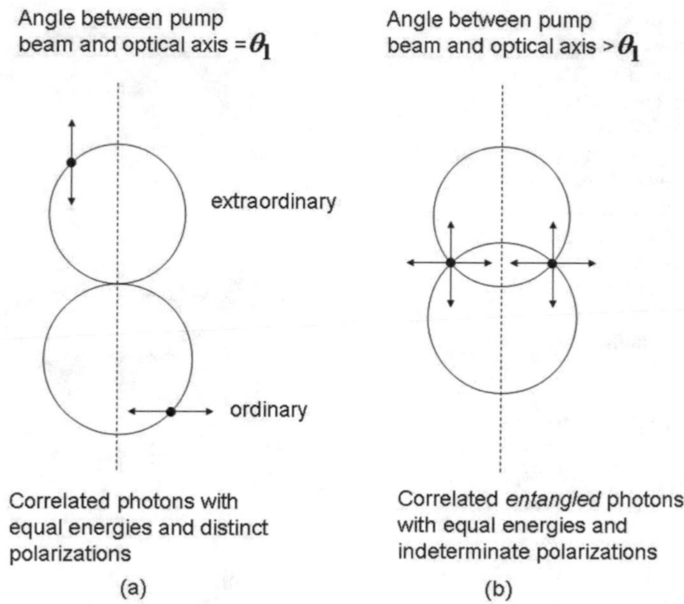

Fig. 5.4 A diagram of the cross-section of the cones along which the down-converted photons propagate. **a** A photon can be emitted along any path along the cone surface, but the position of its pair is correlated so that momentum is conserved. The polarization of the photons is mutually orthogonal. **b** When the angle between the pump beam and the optic axis increases, the beam experiences a smaller index of refraction in BBO, and its wavevector increases. To achieve momentum conservation, the angles α_1 and α_2 must decrease. This causes the two emission cones to intersect, creating pairs of polarization-entangled photons

Technologies at the National University of Singapore. You can view this on-line at: http://qolah.org/research/hqpdc/rings.avi.

Patrick Bronner of the University Erlangen has created a sequence of on-line experiments "Quantum Lab" on quantum photon behavior (Bronner 2009). This presentation includes a video of spontaneous parametric down-conversion as well as experiments using entangled photons prepared by this method. These can be viewed at: http://www.didaktik.physik.uni-erlangen.de/quantumlab/english/index.html "Entanglement".

The trajectories of the photon pairs may exist simultaneously along the two lines where the surface of the cones intersects. This effect can be captured, and a photographic image created by down-converted photons is shown in Fig. 5.5 (Reck 1996). For photons propagating along this line, the polarization is indeterminate, although the net polarization of the pair remains zero. This results in entanglement of the photon pairs (Catalano 2014). Because of this simple symmetrical geometric arrangement, it is straightforward to harvest these entangled pairs.

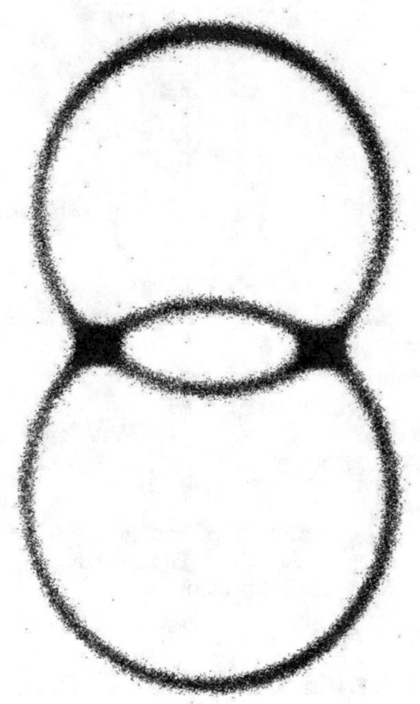

Fig. 5.5 Photographic image of down-converted photon pairs in BBO. The two emission cones have been made to intersect in the experiment creating polarization entanglement. From experiments carried out by P.G. Kwiat and M. Reck. (Photograph courtesy of Prof. Paul Kwiat, Department of Physics, University of Illinois, Urbana-Champaign, reproduced by the kind permission of the authors)

5.6 Two-Photon Interference

Interference of optical beams has been studied at least since the observations of Francesco-Maria Grimaldi in the 17th century. Single photons also show interference. The experiments of Dykstra et al. and Grangier et al., discussed in Chap. 2 show that the wavefunction of a single photon interferes with itself, and produces an interference pattern that is characteristic of the structures encountered during propagation. Self-interference is not restricted to photons, it is a general property of single quanta. The electron interference pattern obtained by Tonomura (Fig. 2.13) is an example of self interference of single electrons. Experiments such as these demonstrate the extended and non-localised nature of the wavefunction of quantum particle-waves.

On this topic, Dirac has stated on page 9 of his text on quantum mechanics (see reference Dirac 1958) "Each photon interferes with itself. Interference between two different photons never occurs".

This is perhaps an overstatement of the situation. To be sure, two photons do not interact with each other, but interference can occur if the two photons are indistinguishable. Indistinguishable photons have the same defining quantum states of frequency and polarization. For interference to occur, the wave functions of two indistinguishable photons must also overlap in space and time. On the other hand, such interference between two different electrons, and in general between fermions is forbidden by the Pauli exclusion principle.

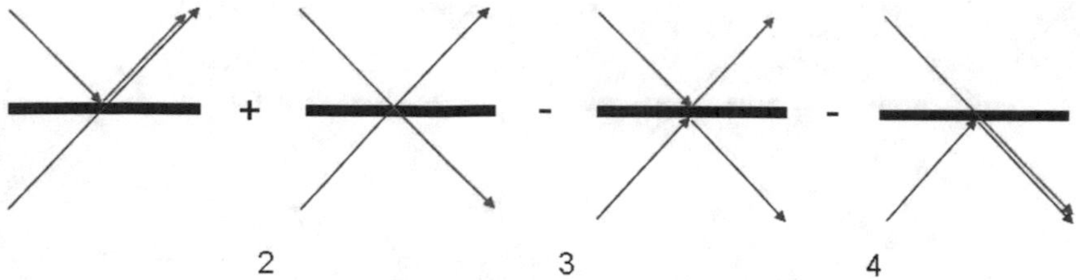

Fig. 5.6 Two photons can intersect a beam splitter in a number of ways. If the photons are distinguishable (one in *red*, the other in *blue*) there are four distinguishable possibilities. There is a relative phase shift of π radians between photons reflected from one surface relative to those reflected from the other surface, represented here by the change in sign of the events 3 and 4 compared to events 1 and 2

We could restate Dirac's claim as follows: "Interference between two distinguishable photons does not occur".

The interference of a photon with itself, or the interference of an electron with itself, is the degree zero of indistinguishability. Although the wavefunction may be reshaped by passage through an object, such as a tunneling barrier or a beam splitter, or a pair of slits, the quantum properties of each part of the wavefunction of a single quantum are identical, and when these components can be made to recombine at the same place at the same time, quantum interference is inevitable (Ou 2007).

In this section, we will present some important experiments that show quantum interference between two photons, demonstrating what is meant by "indistinguishable", and "in the same place at the same time".

A photon has two possibilities when it encounters a beam splitter: reflection or transmission (ignoring absorption or inelastic scattering). When two photons are incident on a beam splitter, one on each side, there are four possible outcomes. These are indicated schematically in Fig. 5.6. The two photons are colored differently making them, as well as the outcomes distinguishable. Note that the "blue" photon experiences a π-phase shift on reflection, relative to the behavior of the "red" photon, following the presentation of the beam splitter in Chap. 2. These photons could be distinguishable for several reasons: different frequencies, different polarizations or different times of passage through the beam splitter.

The two photons are indistinguishable if they have the same frequency and polarization and they arrive at the beam splitter at the same time. This situation is diagrammed in Fig. 5.7.

There are 4 cases as before, but only three of these cases are distinguishable. Cases 2 and 3 cannot be distinguished. In a 50–50% beamsplitter, the amplitudes of the wavefunctions representing these two cases are equal. Because they are out of phase by π radians, they will cancel each other. Only states 1 and 4 remain. This is an example of two-photon interference.

Using the operators of the quantized electromagnetic field developed above, we can develop the quantum mechanical description of this two-photon interference.

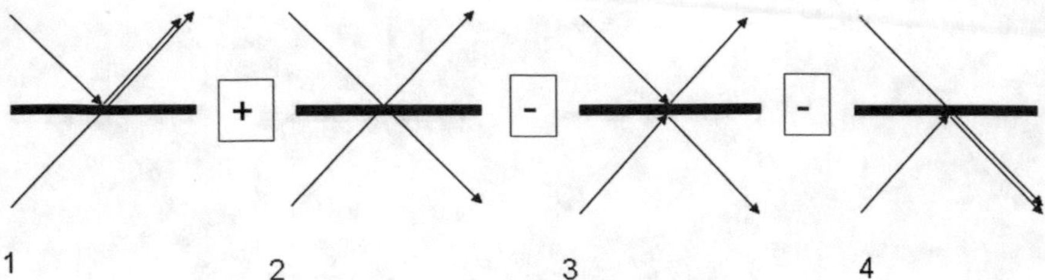

Fig. 5.7 When the two photons are indistinguishable, then the number of possibilities is reduced to 3 events. However, events 2 and 3 are indistinguishable, but have different signs. The wavefunctions corresponding to these two events interfere and cancel each other

In Fig. 5.8 we show a schematic diagram of a beam splitter with input ports **a** and **b** and output ports **c** and **d**.

Using the creation operators a_l^+ and b_l^+, one photon is created at each input simultaneously:

$$a_l^+ b_l^+ |0, 0\rangle_{ab} = |1, 1\rangle_{ab} \tag{5.45}$$

In the experiment to be described below, these photons are created by spontaneous parametric down-conversion with the same frequency and polarization. Subsequently, they propagate to the input ports of the beam splitter. The beam splitter mixes the modes at the input ports and turns them into new modes c and d. The creation operators a_l^+ and b_l^+ are also transformed by the beam splitter. Assuming a 50–50% beam splitter, thus

$$\begin{aligned} a_l^+ &\rightarrow \frac{1}{\sqrt{2}}\left(c_l^+ + d_l^+\right) \\ b_l^+ &\rightarrow \frac{1}{\sqrt{2}}\left(c_l^+ - d_l^+\right) \end{aligned} \tag{5.46}$$

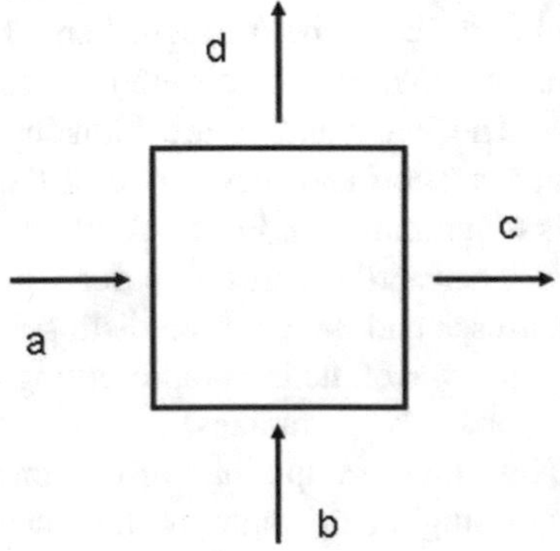

Fig. 5.8 A simplified diagram of *a* beam splitter with input ports *a* and *b* and two output ports *c* and *d*

where the minus sign in (5.46) reflects the relative π phase-shift between the two reflected modes exiting the beam splitter. This ensures that the beam splitter conserves energy; that is, that the transformation is unitary. Arranging in matrix form:

$$\begin{pmatrix} a_l^+ \\ b_l^+ \end{pmatrix} = \frac{1}{\sqrt{2}} \begin{pmatrix} 1 & 1 \\ 1 & -1 \end{pmatrix} \begin{pmatrix} c_l^+ \\ d_l^+ \end{pmatrix} \tag{5.48}$$

Developing (5.45):

$$\begin{aligned} |1,1\rangle_{ab} = a_l^+ b_l^+ |0,0\rangle_{ab} &\rightarrow \frac{1}{2} \left(c_l^+ + d_l^+\right) \left(c_l^+ - d_l^+\right) |0,0\rangle_{cd} \\ &= \frac{1}{2} \{ (c_l^+ c_l^+ - d_l^+ d_l^+) |0,0\rangle_{cd} + [c_l^+ d_l^+ - d_l^+ c_l^+] |0,0\rangle_{cd} \} \end{aligned} \tag{5.49}$$

We recognize the last term as the commutator of the two creation operators; $\lfloor c_l^+, d_l^+ \rfloor$, which vanishes by orthogonality (5.30).

$$|1,1\rangle_{ab} \rightarrow \frac{1}{2} \left\{ \sqrt{2}|2,0\rangle_{cd} - \sqrt{2}|0,2\rangle_{cd} \right\} \tag{5.50}$$

Thus, two identical photons entering a 50–50% beam splitter will appear only as photon pairs at one or the other of the exit ports. As a consequence the photons will not appear coincidentally, one at each of the exit ports. The result that coincidence vanishes when the two identical photons enter the beam splitter is an example of interference of photon wave functions at the quantum level.

The measurement that demonstrates this result is the Hong-Ou-Mandel experiment. This effect is a cornerstone for studies and applications of photon correlation and interference, including quantum communication and quantum computing. Leonard Mandel was a distinguished scientist and teacher whose specialty was quantum optics. The Hong-Ou-Mandel experiment, published in 1987 (see references), marks a major achievement in his lifelong studies of two-photon interference.

In Fig. 5.9, we show a schematic diagram of a Hong-Ou-Mandel measurement.

The measurement proceeds as follows. A UV photon beam (argon laser: $\lambda =$ 351.1 nm) enters a non-linear optical crystal of Potassium di-Hydrogen Phosphate, or KDP. KDP can be used to produce photon pairs by spontaneous parametric down-conversion. The crystal is cut and aligned to produce two photons of the same energy, (degenerate SPDC). Most of the UV beam passes straight through and is absorbed in the beam stop as shown. A small fraction ($\approx 10^{-12}$) of the beam is transformed by SPDC into photon pairs each having one-half the frequency of the photons in the pump beam. These photons have the same frequency and the same polarization and are emitted along the surface of a cone determined by the type-I phase-matching conditions for KDP. Note that the photons have different momenta and propagate in different directions. This does not affect indistinguishability. Mirrors are used to cause the photons to intersect the beam splitter, as shown in Fig. 5.9. From (5.49) we know that two photons will

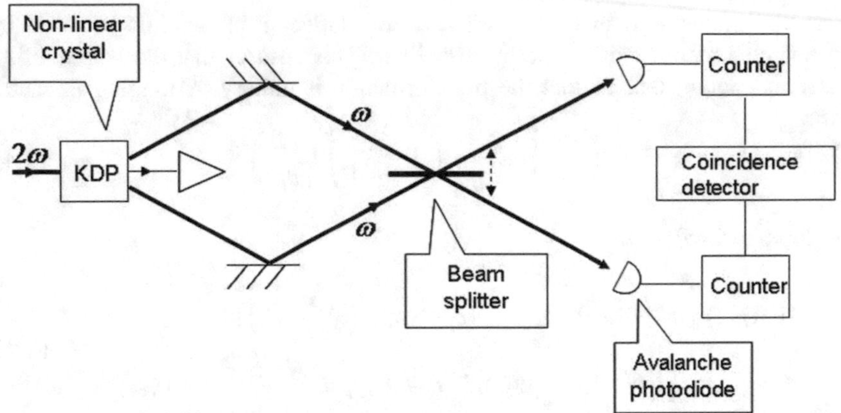

Fig. 5.9 A schematic diagram of the experimental set-up used by Hong, Ou and Mandel for measurement of 2-photon interference (Hong et al. 1987). Spontaneous parametric down-conversion. Type-I phase-matching applies in KDP so the photons are emitted along the surface of a single cone centered on the pump beam axis. The polarizations of the two photons are the same. The relative arrival times of the two photons can be varied by displacement of the beam splitter

experience quantum interference if they are indistinguishable. By design, the two photons in this experiment have the same frequency and the same polarization. The question is: do they intersect the beam splitter at the same place at the same time? In this experiment the beam splitter is mounted on a precise displacement stage and scanned over a range of distances as indicated schematically in Fig. 5.9. The scanner changes the relative distances the photons have to travel in order to reach the beam splitter. In this range is the point in space where the two photons reach the beam splitter and overlap at the same time.

Photon arrival at the two exit paths is recorded by the detectors. A coincidence detector determines whether photons are seen at both detectors at the same time (within a time interval much shorter than the time between detection events. Hong, Ou and Mandel chose this time interval to be 7.5 ns.). Since the creation rate of SPDC photons is in general quite low, (several 1000 per second) it is relatively easy to separate the detection events in time. The measurement of coincidence events published by Hong, Ou and Mandel is shown in Fig. 5.10 and the message is unambiguous. At the left of the figure, the photons do not arrive at the beam splitter at the same time, so they are distinguishable, and no interference occurs. The maximum coincidence rate is about 10 events per second. On the right hand side, the same situation obtains. However, in the center of the figure, the photons arrive at the beam splitter at the same point in space at the same time, and quantum interference reduces the coincidence rate to near zero.

Two photons must have the same polarization in order to interfere. This requirement has been quantified by Kwiat, Steinberg and Chiao in a subsequent publication (Kwiat et al. 1992). They demonstrated how the relative polarization of two photons can be used to determine whether or not two-photon interference takes

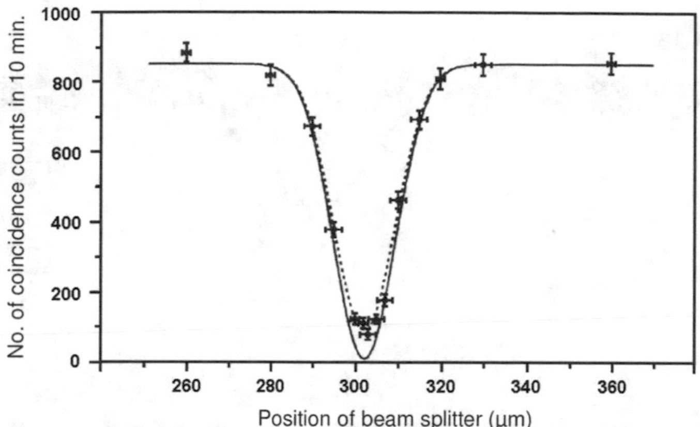

Fig. 5.10 Result of the Hong-Ou-Mandel measurement of 2-photon interference. On the left (beam splitter < 280 μm) and on the right (beam splitter > 330 μm) the two photons pass through the beam splitter at separate times. There is no interference, and detection of a photon in each detector within the same time interval can be recorded as a coincidence. When the beam splitter is positioned in between, the photon wavefunctions overlap in space and time, causing the probability for coincident events to vanish. This figure courtesy of Z-Y Ou. Reproduced by permission of the American Physical Society from Hong et al. (1987)

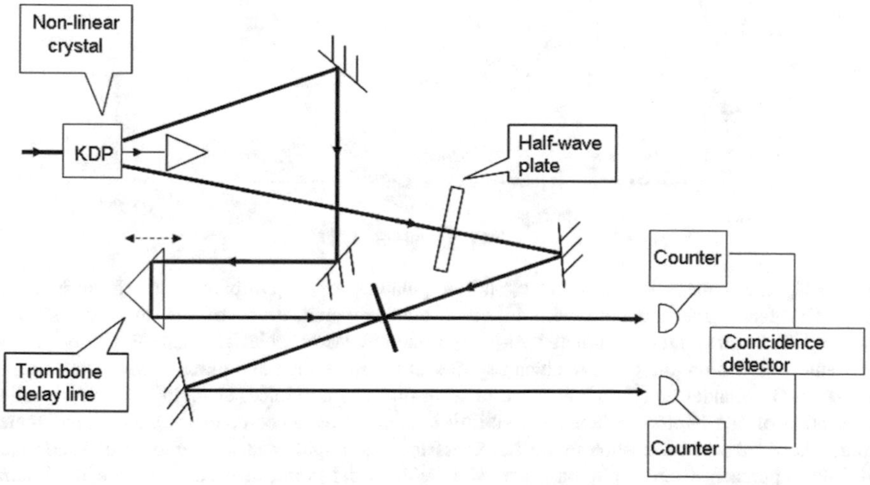

Fig. 5.11 The measurement set-up used by Kwiat, Steinberg and Chiao is similar to the Hong, Ou and Mandel experiment (Kwiat et al. 1992). The superposition of the single photons on the beam splitter is achieved by a "trombone-prism" variable delay line instead of moving the beam splitter directly. A half-wave plate is added in the path of one photon in order to rotate its polarization relative to that of the other photon

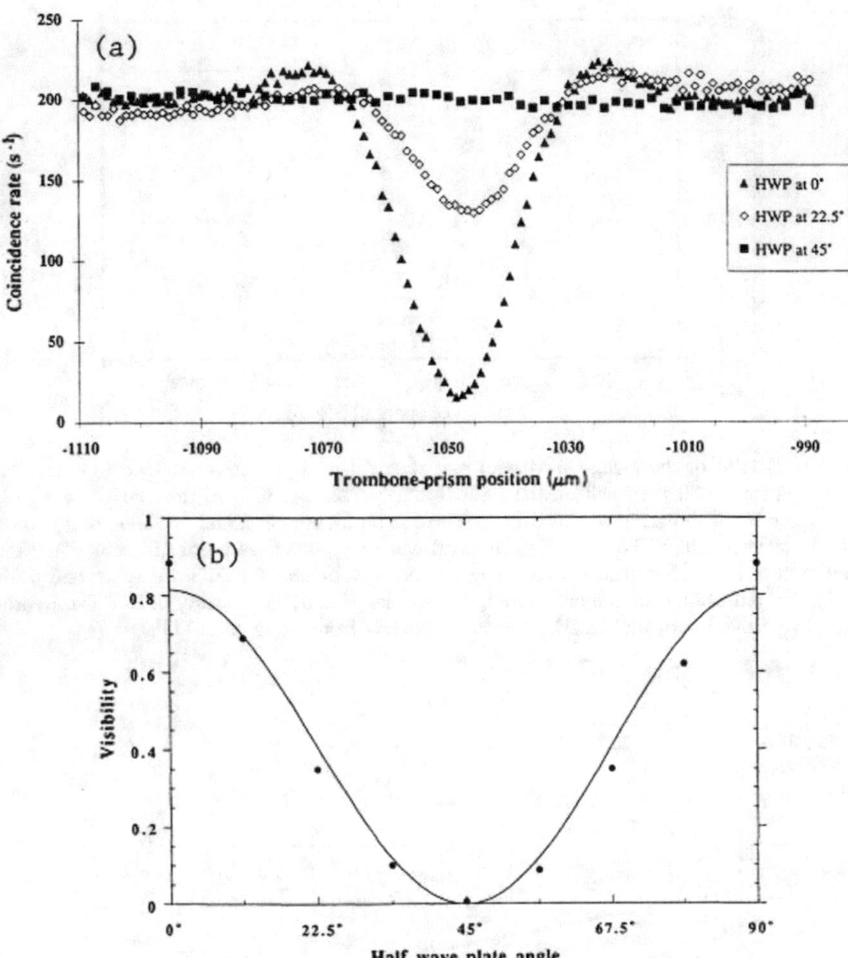

Fig. 5.12 The importance of relative photon polarization in two-photon interference. **a** A Hong-Ou-Mandel measurement with different relative polarizations of the two photons on entering the beam splitter. Complete two-photon interference is achieved when the photons have the same polarization and cross the beam splitter at the same time and place. **b** Measurement of the Hong-Ou-Mandel dip. The half wave plate in one arm is rotated, changing the polarization orientation of one photon. When the visibility is zero, coincidence counts remain at the same values recorded when the photons do not cross the beam splitter at the same time. When the visibility approaches unity, photon correlation reduces the coincidence counts to a minimum level. Reproduced by permission of the American Physical Society from Kwiat et al. (1992)

place. Experiments where polarization rotation is imposed are different from those in which photons are detected. Detection destroys the photon, whereas polarization rotation does not.

The half wave plate shown in Fig. 5.11 is used to rotate the polarization of one of the photons entering the beam splitter. A rotation of the half wave plate by $\theta = 45°$ rotates the polarization of the photon by $2\theta = 90°$. First of all, the trombone delay line is adjusted so that the two photons arrive at the beam splitter at the same time. When the polarizations of both photons are the same, 2-photon interference can occur. When the polarizations are oriented at 90° to each other, the photons are distinguishable and there should be no interference. Thus the Hong-Ou-Mandel measurement should show a dip in the coincidence count when the polarizations are aligned and absence of a dip when the polarizations are orthogonal. This result is demonstrated in Fig. 5.12a.

Kwiat et al. (1992) have demonstrated that 2-photon interference can be restored to a pair of distinguishable photons, if their polarizations can be made identical, thus making the pair indistinguishable. This manipulation is known the quantum eraser effect (Hillmer and Kwiat 2007). A quantum eraser can be created by placing a polarizer set at 45° relative to both the vertical and horizontal in each photon path directly in front of each detector (and far away from the beam splitter). The effect of these polarizers is to erase the information about the polarization labeling and thus the origin of either photon prior to passage through the polarizers, by putting each photon into a superposed state of horizontal and vertical polarizations. Since the photon wavefunctions are now made indistinguishable, two-photon interference will be observed by the absence of coincident detection events. The quantum eraser is an important tool for editing quantum states in quantum communication applications.

5.7 Summary

In this chapter we have investigated the description of the electron-magnetic fields in terms of the photons that constitute them. The Hamiltonian operator for the total energy of the electro-magnetic fields consists of two terms representing the energy stored in the electric and magnetic fields respectively. Propagation of the fields is accompanied by an oscillation of the energy between the electric and magnetic fields with a frequency ω. The quantum of this oscillation is the photon. The description of the electro-magnetic fields in terms of the quanta is formally equivalent to that of the quantum mechanical harmonic oscillator. To quote Dirac (5.1) who detailed this analysis in 1930, as the Schrödinger wave equation was also being developed:

> The dynamical system consisting of an assembly of similar bosons (e.g. photons) is equivalent to the dynamical system of a set of oscillators—the two systems are just the same system looked at from two different points of view.... This is one of the most fundamental results of quantum mechanics, which enables a unification of the wave and corpuscular theories of light to be effected.

The procedure of second quantization, which is the introduction of creation and annihilation operators, enables the analysis of systems of bosons to extract the energies and occupation numbers of the stable states of such systems without having to treat explicitly the eigenvectors of the system. In addition to the fundamental unification of the wave and particle viewpoints, this procedure also contributes a considerable simplification to the analysis of quantized energy levels.

Thus, an electro-magnetic oscillation having a frequency ω can be described at the quantum level as having an energy $E = \hbar\omega(n + \frac{1}{2})$, where n is the number of photons comprising the oscillation. The total energy of a wave consisting of multiple frequencies is just the arithmetic sum of the contribution of each frequency present. Thus treatment of the electro-magnetic field from the quantum point of view is formally equivalent to that of a set of independent harmonic oscillators. The relevant quantum variables are the occupation number n of each frequency mode and the energy of that mode.

The creation and annihilation operators can be used to describe the emission and absorption of photons in a direct and transparent way, without having to derive the quantum wavefunction in explicit form. This is a key practical result of this chapter. We can use directly the results to analyze stimulated and spontaneous emission (and absorption) of photons, and thus the performance of Lasers, LEDs and photodiodes.

The method of quantization of the electromagnetic field is complementary to the analysis of electromagnetic behavior using Maxwell's field equations. There is a range of situations where both methods can be used in an equivalent manner. However, there are limits. When photons or electrons interact with vacuum fluctuations, then the quantized electromagnetic method must be used. This is the case for the description of laser or LED emission, and as we showed in the previous section, for spontaneous parametric down-conversion. Analysis using Maxwell's equations fails to account for experimental observation. On the other hand, interactions of photon flux or electron current with the macroscopic world is much easier using Maxwell's equations which describe the coupling and propagation of the electric and magnetic fields belonging to these particles.

5.8 Exercises

5.1 Use the Heisenberg uncertainty principle to show that the minimum energy of the quantum harmonic oscillator is $E_{\text{min}} = \frac{\hbar\omega}{2}$.

5.2 The wave function for the $n = 1$ state of the quantum harmonic oscillator is;

$$\phi_1(x) = \left(\frac{\alpha}{\pi^{\frac{1}{2}} n! 2^n}\right)^{\frac{1}{2}} (2\alpha x) e^{-\frac{\alpha^2 x^2}{2}}, \quad \text{where } n = 1$$

And the wave function for the $n = 2$ state of the quantum harmonic oscillator is;

$$\phi_2(x) = \left(\frac{\alpha}{\pi^{\frac{1}{2}} n! 2^n}\right)^{\frac{1}{2}} (4\alpha^2 x^2 - 2) e^{-\frac{\alpha^2 x^2}{2}}, \quad \text{where } n = 2$$

Show by algebraic derivation that $a^+|\phi_1\rangle = \sqrt{2}|\phi_2\rangle$

5.3 Prove that the commutator $[a, a^+] = 1$

5.4 Show that $a^+(a\phi_n) = n\phi_n$, where ϕ_n is the wavefunction describing n photons of frequency ω

5.5 Derive the commutator $[a_l(t), a_l^+(t)]$ for the creation and annihilation operators

$$a_l^+(t) = \sqrt{\frac{1}{2\hbar\omega_l}}[\omega_n q_l(t) - ip_l(t)] \text{ and } a_l(t) = \sqrt{\frac{1}{2\hbar\omega_l}}[\omega_n q_l(t) + ip_l(t)]$$

5.6 Two-photon interference

(a) Using data from the Hong-Ou-Mandel experiment shown in Fig. 5.10, estimate the width in time of the wavepacket of the single photon created by Spontaneous Parametric Down-Conversion.

(b) What is the approximate real-space size of the single photon wave-packet when it intersects the beamsplitter? Comment on possible procedures to align two wavepackets so that they overlap?

(c) What happens to the wavepacket alignment and the measurement when the beamsplitter position in Fig. 5.9 is scanned? How is this "walk-off" effect resolved in Fig. 5.11?

References

P. Bronner, J.P. Meyn, Quantum-lab, an on-line interactive experimental lab in in quantum optics (2009). http://www.didaktik.physik.uni-erlangen.de/quantumlab/english/index.html?/quantumlab/english/imprint/index.html

J.L. Catalano, Spontaneous parametric down-conversion and quantum entanglement Thesis, Portland State University (2014). http://pdxscholar.library.pdx.edu/cgi/viewcontent.cgi?article=1088&context=honorstheses

P.A.M. Dirac, *The Principles of Quantum Mechanics*, 4th ed. (Oxford University Press, Oxford, 1958), P. 27. ISBN 0-19-852011-5

G. Grynberg, A. Aspect, C. Fabre, *Introduction to Quantum Optics* (Cambridge University Press, Cambridge, 2010). ISBN 978-0-521-55112-0

R. Hillmer, P. Kwiat, A do-it-yourself quantum eraser. Sci. Am. 90–95 (2007). http://www.arturekert.org/miscellaneous/quantum-eraser.pdf

C.K. Hong, Z.Y. Ou, L. Mandel, Measurement of subpicosecond time intervals between two photons by interference. Phys. Rev. Lett. **59**, 2044–2046 (1987). http://quantumagic.narod.ru/Articles/Mandel_1987.pdf

P.G. Kwiat, K. Mattle, H. Weinfurter, A. Zeilinger, New high-intensity source of polarization-entangled photon pairs. Phys. Rev. Lett. **75**, 4337–4341 (1995). http://journals.aps.org/prl/abstract/10.1103/PhysRevLett.75.4337

P.G. Kwiat, A.M. Steinberg, R.Y. Chiao, Observation of a “quantum eraser”: a revival of coherence in a two-photon interference experiment. Phys. Rev. A **45**, 7729–7739 (1992). http://research.physics.illinois.edu/QI/Photonics/papers/My%20Collection.Data/PDF/Observation%20of%20a%20quantum%20eraser%20A%20revival%20of%20coherence%20in%20a%20two-photon%20interference%20experiment.pdf

L.D. Landau, E.M. Lifshitz, *Quantum Mechanics*, (Reading Addison-Wesley, 1958). ISBN 0-08-020940-8

A. Ling, Entangled state preparation for optical quantum communication: creating and characterizing photon pairs from spontaneous parametric down conversion inside bulk uniaxial crystals, Thesis, Department of Physics, National University of Singapore, 2008. https://www.quantumlah.org/media/thesis/thesis-alex.pdf. This work is an clear and comprehensive treatment of spontaneous parametric down-conversion for creation of entangled photon states

A. Ling, Center for quantum technologies national university of singapore, Video recordings of experiments using spontaneous parametric down-conversion. http://qolah.org/research/hqpdc/hqpdc.htm

E. Merzbacher, *Quantum Mechanics* (Wiley, New York, 1960). ISBN-13: 978-0471887027

A. Migdall, *Quantum Optics Parametric Down-Conversion* (One photon in, Two Photons Out, YouTube, 2010). https://www.youtube.com/watch?v=1MaOqvnkBxk

Z.Y.J. Ou, *Multi-photon Quantum Interferences* (Springer, New York, 2007). ISBN 978-0-187-25532-3

Michael H. A. Reck, *Quantum interferometry with multiports: entangled photons in optical fibers* (Institut für Experimentalphysik, Leopold-Franzens Universität, Innsbruck, Thesis, 1996). http://www.univie.ac.at/qfp/publications/thesis/mrdiss.pdf

Ö. Süzer, T.G. Goodson III, Does pump beam intensity affect the efficiency of spontaneous parametric down conversion? Opt. Express **16**, (2008). https://www.osapublishing.org/oe/fulltext.cfm?uri=oe-16-25-20166&id=174994

A. Yariv, *Quantum Electronics*, 2nd edn. (Wiley, New York, 1975). ISBN : 0-471-97176-6

Chapter 6
Entanglement and Non-locality of Quantum Photonics

Abstract Entanglement describes a situation in which two entities share a common attribute. The Einstein–Podolsky–Rosen (*EPR*) paradox concerns the behavior of entangled states. Their 1935 publication in Physical Review considered the measurements of properties on two systems which are first allowed to interact, and therefore become entangled, and which are subsequently separated spatially so that they appear no longer to interact with each other. In proposing this condition, *EPR* supposed locality. Bell's theorem analyzes the outcomes of measurements at two different locations, A and B under the assumption of locality. Bell's theorem allows for the presence of a "local hidden variable" that accounts for some or all of the apparent statistical nature of the measurements at locations A or B. The analysis places upper limits on the correlations that can be measured between results recorded at locations A and B, regardless of the nature or form of the hidden variable. For more than 2 decades, careful measurements of entangled photon pairs show significant violations of Bell's limit. This result is widely accepted as demonstration that measurement of an entangled quantum does not depend entirely on local conditions where the measurement is made. The reality of the non-local behavior of entangled photon pairs is a foundation stone for quantum cryptography. This technology of ultra-secure communication is based on the distribution to authorized receivers of a quantum key that enables reading of encrypted messages. The quantum key is determined by the detection of photons correlated by entanglement at two different locations.

6.1 Introduction

Why did Marguerite fall in love with Faust? Faust, according to legend, was a scientist and philosopher in full mid-life crisis. His story has been treated by many playwrights and composers. However, Goethe's early nineteenth-century treatment stands out among all others. Johann Wolfgang von Goethe is a colossus of German literature, and his impact on western European culture is truly profound. His literary output of plays, poetry, novels, essays, travelogues, and scientific texts fills

© Springer Nature Switzerland AG 2020

T. P. Pearsall, *Quantum Photonics*, Graduate Texts in Physics,
https://doi.org/10.1007/978-3-030-47325-9_6

more than 100 volumes, including **Theory of Colours**, published in 1810 and still in print today (von Goethe 2006).

In Goethe's *Faust*, Marguerite, a chaste, devout and beautiful young woman falls hopelessly in love with Dr. Faust, an aging and bitter scientist seeking to recover his lost years. Their lives become entwined, and things go rapidly downhill as a result. Why would Marguerite fall for such a reprobate? She does not understand herself. However the audience knows that there is a *hidden variable*: Mephistopheles, a representative of Satan with supernatural powers. Mephistopheles controls the relationship between Marguerite and Faust. As a result Marguerite's free will to decide is sacrificed as part of a deal between Faust and the devil. An important theme in Faust concerns free choice.

Consider the following simple experiment: There are 2 subjects and a bag with an equal number of black and white balls inside. The balls are wrapped so their color cannot be seen. Subject no. A reaches inside the bag and picks a ball. His chosen ball has a definite color, but he does not know what it is. The bag is then taken to another room where subject no. B is waiting. She reaches into the bag and takes a second ball. After the selection is complete, subject no. A first unwraps his ball and notes the color. Then without any communication, subject no. B does the same.

After several repetitions of this experiment, analysis shows that subject no. A chooses a black ball 50% of the time, as expected. The same is observed for subject no. B. However, when taken together, there is a strong correlation between the results of the two subjects. When subject no. A chooses a black ball, the chances are 2 to 1 that subject no. B chooses a white one. And, the reverse is also true. When subject no. A chooses a white ball, then the chances are 2 to 1 that subject no. B will choose a black ball.

How can the choice made by subject no. A so strongly influence the choice made by subject no. B when he does not know what color of ball he has chosen, and there is no path for communication between the two subjects?

The explanation is that there is *hidden information*. There are only 4 balls in the bag: two black and two white. If we complete the description of the experiment with this additional information, then the statistical behavior of observed results is apparent. In quantum mechanics, the situation can be more mysterious. Two particles which are entangled quantum mechanically can be separated by significant spatial distance. A measurement (equivalent to looking at the color of a ball) made on one particle, forces the second particle with 100% certainty to adopt a specific state. The question is whether a hidden variable causes this reaction to occur. Or, on the other hand, does the entangled pair behave like a quantum with a single wavefunction, so that wavefunction collapse on measurement of one of the particles forces the quantum into an eigenstate, determining instantaneously the states of both particles.

6.2 The Einstein, Podolsky Rosen Paradox: Is the Description of Reality by Quantum Mechanics Complete?

6.2.1 Entangled States in Quantum Mechanics

In quantum mechanics two particles become entangled when they share a common measureable attribute like spin or polarization (see Sect. 2.5). Although the entangled pair consists of two particles, under certain measurements they behave like a single entity. Suppose we have an entangled electron-positron pair. The measurement of the spin of the electron, by having it pass between the poles of a magnet, will also determine the spin of the positron, even though it may be spatially far away. After the measurement, the particles are no longer entangled, because each is now in a well-defined spin state.

We can express the wavefunction of a pair of particles as $|\Psi_{1+2}\rangle$. In certain cases, we can write this wavefunction as the tensor product of the wavefunctions ψ_1 and ψ_2 of the two particles that make up the pair. That is:

$$|\Psi_{1+2}\rangle = |\psi_1\rangle \otimes |\psi_2\rangle \tag{6.1}$$

In this case the wavefunction of the pair can be factored into 2 pure states. Measurements on particle 1 can be made without affecting particle 2.

However, (6.1) is only one possibility. In general, the wavefunction of a pair of particles is a superposition of states which can be factored only under special circumstances. For example, the state:

$$|\Psi_{1+2}\rangle = \frac{1}{\sqrt{2}}\left(|+\rangle_1\,|+\rangle_2 + |-\rangle_1\,|-\rangle_2\right), \tag{6.2}$$

which could represent the spin of an electron-positron pair, cannot be factored into 2 pure states. As a result this is called an entangled state.

On the other hand, the state:

$$|\Psi_{1+2}\rangle = \frac{1}{\sqrt{2}}(|+\rangle_1|-\rangle_2 - |-\rangle_1|-\rangle_2) = \frac{1}{\sqrt{2}}(|+\rangle_1 - |-\rangle_1) \otimes |-\rangle_2 \tag{6.3}$$

can be factored into pure states of the individual particles, so it is not entangled.

The Einstein–Podolsky–Rosen (*EPR*) paradox concerns the behavior of entangled states (Fine 2015). Their 1935 publication in Physical Review considered the measurements of properties on two systems which are first allowed to interact, and therefore become entangled, and which are subsequently separated so that they are no longer interacting with each other (Einstein et al. 1935). In proposing this condition, *EPR* supposed locality. That is the measurements on system I cannot affect the state of system II. In particular, the state of system II cannot be affected simultaneously by measurements on system I.

> *EPR* states: In attempting to judge the success of a physical theory, we may ask ourselves two questions: (1) "Is the theory correct?" and (2) "Is the description given by the theory complete?" It is only in the case in which positive answers may be given to both of these questions, that the concepts of the theory may be said to be satisfactory (Einstein et al. 1935).

This statement is the equivalent of the concept of logical proof, **both necessary and sufficient**.

> *EPR* states next: Every element of the physical reality must have a counterpart in the physical theory. We shall call this the condition of completeness. The elements of the physical reality must be found by an appeal to results of experiments and measurements.

The analysis of *EPR* considers a wavefunction that describes the entangled state of the two systems:

$$\Psi(x_1, x_2) = \sum_{n=1}^{\infty} \psi_n(x_2)u_n(x_1) \tag{6.4}$$

where x_1 are the variables that describe system I and x_2 are the variables to describe system II. The $u_n(x_1)$ are the eigenfunctions that correspond to the measurement of some property A of system I. After a measurement of A, eigenvalue a_k is found and system I collapses to the single state u_k. The expression for $\Psi(x_1, x_2) = \psi_k(x_2)u_k(x_1)$ has simplified to a single term and the wavefunction for system II is $\psi_k(x_2)$. If we now make a second measurement on system I of a different property B, having a different set of eigenfunctions $v_n(x_1)$ and eigenvalues b_n, then we can write

$$\Psi(x_1, x_2) = \sum_{n=1}^{\infty} \varphi_n(x_2)v_n(x_1) \tag{6.5}$$

After making the measurement on system I, we find the eigenvalue b_τ, and the system is in the eigenstate $v_\tau(x_1)$. As a result, system II is now found to be in a single state $\varphi_\tau(x_2)$. *EPR* note the contradiction, that a measurement of different properties of system I results in changing the state of system II, even though it is postulated that these systems are not interacting. They conclude that it is possible to assign two different wavefunctions to describe the same reality (system II), and that this result suggests that the quantum mechanical description is not complete. A better conclusion would be that analysis of *EPR* is rather an elegant demonstration of the non-local nature of entangled systems.

Fifteen years later, David Bohm illustrated the situation by a simple, but concrete model experiment involving an entangled electron-positron pair (Bohm 1951, 1989). In this model a source emits an electron-positron pair whose global magnetic moment is zero. The electron proceeds from the source towards measurement station A where there is an analyzing magnet and detector. The positron proceeds in the opposite direction toward station B where there is a similar measurement apparatus. This model is shown schematically in Fig. 6.1.

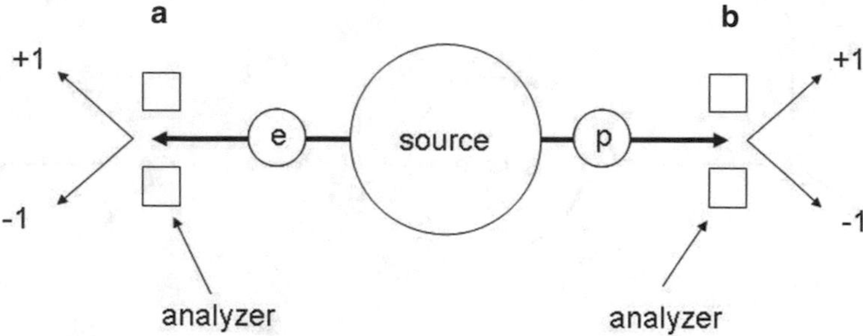

Fig. 6.1 Schematic diagram of the *EPR*-Bohm model experiment that illustrates non-local behavior of an entangled pair of particles. The source emits entangled electron-positron pairs. The spin of the particle pair is zero. While such an experiment is simple to imagine, its actual realization is difficult, requiring high vacuum over substantial distances and other practical difficulties

In the absence of a measurement, there is no way to know what the orientation of the electron spin is. In the Copenhagen interpretation of quantum mechanics the orientation of the electron spin is in a superposition of the 2π possibilities. The same statement can be made about the positron. However we do know with certainty in this case that the magnetic moment of the electron is equal and opposite to the magnetic moment of the positron.

When the electron passes through the magnetic field of the analyzing magnet at A, it is forced into an eigenstate with a distinct eigenvalue: either spin up $(= |+1\rangle)$ or spin down $(= |-1\rangle$. As a result of the wavefunction collapse, the positron must adopt a distinct eigenstate with its spin oriented opposite to that of the electron, i.e., spin down $(= |-1\rangle$ or spin up $(= |+1\rangle)$ respectively. There may appear to be a cause and effect relationship between the measurement of electron spin at A and the determination of the positron spin at B. The causality requirement of special relativity would require that the measurement of electron spin at A could affect the positron spin at B no faster than the speed of light. However, wavefunction collapse is not a cause and effect relationship. It is an operation that affects the whole wavefunction, and not just the part that is localised at A (or B). Thus the collapse happens everywhere in space where the wavefunction is defined.

The collapse of the wavefunction accomplishes the transition from the quantum environment (wavefunction) to the world of classical reality (eigenvalues). The collapse does not transmit information. Before the collapse we know already that the spin of the electron is opposite to the spin of the positron. At the moment that the electron spin is determined, nothing additional is learned about the spin of the positron, and so no information is transmitted. The correlation of the spin orientation between these two particles is not limited in quantum mechanical theory by the distance of their physical separation. After a short reflection, it must be the case that the wavefunction collapse is instantaneous. If it were any other way there could be serious problems of conservation of energy, momentum, and in some events like electron tunneling, even conservation of matter.

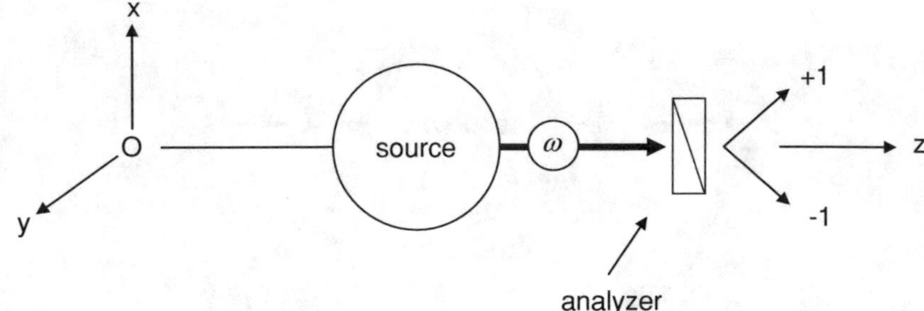

Fig. 6.2 Measurement of photon polarization. One of the two detectors at the output of the polarising beam-splitter will measure a signal: "+1" is recorded when the polarization lies along the axis of the polarizer and "−1" is recorded when the polarization is perpendicular to the axis of the polarizer. As demonstrated in Chap. 2, for a single photon, no other results are possible. Thus the measurement forces the photon to adopt a polarization along one of the two directions

It is apparent that the *EPR* situation can be compared directly to the experiment of electron tunneling through a potential barrier. In the case of *EPR* the entangled quantum pair behaves like a single entity with a single wavefunction having two spatially separated components. In the case of electron tunneling, the wavefunction of the electron also has two spatially separated components representing the probability densities for transmission and reflection. If electron transmission is detected, then we know instantly that the probability density for reflection is zero.

For purely practical reasons, the *EPR*-Bohm experiment is difficult to carry out. One significant difficulty arises from having to put the apparatus in high vacuum to allow the particles to propagate.[1] There is a simple analogue to this idea using photons. A photon has two states of polarization (parallel and perpendicular) and an entangled pair of photons sharing an overall polarization can by created. In such a system, the determination through measurement of the polarization of one of the photons will instantaneously determine the polarization of the other so that the overall polarization of the pair remains unchanged.

6.2.2 Measurements of Photon Polarization

Suppose that we have 2 photon modes l' and l'' characterized by the same wavevector $\mathbf{k}$ and frequency ω, but with polarizations ε' and ε''. The photon states are defined by the basis $|X\rangle; |Y\rangle$. We can define a coordinate system as shown in Fig. 6.2, with photon propagation along the O-z direction.[2]

[1]For example, in 2015, a group led by R. Hensen at the Delft University of Technology achieved an elegant demonstration using a combination of entangled electron spins and photon polarisations (2015). This is an area of active research and discovery.

[2]This section is adapted from, "The Naive View of an Experimentalist" by Aspect (2002).

We will define θ as the angle between the O-x axis and the axis of the polarizer. Next, introduce $\widehat{E}(\theta)$ as the operator for the polarization measurements along θ and $\theta + \frac{\pi}{2}$. When the polarization analyser axis is aligned along O-x,

$$\widehat{E}(0) = \begin{pmatrix} 1 & 0 \\ 0 & -1 \end{pmatrix} \tag{6.6}$$

The polarizer can be rotated freely around the z-axis. When the polarization analyser is oriented at θ from O-x,

$$\widehat{E}(\theta) = \begin{pmatrix} \cos 2\theta & \sin 2\theta \\ \sin 2\theta & -\cos 2\theta \end{pmatrix} \tag{6.7}$$

The eigenvectors of $\widehat{E}$ are:

$$|^{+}\theta\rangle = \cos\theta|X\rangle + \sin\theta|Y\rangle \tag{6.8}$$

and

$$|^{-}\theta\rangle = -\sin\theta|X\rangle - \cos\theta|Y\rangle$$

Now assume a photon with a polarization at λ relative to O-x. Its wavefunction can be written:

$$|\psi\rangle = \cos\lambda|X\rangle + \sin\lambda|Y\rangle \tag{6.9}$$

The probability of detecting a "+1" is expressed

$$\begin{aligned} P_{+}(\theta,\lambda) &= |\langle^{+}\theta|\psi\rangle|^2 = |\cos\theta\cos\lambda\langle X|X\rangle + \sin\theta\sin\lambda\langle Y|Y\rangle|^2 \\ &= |\cos\theta\cos\lambda + \sin\theta\sin\lambda|^2 \\ P_{+}(\theta,\lambda) &= \cos^2(\theta-\lambda) \end{aligned} \tag{6.10}$$

Similarly, the probability for detecting a "−1" is

$$P_{-}(\theta,\lambda) = \sin^2(\theta-\lambda) \tag{6.11}$$

Equations (6.10) and (6.11) are known as Malus' Law (https://en.wikipedia.org/wiki/Polarizer).

6.2.3 *Joint Polarization Measurements*

Next we consider the ensemble polarization of a pair of photons. We can express the wavefunctions as:

$$\psi_1 = |X_1\rangle + |Y_1\rangle$$
$$\psi_2 = |X_2\rangle + |Y_2\rangle$$

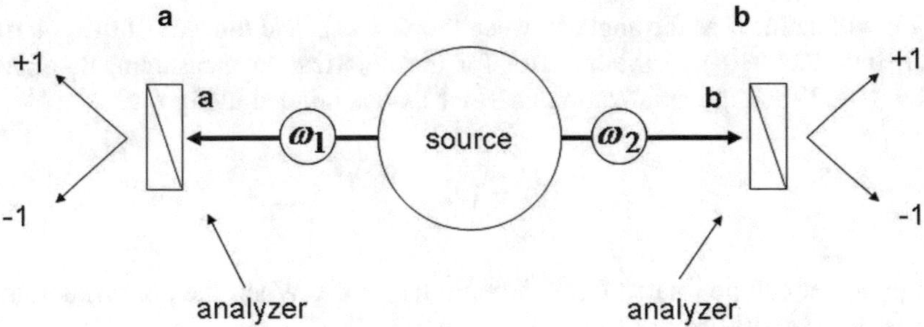

Fig. 6.3 A photon source emits pairs of entangled photons which propagate in different directions. They are detected at measurement stations *A* and *B*. Whenever a photon is detected at *A*, the other is detected at *B*. At station *A* there is a polarization analyzer, the axis of which is oriented at an angle **a** with respect to the axis *O-x*. At station *B* the orientation of the analyser is set at angle **b** with respect to the axis *O-x*. The measurement of polarization at each station gives either +1 (parallel to the polarizer axis) or −1 (perpendicular to the polarizer axis)

The joint polarization can be represented by a 4 × 4 matrix:

$$\varepsilon = \varepsilon_1 \otimes \varepsilon_2 = [|X_1, X_2\rangle; |X_1, Y_2\rangle; |Y_1, X_2\rangle; |Y_1, Y_2\rangle] \tag{6.12}$$

The joint probabilities are expressed:

For polarization measuring parallel relative to *O-x* (see Figs. 6.1 and 6.3) at both A and B:

$$(+1, +1) = P_{++}(\mathbf{a}, \mathbf{b}) = |\langle^+\mathbf{a}, {}^+\mathbf{b}|\psi\rangle|^2 \tag{6.13}$$

and for measuring polarization perpendicular to *O-x* at both A and B,

$$(-1, -1) = P_{--}(\mathbf{a}, \mathbf{b}) = |\langle^-\mathbf{a}, {}^-\mathbf{b}|\psi\rangle|^2 \tag{6.14}$$

The probability for measuring parallel polarization at A alone is:

$$P_+ = P_{++}(\mathbf{a}, \mathbf{b}) + P_{+-}(\mathbf{a}, \mathbf{b}) \tag{6.15}$$

The wavefunction

$$|\psi_{EPR}\rangle = \frac{1}{\sqrt{2}}(|X, X\rangle + |Y, Y\rangle) \tag{6.16}$$

represents an entangled pair of photons with frequencies ω_1 and ω_2. The state $|X, X\rangle$ represents a photon ω_1 polarised in the *O-x* direction and a photon ω_2 also polarised in the *O-x* direction. The 2-photon state given by (6.16) is entangled because it cannot be factored into 2 terms, one associated with ω_1 and the other associated with ω_2.

Now let us proceed to measurements on this state. The polarizer at A is set to **a** degrees relative to *O-x*, and the polarizer at *B* is set to **b** degrees relative to *O-x*.

The joint probability of measuring a "+1" at both A and *B* is

$$P_{++}(\mathbf{a},\mathbf{b}) = |\langle {}^{+}\mathbf{a}, {}^{+}\mathbf{b}|\psi_{EPR}\rangle|^2 = \frac{1}{2}\cos^2(\theta_{\mathbf{a}} - \theta_{\mathbf{b}}) = \frac{1}{2}\cos^2(\mathbf{a},\mathbf{b}) \tag{6.17}$$

and

$$P_{+-}(\mathbf{a},\mathbf{b}) = |\langle {}^{+}\mathbf{a}, {}^{-}\mathbf{b}|\psi_{EPR}\rangle|^2 = \frac{1}{2}\sin^2(\theta_{\mathbf{a}} - \theta_{\mathbf{b}}) = \frac{1}{2}\sin^2(\mathbf{a},\mathbf{b}) \tag{6.18}$$

These results depend only on the difference between the angle of the axes of the 2 polarizers and is invariable in rotation about the z-axis.

The probability of measuring parallel polarization at A (“1”) regardless of what happens at B is:

$$P_{+}(\mathbf{a},\mathbf{b}) = P_{++}(\mathbf{a},\mathbf{b}) + P_{+-}(\mathbf{a},\mathbf{b}) = \frac{1}{2}(\cos^2(\mathbf{a},\mathbf{b}) + \sin^2(\mathbf{a},\mathbf{b})) = \frac{1}{2} \tag{6.19}$$

Similarly the probability of measuring a $\perp$-polarization at A is also $\frac{1}{2}$. That is, taken separately, each set of polarization measurements at A and at B gives the expected result for a random variable. The results appear to be independent, yet there is strong correlation, as we will now see.

For this *EPR* state, the conditional probability of finding “+1” at B if “+1” has been found at A is:

$$\text{pr}(B \text{ given } A) = \left(\frac{\text{pr}(B \text{ and } A)}{\text{pr}(A)}\right) = \frac{P_{++}(\mathbf{a},\mathbf{a})}{P_{+}(\mathbf{a})} = \frac{\frac{1}{2}\cos^2(\theta = 0)}{1/2} = \frac{1/2}{1/2} = 1 \tag{6.20}$$

These results reveal a fundamental property about measurements of correlated particles. Suppose you are operating a laboratory measuring photons reaching you from a secure communication channel. You are told that some photons have random polarization and others are entangled photon pairs with polarization predetermined by a measurement at the other end of the channel. Equation (6.19) says you that you cannot do a measurement on the photons that distinguishes one from the other. On the other hand, (6.20) says that you can confirm that a photon pair is correlated, but you must have access to measurements performed on each photon.

6.3 Bell's Theorem

6.3.1 *John S. Bell and the Completeness of Quantum Mechanics*

So far we have examined the behavior of entangled particles. Measurements[3] on entangled pairs of particles show that the result of measurements taken on one particle is determined by the state of the second particle, even though it may be

[3] This section is based on the demonstration and discussion by Bell (1987).

too far distant to communicate that state at the speed of light. Einstein referred to this situation as “Spooky action at a distance”—apparently in violation of the theory of special relativity. For several decades, the *EPR* paradox gathered dust because the apparatus needed to demonstrate the experiments did not exist, and there was much more exciting work to be done on elementary particle physics. John Bell changed all this.

John S. Bell was a physicist, born in Belfast and educated in the UK, earning a Ph.D. from the University of Birmingham. He spent most of his career working at CERN on particle accelerator design and theoretical particle physics. One of his hobbies was a better understanding of the basics of quantum theory. You might compare his situation to that of Einstein working for the Swiss patent office, and doing special relativity on the side. In 1964, he took a year-long sabbatical, the result of which was the development of an inequality relationship now known as Bell’s theorem. Bell’s analysis considers measurements having a statistical result. These measurements could be on classical macroscopic systems or on quantum mechanical particles. The analysis compares measurements made at two different locations: A and B (cf. Fig. 6.4). Of course, the context that interested Bell is quantum mechanical measurements, because in specific circumstances, measurements on quanta can show correlations that exceed the upper limit of Bell’s inequality.

Bell was also interested by the probabilistic characteristic of quantum measurements. A hidden variable (a phenomenon, another particle, or information like the number of balls in the bag) could be responsible for some or all of the seemingly statistical nature of the measurement results. (Recall from Chap. 1 that Pauli proposed the existence of a nearly invisible particle, the neutrino, to account for seemingly random fluctuations in energy and momentum during beta decay.) Bell

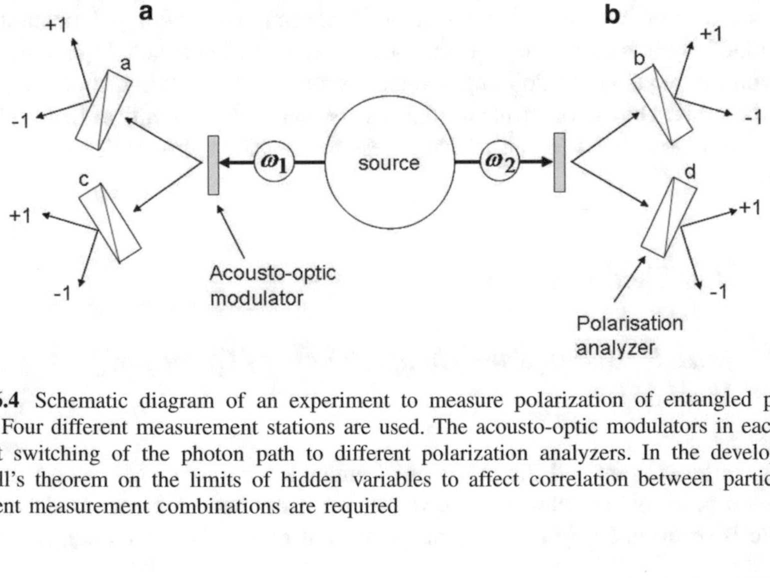

Fig. 6.4 Schematic diagram of an experiment to measure polarization of entangled photon pairs. Four different measurement stations are used. The acousto-optic modulators in each arm permit switching of the photon path to different polarization analyzers. In the development of Bell’s theorem on the limits of hidden variables to affect correlation between particles, 4 different measurement combinations are required

shows us that statistical analysis of measurements carried out independently in different locations could be correlated, but that there are upper limits on the possible effect of the correlation. Bell's theorem shows that this limit exists for **any** hidden variable. This result is an awe-inspiring intellectual achievement.

Bell's theorem states that any physical theory based on local reality cannot reproduce all the predictions of quantum mechanical theory.

Thus, completing quantum mechanics by addition of more detailed theory cannot explain the *EPR* paradox if we insist that each measurement **depend only on local settings** of the measuring equipment, and be thus independent of the situation where the second particle is being measured. Bell's theorem sparked renewed interest in the *EPR* paradox, and five years later in 1969 Clauser, Horne, Shimony and Holt (CHSH) (Clauser et al. 1969) published a proposal to make the measurements by photonics, using entanglement of the polarization of photon pairs emitted in an atomic cascade. Implementation of the measurements requires sensitive equipment, like single-photon counters that were still under development. Initial attempts showed promising results. More than 10 years later, Aspect et al. (1981) carried out experiments based on the proposal of CHSH which showed strong violation of Bell's inequalities, thus demonstrating that in specific experiments, behavior of entangled quanta cannot be explained by theories of local interactions, even those incorporating local hidden variables.

6.3.2 *Hidden Variables in Quantum Mechanics*

We will consider measurements made on pairs of entangled quantum particles, such as photons which have been separated spatially.

As we have shown in Chap. 2, when a single photon passes through a beam-splitting polarizing cube, because it is a quantum, there are only 2 possible results: parallel polarization or perpendicular polarization.

Let $A(\mathbf{a})$ be a function that gives the result of a measurement made at station A in Fig. 6.4 using the polarizing cube with its axis oriented in direction $\mathbf{a}$ with respect to a common reference. Then $A(\mathbf{a}) = \pm 1$ for the orientation parallel or perpendicular to $\mathbf{a}$.

First of all, we assume that the experimenter has the freedom to choose polarizer orientation, and other experimental conditions that prevail at A, independently from those at location B. The key element of Bell's analysis is the assumption of locality. Thus, measurements $A(\mathbf{a})$ are independent of $B(\mathbf{b})$, and the result $E(\mathbf{a}, \mathbf{b})$ of a joint measurement is the simple product:

$$-1 \leq E(\mathbf{a}, \mathbf{b}) = A(\mathbf{a})B(\mathbf{b}) \leq 1 \tag{6.21}$$

Now, let us assume that there is a *local* hidden variable λ—a parameter or function or operator for which the only function is to give preferential weight to particular values of orientation $\mathbf{a}$ or $\mathbf{b}$. λ can therefore change the conditions under which

$A(\mathbf{a}, \lambda) = 1$ or -1, but not those values which are imposed by the quantum nature of the particles. That is, relationship (6.21) is not changed:

$$-1 \le E(\mathbf{a}, \mathbf{b}) = A(\mathbf{a}, \lambda)B(\mathbf{b}, \lambda) \le 1 \tag{6.22}$$

The result of the joint measurement, taking into account the effect of the hidden variable, is

$$E(\mathbf{a}, \mathbf{b}) = \int_\lambda d\lambda\rho(\lambda)A(\mathbf{a}, \lambda)B(\mathbf{b}, \lambda)$$

where:

$$\begin{aligned} &-1 \le E \le 1 \\ &A(\mathbf{a}, \lambda) = \pm 1 \\ &B(\mathbf{b}, \lambda) = \pm 1 \end{aligned} \tag{6.23}$$

and

$$\int_\lambda d\lambda\rho(\lambda) = 1$$

Each measurement can be made only once on a given pair of particles, and $E(\mathbf{a}, \mathbf{b})$ represents a single experiment with orientations $\mathbf{a}, \mathbf{b}$, while $E(\mathbf{a}, \mathbf{c})$ represents a different experiment using a different pair of particles, totally independent of the previous pair, and using a different choice for the orientation of the polarising cube at B.

Let's look at the difference in the results between these two experiments:

$$E(\mathbf{a}, \mathbf{b}) - E(\mathbf{a}, \mathbf{c}) = \int_\lambda d\lambda\rho(\lambda)[A(\mathbf{a}, \lambda)B(\mathbf{b}, \lambda) - A(\mathbf{a}, \lambda)B(\mathbf{c}, \lambda)] \tag{6.24}$$

$$E(\mathbf{a}, \mathbf{b}) - E(\mathbf{a}, \mathbf{c}) = \int_\lambda d\lambda\rho(\lambda)[A(\mathbf{a}, \lambda)B(\mathbf{b}, \lambda)] - \int_\lambda d\lambda\rho(\lambda)[A(\mathbf{a}, \lambda)B(\mathbf{c}, \lambda)] \tag{6.25}$$

Next we can expand this integral by adding and subtracting the same term: $A(\mathbf{a}, \lambda)B(\mathbf{b}, \lambda)A(\mathbf{d}, \lambda)B(\mathbf{c}, \lambda)$. This enables the factoring out of some common elements.

$$\begin{aligned} E(\mathbf{a}, \mathbf{b}) - E(\mathbf{a}, \mathbf{c}) = &\int_\lambda d\lambda\rho(\lambda)[A(\mathbf{a}, \lambda)B(\mathbf{b}, \lambda) \pm A(\mathbf{a}, \lambda)B(\mathbf{b}, \lambda)A(\mathbf{d}, \lambda)B(\mathbf{c}, \lambda)] \\ &- \int_\lambda d\lambda\rho(\lambda)[A(\mathbf{a}, \lambda)B(\mathbf{c}, \lambda) \pm A(\mathbf{a}, \lambda)B(\mathbf{b}, \lambda)A(\mathbf{d}, \lambda)B(\mathbf{c}, \lambda)] \end{aligned} \tag{6.26}$$

$$E(\mathbf{a},\mathbf{b}) - E(\mathbf{a},\mathbf{c}) = \int_{\lambda} d\lambda \rho(\lambda)[A(\mathbf{a},\lambda)B(\mathbf{b},\lambda)][1 \pm A(\mathbf{d},\lambda)B(\mathbf{c},\lambda)] - \int_{\lambda} d\lambda \rho(\lambda)[A(\mathbf{a},\lambda)B(\mathbf{c},\lambda)][1 \pm A(\mathbf{d},\lambda)B(\mathbf{b},\lambda)] \quad (6.27)$$

By incorporation of an additional measurement $A(\mathbf{d},\lambda)$, we introduced 2 additional experiments, and thus two additional pairs of particles, in order to take into account all the possible permutations of the orientations of the polarising cubes at A and B.

Next, take the absolute value of each side.

$$|E(\mathbf{a},\mathbf{b}) - E(\mathbf{a},\mathbf{c})| = \left| \begin{matrix} \int_{\lambda} d\lambda \rho(\lambda)[A(\mathbf{a},\lambda)B(\mathbf{b},\lambda)][1 \pm A(\mathbf{d},\lambda)B(\mathbf{c},\lambda)] \\ -\int_{\lambda} d\lambda \rho(\lambda)[A(\mathbf{a},\lambda)B(\mathbf{c},\lambda)][1 \pm A(\mathbf{d},\lambda)B(\mathbf{b},\lambda)] \end{matrix} \right| \quad (6.28)$$

Use the "triangle inequality" to separate the RHS of the equation into two terms. Notice that the "–" sign changes to a "+" sign.

$$|E(\mathbf{a},\mathbf{b}) - E(\mathbf{a},\mathbf{c})| \leq \left| \int_{\lambda} d\lambda \rho(\lambda)[A(\mathbf{a},\lambda)B(\mathbf{b},\lambda)][1 \pm A(\mathbf{d},\lambda)B(\mathbf{c},\lambda)] \right| + \left| \int_{\lambda} d\lambda \rho(\lambda)[A(\mathbf{a},\lambda)B(\mathbf{c},\lambda)][1 \pm A(\mathbf{d},\lambda)B(\mathbf{b},\lambda)] \right| \quad (6.29)$$

Replace the products $A(\mathbf{a},\lambda)B(\mathbf{b},\lambda)$ and $A(\mathbf{a},\lambda)B(\mathbf{c},\lambda)$ by their maximum values, i.e.: 1. This substitution does not change the inequality

$$|E(\mathbf{a},\mathbf{b}) - E(\mathbf{a},\mathbf{c})| \leq \left| \int_{\lambda} d\lambda \rho(\lambda)[1 \pm A(\mathbf{d},\lambda)B(\mathbf{c},\lambda)] \right| + \left| \int_{\lambda} d\lambda \rho(\lambda)[1 \pm A(\mathbf{d},\lambda)B(\mathbf{b},\lambda)] \right| \quad (6.30)$$

It can be seen that $[1 \pm A(\mathbf{d},\lambda)B(\mathbf{c},\lambda)]$ is therefore always positive, lying between 0 and 2. The same goes for $[1 \pm A(\mathbf{d},\lambda)B(\mathbf{b},\lambda)]$.

As a result, we can take away the two absolute value operations on the RHS:

$$|E(\mathbf{a},\mathbf{b}) - E(\mathbf{a},\mathbf{c})| \leq \int_{\lambda} d\lambda \rho(\lambda)[1 \pm A(\mathbf{d},\lambda)B(\mathbf{c},\lambda)] + \int_{\lambda} d\lambda \rho(\lambda)[1 \pm A(\mathbf{d},\lambda)B(\mathbf{b},\lambda)] \quad (6.31)$$

Now we can perform the integrations, using $\int_{\lambda} d\lambda \rho(\lambda) = 1$.

$$|E(\mathbf{a},\mathbf{b}) - E(\mathbf{a},\mathbf{c})| \leq 2 \pm (E(\mathbf{d},\mathbf{c}) + E(\mathbf{d},\mathbf{b})) \tag{6.32}$$

And finally;

$$|E(\mathbf{a},\mathbf{b}) - E(\mathbf{a},\mathbf{c})| \pm (E(\mathbf{d},\mathbf{c}) + E(\mathbf{d},\mathbf{b})) \leq 2 \tag{6.33}$$

We can rearrange and remove the absolute value operator

$$-2 \leq E(\mathbf{a},\mathbf{b}) - E(\mathbf{a},\mathbf{c}) + E(\mathbf{d},\mathbf{c}) + E(\mathbf{d},\mathbf{b}) \leq 2 \tag{6.34}$$

This proves Bell's theorem, in the form developed by Clauser et al. (1969). It describes the results of four measurements, each made on a different pair of particles using different orientations of the polarizing cubes at A: (**a** and **d**) and at B: (**b** and **c**), under the assumption of locality, that is: the measurements at A are independent of the measurements at B. It is apparent that the inequality in (6.34) can be evaluated in experiment. We will refer to the expression:

$$S = E(\mathbf{a},\mathbf{b}) - E(\mathbf{a},\mathbf{c}) + E(\mathbf{d},\mathbf{c}) + E(\mathbf{d},\mathbf{b}) \tag{6.35}$$

as the Bell function.

6.3.3 Experiments

In order to perform experiments that evaluate (6.35) we need to specify the settings of the polarizer axes. Typically the polarizers are each set at equal angles one with respect to the other, all relative to the O-x axis

$$\begin{aligned}\theta_{\mathbf{a}} &= 0\\ \theta_{\mathbf{b}} &= \phi\\ \theta_{\mathbf{d}} &= 2\phi\\ \theta_{\mathbf{c}} &= 3\phi\end{aligned}$$

Then it follows that

$$\begin{aligned}\angle(\mathbf{a},\mathbf{b}) &= \phi\\ \angle(\mathbf{d},\mathbf{c}) &= \phi\\ \angle(\mathbf{d},\mathbf{b}) &= \phi\\ \angle(\mathbf{a},\mathbf{c}) &= 3\phi\end{aligned} \tag{6.36}$$

The next step is to specify the form of $E(\mathbf{a},\mathbf{b})$.

We will first consider Bell's own hidden variable model.

In this case, the polarization of the two photons in a given pair depends on a parameter λ and the measurement of polarization depends on this common parameter. Then it is easy to see that there will be a correlation between the measurements at A and B. We suppose that the 2 photons have a well defined polarization $\boldsymbol{\varepsilon}$ in the x-y plane that is specified as the angle λ between the O-x axis and the polarization direction. The orientation of $\boldsymbol{\varepsilon}$ changes from pair to pair in a random manner, and so λ is a random variable equally distributed over all angles: $0 \le \lambda \le 2\pi$; that is, $\rho(\lambda) = \frac{1}{2\pi}$.

The polarizer at A is oriented at angle $\mathbf{a}$ with respect to the O-x axis. If the absolute value of the angle between the polarization and the polarizer $= |(\boldsymbol{\varepsilon}, \mathbf{a})|$ is less than or equal to $\frac{\pi}{4}$, the polarizer measurement gives "+1". If $\frac{\pi}{4} < |(\boldsymbol{\varepsilon}, \mathbf{a})| < \frac{\pi}{2}$, the polarizer gives "−1". We can model the polarizer response at A using the sign function:

$$A(\lambda, \mathbf{a}) = \text{sign}[\cos(2(\theta_{\mathbf{a}} - \lambda))] \tag{6.37}$$

where $\text{sign}(x) = +1$ if $x \ge 0$ and $\text{sign}(x) = -1$ if $x < 0$.

A similar argument applies at B:

$$B(\lambda, \mathbf{b}) = \text{sign}[\cos(2(\theta_{\mathbf{b}} - \lambda))]$$

Then we can express the expectation value of a joint measurement as

$$E_{\text{Hidden-Value}}(\mathbf{a}, \mathbf{b}) = \int_0^{2\pi} d\lambda \rho(\lambda) A(\lambda, \mathbf{a}) B(\lambda, \mathbf{b}) = 1 - 4\frac{|\theta_{\mathbf{a}} - \theta_{\mathbf{b}}|}{\pi} \tag{6.38}$$

For $-\frac{\pi}{2} \le \theta_{\mathbf{a}} - \theta_{\mathbf{b}} \le \frac{\pi}{2}$.

We show the expectation value for the result of a joint measurement in Fig. 6.5.

The Bell function for this model can be evaluated assuming the equal angle settings of the polarizer-analyser:

$$S_{\text{Hidden-Value}} = E(\mathbf{a}, \mathbf{b}) - E(\mathbf{a}, \mathbf{c}) + E(\mathbf{d}, \mathbf{c}) + E(\mathbf{d}, \mathbf{b}) = 2. \tag{6.39}$$

independent of the setting angle of the polarizers, indicating a high degree of correlation between the measurements at A and B. This is the absolute maximum value of the parameter that is allowed by Bell's Theorem. While Bell's hidden parameter model may seem crude, Bell's theorem shows that one cannot do better.

Next we consider the quantum mechanical case. The polarization is not determined until a detection takes place. The probabilities for detection are given by Malus' Law (6.10) and (6.11). The expectation value for a join measurement is expressed in the usual way:

$$E_{Q-M}(\mathbf{a}, \mathbf{b}) = (+1)P_{+}(\mathbf{a}, \mathbf{b}) + (-1)P_{-}(\mathbf{a}, \mathbf{b}) = \cos^2(\theta) - \sin^2(\theta) = \cos(2\theta) \tag{6.40}$$

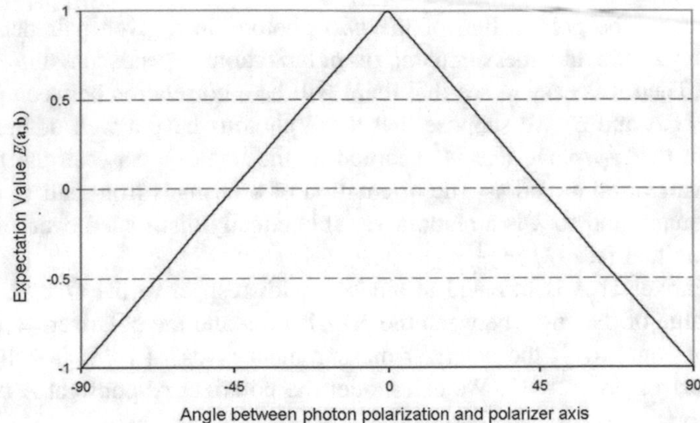

Fig. 6.5 Expected value of a joint measurement of polarizations of a photon pair. Measurement results at *A* are completely locally determined, but dependent on a hidden parameter λ. A similar situation prevails at *B*

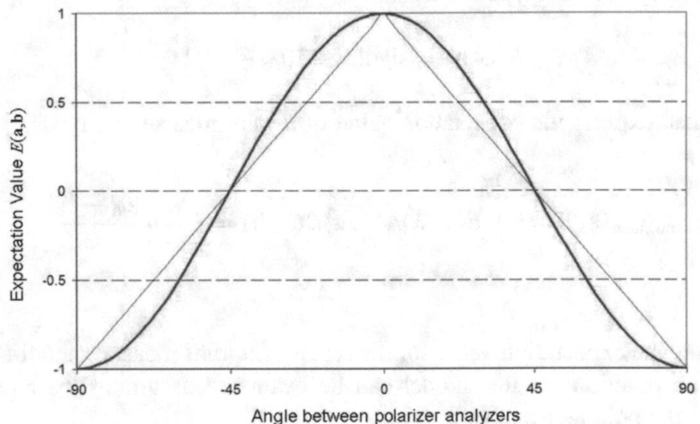

Fig. 6.6 Expectation value for the joint polarization measurement based on quantum-mechanical behavior of the photons in the pair. The result for Bell's hidden variable model is shown for comparison. At certain angles between the polarizers, $(0°, 45°, 90°$, etc.) the two models give the same result, but elsewhere, the quantum mechanical model gives a larger expectation value

In Fig. 6.6 we show how the expectation value for joint detection varies as a function of angles between the axes of the two polarization analysers.

In order to evaluate the Bell function, we must specify the angle between the polarizers. This choice is up to the experimenter. We will use the convention of equal angles as discussed earlier. The behavior of the Bell function is shown in Fig. 6.7.

As an example, we make the choice that gives the most significant departure from the limits imposed by Bell's Theorem.

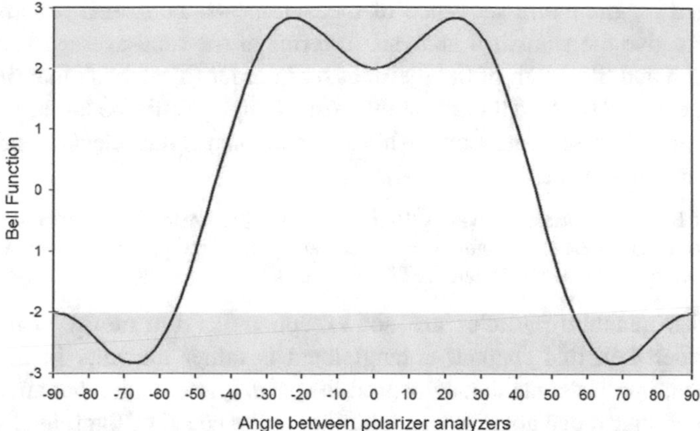

Fig. 6.7 Evaluation of the Bell function in the quantum mechanical case using equal angles between the polarizers. The Bell function varies between 0 and $\pm 2\sqrt{2}$. Any excursion beyond the Bell limit of ± 2 (shown as the *shaded regions*) implies that non-local interactions are influencing results. Note that there are many choices for the angle separating the polarizers that do not result in a violation of Bell's Theorem. The setting that gives the largest deviation is $\theta = \frac{\pi}{8}$

$$(\mathbf{a}, \mathbf{b}) = (\mathbf{d}, \mathbf{c}) = (\mathbf{d}, \mathbf{b}) = \frac{\pi}{8} \tag{6.41}$$

and, of course: $(\mathbf{a}, \mathbf{c}) = \frac{3\pi}{8}$

$$\begin{aligned} S_{Q-M} &= E(\mathbf{a}, \mathbf{b}) - E(\mathbf{a}, \mathbf{c}) + E(\mathbf{d}, \mathbf{c}) + E(\mathbf{d}, \mathbf{b}) \\ &= \frac{1}{2}\sqrt{2} + \frac{1}{2}\sqrt{2} + \frac{1}{2}\sqrt{2} + \frac{1}{2}\sqrt{2} = 2\sqrt{2} \end{aligned} \tag{6.42}$$

6.4 More About Bell's Theorem—Loopholes, Bohm—deBroglie and Free-Will

Bell's Theorem is remarkable and occupies perhaps a unique place in physics because it does not describe the behavior of a particle or even a pair of particles. It describes an environment, and a means to test if one is working within this environment. In order to evaluate (6.34), at least four separate independent experiments should be carried out. The result of such a test does not describe the nature of the entanglement of individual photon pairs. Indeed, a typical experiment to test of Bell's inequality involves millions of measurements on pairs emitted by a suitable source. The results are combined using the recorded orientation of the polarizer in the arm where the photon is detected, and the evaluation of (6.34) is calculated for the entire sequence of measurements. Published reports of such experiments give the statistical analysis in terms of the total average value of the Bell function and the width of the distribution in order to establish margin of error (see Fuwa et al. 2015; Kwiat et al. 1999). An example would be the report photon pairs transmitted in separate optical fibre circuits using the telecoms network of the Swiss postal service:

> A test of Bell inequalities by photons 10.9 km apart is presented. Energy-time entangled photon pairs are measured using two-channel analyzers, leading to a violation of the inequalities by 16 standard deviations (Tittel et al. 1998).

Entangled quantum particles are not exceptional. From atoms to the photon pairs considered in this chapter, entanglement is rather the rule. In most cases, entanglement implies some delocalization in space. Photon pairs that are separated by a beam-splitter move apart from each other at the speed of light, leading to significant and macroscopic spatial separation.

Detection of one of the quanta forces it to adopt an eigenstate with measurable eigenvalues. However, in the case of an entangled state, the measurement made on one quantum is actually made on both quanta that constitute the state. Both photons are thus forced into eigenstates with eigenvalues, simultaneous with the measurement. If it happened any other way, it would not be possible to normalize the wavefunction during the time necessary for both particles to adopt their appropriate eigenstate. The *EPR* paradox is not a paradox if one accepts that entanglement of quanta introduces a wavefunction for the corresponding state that is distributed in space, and that wavefunction collapse is necessary to conserve the normalization of that wavefunction.

Bell's theorem shows that a local hidden variable theory might explain some, **but not all** the behavior of entangled quantum particles. Figure 6.7 shows that there is a restricted set of experimental conditions which identifies the presence of entanglement. Quantum mechanics, on the other hand successfully predicts observations, but takes into account the non-local behavior of entangled quanta. Experiments that violate Bell's inequality reveal that entangled pairs of quanta are described by a wavefunction that is distributed in space. Thus, analysis based only on local settings will not be able to describe all the observations.

6.4.1 Bell Test Experiments—Loopholes

With local hidden variables relegated to the sidelines, there remains a legitimate question about the effect of *non-local* hidden variables and their effect on experiments. This would be a different situation that lies outside the basic premise underlying Bell's theorem. An example of a non-local hidden variable would be communication ($\leq$speed of light) of polarizer settings between stations A and B (see Larsson 2014).

Looking in more detail at an actual experiment shows important concerns, sometimes called loopholes, that need to be addressed if the results are to be considered a fair test of Bell's inequalities. Aspect et al. (1982) performed one of the first measurements of Bell inequalities (http://journals.aps.org/prl/pdf/10.1103/PhysRevLett.49.1804) with close attention paid to closing the "locality loophole". Referring to Fig. 6.4, photon pairs are spatially separated and sent to the 2 analysis stations A and B. As in the experiments described in Chap. 2, The photons arrive at the detectors nearly simultaneously; Coincident detection at A and B within a chosen time interval signals the presence of entangled pairs.

The **locality loophole** consists of communication of settings and results between stations A and B prior to or during the measurement To close the loophole, it is necessary to set the two detection stations far enough apart that a signal can not be transmitted between them during the detection interval. In the case of Aspect, Dalibard and Roger, that distance exceeds 13 m, corresponding to a photon time-of-flight of 43 ns. The acousto-optic modulators switch the photon pairs between the polarization detection stations **a** and **d** at A and **b** and **c** at B at a time interval that is shorter than the photon transit time. These modulators are unsynchronized so that the choice of which polarization detection station actually receives the photon is close to random. Under these conditions, the communication of the detection settings between each other or with the apparatus that generates the photons is reduced to a minimum. On each side, the orientation of the polarizer as well as the results of polarization measurements, are recorded as a function of time. After a sequence of measurements, the data from the two sides are integrated in order to calculate the Bell function for each quartet of polarization measurements.

In any experiment, the detection efficiency is less than 100% efficient. Photon pairs at A and B are only identified as belonging to a single pair after the experiment is performed, by judging whether or not their detection times are close enough to one another. If the signal to noise is not too different from unity, detection of a "noise" photon will occur in the coincidence window, leading to an ambiguous result. During post-experiment analysis, such results may be discarded from consideration, creating a **coincidence loophole**. This loophole can be minimized by narrowing the coincidence window and simultaneously increasing signal to noise by using a brighter photon source. The adoption of parametric down-conversion using BBO, which we introduced in Chap. 2 (see Fig. 2.18), has been an important contribution toward elimination of the coincidence loophole. A good demonstration is that of Kwiat and co-workers (http://arxiv.org/abs/quant-ph/9810003) (Kwiat et al. 1999). They have measured a Bell function (6.42) of $S = 2.7$, proving a violation of the Bell limit by over 200 standard deviations. They generated entangled pairs of photons at 702 nm using parametric down-conversion in BBO. Nonetheless, their experiment did not close the locality loophole. The detection stations were not separated far enough spatially to preclude communication between detectors, and fast switching between polarizers was not used to ensure random selection of polarization orientation.

Experiments that test Bell's equalities are relatively simple to describe, but the devil is in the details. There are additional important loopholes than those described here, and experiments continue to demonstrate a loophole-free test of Bell's inequality. The work of Hensen et al. (2015) cited earlier in this chapter has succeeded on closing several important loopholes: locality, detection efficiencies and fair sampling. This is an excellent achievement, but more work remains to be done.

6.4.2 Locality, Reality and Free-Choice

Equations (6.18)–(6.20) show that an observer at A cannot distinguish by a measurement on a single photon whether it belongs to an entangled pair or whether it is an uncorrelated quantum with random polarization. This photon will have a determined polarization if a measurement has previously been made on the other photon of the pair. This mental exercise suggests that there is a level of reality that quantum mechanics cannot describe. This situation poses a question: does quantum mechanics describe reality, as Einstein required, or does quantum mechanics describe the observer's state of knowledge about reality? Is there a deterministic reality that we can't see, with standard quantum mechanics describing our limited knowledge?

A theory developed by David Bohm and Louis de Broglie treats the quantum as both a particle and a wave. The particle has well-defined eigenvalues: energy, momentum, position, polarization, spin, etc., just like a classical object, and is controlled by a wavefunction that determines how it can be observed. This theory reproduces all the results of standard quantum mechanics, including the fundamental uncertainty relationships. There is an important difference with respect to standard quantum mechanics. In the Bohm-de Broglie theory, the quantum does not reside in a "ghostly" superposition of eigenstates prior to measurement, with wavefunction collapse on measurement as postulated by the Copenhagen interpretation. Instead, an observable, such as polarization, always has a defined value. Its evolution in space and time is governed by a wavefunction, called a pilot wave. A good number of serious physicists, including John Bell himself, have supported this theory as a viable alternative to conventional quantum mechanics.

There are three key concepts that are present in the EPR question: *Locality, Reality* and *Free-choice*. From EPR to the development of a rigorous experiment by Aspect, the question of locality in quantum mechanics has been deeply scrutinized. Quantum entanglement introduces non-locality and the effect of this non-locality can be measured. To date all experiments show that the same particular experimental conditions demonstrate experimental results that cannot be explained by local influences alone.

6.4.3 Free-Will

Bell's theorem and subsequent experiments show that there is no local hidden-variable that can reproduce all the results of quantum mechanical behavior. However, a theory based on *non-local* hidden variables does not have this limitation. The Bohm-de Broglie is a step in this direction. The concept of *non-local* hidden variables leads quickly to the idea of pre-determined outcomes, not in an individual laboratory, but in the universe at large. This brings the question of free-will to the fore. Is the experimenter truly free to choose the apparatus and experiments to be performed, or are his actions influenced by conditions over which he has no knowledge or control? This is the situation that Goethe imposed on Marguerite.

Bell addressed the dilemma of determinism in an interview with BBC radio in 1985.

> There is a way to escape the inference of superluminal speeds and spooky action at a distance, but it involves absolute determinism in the universe, the complete absence of free will. Suppose the world is super-deterministic, with not just inanimate nature running on behind-the-scenes clockwork, but with our behavior, including our belief that we are free to choose to do one experiment rather than another, absolutely predetermined, including the 'decision' by the experimenter to carry out one set of measurements rather than another, the difficulty disappears. There is no need for a faster-than-light signal to tell particle A what measurement has been carried out on particle *B*, because the universe, including particle *A*, already 'knows' what that measurement, and its outcome, will be.[4]

Shimony et al. (1976) have also addressed this issue:

> In any scientific experiment in which two or more variables are supposed to be randomly selected, one can always conjecture that some factor in the overlap of the backward light cones has controlled the presumably random choices. But, we maintain, skepticism of this sort will essentially dismiss all results of scientific experimentation. Unless we proceed under the assumption that [superdeterminism does not hold], we have abandoned in advance the whole enterprise of discovering the laws of nature by experimentation.[5]

6.5 Summary

Entanglement describes a situation in which two entities share a common attribute. In the case of two photons, this attribute could be polarization. While the net polarization of the pair of entangled photons is well-defined, the polarization of the individual photons is not. If the photons are physically separated by a beam-splitter, they can each be brought to a different measurement apparatus. However, the determination by measurement of the polarization of one photons will fix instantaneously and with certainty the polarization of the second photon so that the net polarization is not altered by measurement.

Measurement causes the entangled state wavefunction to collapse. Once the polarization of one photon is determined, then the polarization of the other is also known. This between two photons has been shown experimentally to exist over

[4]Excerpt from the interview on BBC radio, conducted by Paul Davies. A transcript can be found in the book Davies and Brown (1993).

[5]Shimony et al. (1976), Bell et al. (1985).

significant macroscopic distances, and provides evidence for the non-locality in physical measurements. That is, the measurement of a property of one particle at location A is not completely determined by conditions at location A. The results at location A can also be affected by the results of a measurement of another particle at a different location *B*, provided that these particles are entangled.

Bell's theorem analyzes the outcomes of measurements at two different locations, *A* and *B* under the assumption of locality. That is the outcome at A is determined only by the settings of the measurement equipment at location *A*, and the properties of the object to be measured at A. In this context, the theorem allows for the presence of a "local hidden variable" that accounts for some or all of the apparent statistical nature of the measurements at locations A or B. Bell's analysis places upper limits on the correlations that can be measured between results recorded at locations A and B, regardless of the nature or form of the hidden variable.

For more than 2 decades, careful measurements of entangled photon pairs demonstrate significant violations of Bell's limit. We have shown that the measured results can be explained by quantum mechanical analysis, and that the assumption of locality is not valid. The polarization entanglement of a photon pair can be described by a wavefunction that is non-local, spread over both photons of the pair. Measurement of the polarization one photon forces this wavefunction collapse into two disentangled states with distinct polarizations.

The demonstration of the non-local behavior of entangled photon pairs is a foundation stone for quantum cryptography. This technology of ultra-secure communication is based on the distribution to authorized receivers of a quantum key that enables reading of encrypted messages. The quantum key is determined by the detection of photons correlated by entanglement at two different locations. Any attempt to intercept the photons that constitute the quantum key destroys the correlation, by wavefunction collapse. The cryptographic key is composed only of those detected photons for which the correlation is preserved. Using the quantum key, encrypted communications can be exchanged over conventional communication networks without having to worry about interception.

6.6 Exercises

6.1 Single photon detection can be achieved using a photomultiplier or a silicon avalanche photodiode, both of which work best for visible and near-ir photons. The quantum efficiency η of these detectors is always less than unity.

(a) Show that the expression for the Bell function is modified to:

$$2 - \frac{4}{\eta} \leq E(\mathbf{a}, \mathbf{b}) - E(\mathbf{a}, \mathbf{c}) + E(\mathbf{d}, \mathbf{c}) + E(\mathbf{d}, \mathbf{b}) \leq \frac{4}{\eta} - 2$$

(b) Presuming that an experiment is carried out using entangled photons with polarizers oriented at the optimum settings of $\frac{\pi}{8}, \frac{\pi}{8}, \frac{\pi}{8}, \frac{3\pi}{8}$. At what level of quantum efficiency will it no longer be possible to detect a violation of Bell's inequality?

References

A. Aspect, Bell's theorem: "The Naive view of an experimentalist", in *Quantum [Un] speakables—From Bell to Quantum Information*, ed. by R.A. Bertlmann, A. Zeilinger (Springer, Berlin, 2002). http://arxiv.org/ftp/quant-ph/papers/0402/0402001.pdf

A. Aspect, P. Grangier, G. Roger, Experimental tests of realistic local theories via Bell's theorem. Phys. Rev. Lett. **47**, 460–463 (1981). http://www.physics.drexel.edu/~bob/Entanglement/Aspect82a.pdf

A. Aspect, J. Dalibard, G. Roger, Experimental test of Bell's inequalities using time-varying analyzers. Phys. Rev. Lett. **49**, 1804–1807 (1982). https://doi.org/10.1103/PhysRevLett.49.1804

J.S. Bell, *Speakable and Unspeakable in Quantum Mechanics* (Cambridge University Press, Cambridge, 1987). ISBN 0-521-33495-0. https://is.muni.cz/el/ped/podzim2017/FY2BP_TF1/um/Uceni_text_John_S._Bell_Speakable_and_Unspeakable_in_Quantum_Mechanics_First_Edition.pdf

J.S. Bell, J.F. Clauser, M.A. Horne, A. Shimony, An exchange on local beables. Dialectica **39**, 85–96 (1985). https://www.jstor.org/stable/42970534?seq=1

D. Bohm, *Quantum Theory* (Prentice Hall, New York, 1951) (Reprinted by Dover, New York, 1989). ISBN-13: 978-0486659695

J.F. Clauser, M.A. Horne, A. Shimony, R.A. Holt, Proposed experiment to test local hidden-variable theories. Phys. Rev. Lett. **23**, 880–884 (1969). http://www.sophphx.caltech.edu/Adv_Lab/References/CHCSPhysRevLett.23.880.pdf

P.C.W. Davies, J.R. Brown, *The Ghost in the Atom: A Discussion of the Mysteries of Quantum Physics* (Cambridge University Press, Cambridge, 1993), pp. 45–46. ISBN 0 521 30790 2

A. Einstein, B. Podolsky, N. Rosen, Can quantum-mechanical description of reality be considered complete? Phys. Rev. **47**, 777–780 (1935). Available on-line: http://journals.aps.org/pr/pdf/10.1103/PhysRev.47.777

A. Fine, *Einstein-Podolsky-Rosen Argument in Quantum Theory* (2015) http://plato.stanford.edu/entries/qt-epr/

M. Fuwa, S. Takeda, M. Zwierz, H.M. Wiseman, A. Furusawa, Experimental proof of nonlocal wavefunction collapse for a single particle using homodyne measurements. Nat. Commun. **6** (Article number: 6665, 24 Mar 2015). https://arxiv.org/abs/1412.7790

B. Hensen, H. Bernien, A.E. Dréau, A. Reiserer, N. Kalb, M.S. Blok, J. Ruitenberg, R.F.L. Vermeulen, R.N. Schouten, C. Abellán, W. Amaya, V. Pruneri, M.W. Mitchell, M. Markham, D.J. Twitchen, D. Elkouss, S. Wehner, T.H. Taminiau, R. Hanson, Experimental loophole-free violation of a Bell inequality using entangled electron spins separated by 1.3 km. Nature **526**, 682–686 (2015). http://users.df.uba.ar/bragas/TEP/Loophole-free%20Bell%20inequality%20violation%20using%20electron%20spins%20separated%20by%201kilometers.pdf

P.G. Kwiat, E. Waks, A.G. White, I. Appelbaum, P.H. Eberhard, Ultra-bright source of polarization-entangled photons. Phys. Rev. A. **60**, R773–R776 (1999). http://arxiv.org/abs/quant-ph/9810003

J.-A. Larsson, Loopholes in Bell inequality tests of local realism. J. Phys. A **47**, 424003 (2014). http://arxiv.org/pdf/1407.0363v1.pdf

A. Shimony, M.A. Horne, J.F. Clauser, The theory of local beables. Epistemol. Lett. **13**, 1 (1976)

W. Tittel, J. Brendel, B. Gisin, T. Herzog, H. Zbinden, N. Gisin, Experimental demonstration of quantum correlations over more than 10 km. Phys. Rev. A **57**, 3229–3233 (1998). https://arxiv.org/pdf/quant-ph/9707042.pdf

J.W. von Goethe, *Theory of Colours* (English translation by C.E. Eastlake) (Dover Fine Arts, New York, 2006). ISBN -13 987-0-486-44805-3

Chapter 7
Lasers

Abstract LASER is an acronym for **L**ight **A**mplification by **S**timulated **E**mission of **R**adiation. Laser action is a general principle of the behavior of light absorption and emission by matter. A laser is an amplifier with positive feedback. Amplification is generated by stimulated emission of photons, and positive feedback is achieved using mirrors. The physics of spontaneous and stimulated emission of photons is directly seen though quantization of the electromagnetic fields. The transition rate for stimulated emission depends on the number of photons already present in the electro-magnetic mode. Spontaneous emission can only be understood as a quantum reaction with the vacuum. Photons created by spontaneous emission act as seeds for subsequent stimulated optical transitions. Population inversion is needed to raise the stimulated emission rate above the rate of spontaneous emission. Optical gain will occur if the stimulated emission rate exceeds the stimulated absorption rate. Such a situation cannot occur in a two level system because the same two levels are responsible both for absorption and emission of photons. In a three or four level system, the equilibrium between absorption and emission is maintained, but the absorption takes place between one set of levels and the emission takes place between a different set. A critical parameter for laser performance is the threshold current density. The double-heterostructure laser concept is based on confinement of the optical mode and the recombination of electrons and holes to the same region of space. Application of quantum confinement has dramatically improved this design by separate optimization of the optical confinement region and the electron-hole recombination region. This results in a steep reduction in the threshold current and a gain factor that is independent of operating drive current.

7.1 Introduction

LASER is an acronym for **L**ight **A**mplification by **S**timulated **E**mission of **R**adiation. Laser action is a general principle of the behavior of light absorption and emission by matter. As a result, lasing has been observed in a wide range

© Springer Nature Switzerland AG 2020

T. P. Pearsall, *Quantum Photonics*, Graduate Texts in Physics,
https://doi.org/10.1007/978-3-030-47325-9_7

of materials and conditions where luminescence is generated, including chemical reactions, antifreeze, gases, solids, liquids, and semiconductor p–n junctions. Even water can be made to support lasing in the far infra red. Semiconductor p–n junctions are among the materials in which laser action can be achieved most easily.

Semiconductor lasers cover a very wide range of optical wavelengths. Lasers can be built that span a range from <400 nm to more than 10,000 nm. No other materials system has this flexibility. Semiconductor lasers are relatively inexpensive. The cheapest examples sell for less than a dollar, and the most expensive for less than $10,000. As a result of these and other considerations, the semiconductor laser in by far the industry leader in terms both of the number of units sold and the volume of revenues. Semiconductor laser diodes are the key component in a number of common devices such as a CD-player, scanner, printer, and DVD reader. They are also a key component in optical fiber telecommunications for generating the lightwave that travels down the fiber.

7.2 Spontaneous and Stimulated Emission

In this section we will consider optical absorption and emission in a finite resonant cavity of length L, such as the active region of a semiconductor laser. To a good approximation, the electric and magnetic fields can be considered as plane waves. The direction of propagation is taken along the z-axis, the magnetic field along the x-axis and the electric field along the y-axis.

The allowed modes of this cavity are quantized: $k_l = \frac{l\pi}{L}$, where l is an integer. The cavity modes can be written down:

$$\begin{aligned}\mathbf{H}(\mathbf{r}) &= -\mathbf{i}\sqrt{\frac{2}{V}}\cos(k_l z)\\ \mathbf{E}(\mathbf{r}) &= \mathbf{j}\sqrt{\frac{2}{V}}\sin(k_l z)\end{aligned} \tag{7.1}$$

where V is the mode volume.

Using (5.27) and (5.28), we can express the total resonator fields as:

$$\begin{aligned}\mathbf{H}_l(\mathbf{r}, t) &= -\omega_l\sqrt{\frac{2}{V\mu}}Q_l(t)\cos(k_l z)\\ \mathbf{E}_l(\mathbf{r}, t) &= -\sqrt{\frac{2}{V\varepsilon}}P_l(t)\sin(k_l z)\end{aligned} \tag{7.2}$$

Using (5.33) and (5.34), we can transform the expressions for the magnetic and electric fields directly into quantized form:

$$\mathbf{H}_l(\mathbf{r}, t) = -\sqrt{\frac{\hbar\omega_l}{V\mu}}(a_l^+ + a_l)\cos(k_l z)$$
$$\mathbf{E}_l(\mathbf{r}, t) = -i\sqrt{\frac{\hbar\omega_l}{V\varepsilon}}(a_l^+ - a_l)\sin(k_l z) \tag{7.3}$$

The strength of the emission or absorption of photons can be expressed as the product of the optical transition rate, the probability that the initial state is occupied and the probability that the final state is empty.

The transition rate from state 2 to 1 is given by Fermi's golden rule, and can be derived easily from first-order perturbation theory:

$$W_{2\to1} = \frac{2\pi}{\hbar}|\mathcal{H}_{2\to1}|^2\delta(E_2 - E_1 - \hbar\omega) \tag{7.4}$$

$|\mathcal{H}_{2\to1}|$ is the interaction Hamiltonian operator for the photonic transition:

$$\mathcal{H}_{2\to1} = -q\mathbf{E}_l(\mathbf{r}, t)\cdot\mathbf{r} = -qE_l(z, t)y \tag{7.5}$$

where $q =$ the electronic charge

$$\mathcal{H}_{2\to1} = iqy\sqrt{\frac{\hbar\omega_l}{V\varepsilon}}(a_l^+ - a_l)\sin(k_l z) \tag{7.6}$$

We consider first the transition rate for optical emission. In this case an electron in state 2, with n photons in the field, makes a transition to state 1 accompanied by the emission of a photon so that the final photon state has $n + 1$ photons present. The mode l of the radiation field has frequency ω_l where $\hbar\omega_l = E_2 - E_1 = \Delta E$

$$\mathbf{\Psi}_2 = |2, n\rangle \tag{7.7}$$

$$\mathbf{\Psi}_1 = \langle 1, n + 1| \tag{7.8}$$

The transition rate is:

$$W_{2\to1} = \frac{2\pi q^2\omega_l}{V\varepsilon}|\langle 1, n + 1|ya_l^+|2, n\rangle|^2\sin^2(k_l z)\delta(\Delta E - \hbar\omega) \tag{7.9}$$

Using the property of the creation operator: $\langle n + 1|a^+|n\rangle = \sqrt{n + 1}$,

$$W_{2\to1} = \frac{2\pi q^2\omega_l}{V\varepsilon}(n + 1)|\langle 1|y|2\rangle|^2\sin^2(k_l z)\delta(\Delta E - \hbar\omega) \tag{7.10}$$

The number $|\langle 1|y|2\rangle|^2 \equiv y_{2\to1}^2$ is the dipole transition matrix element between $\mathbf{\Psi}_2$ and $\mathbf{\Psi}_1$. Note that $y_{2\to1}^2 = y_{1\to2}^2$.

Assuming that energy is conserved, we can simplify (7.10) in order to highlight the most important results:

$$W_{2\to1} = W_o(n+1)\sin^2(k_l z) \tag{7.11}$$

The first term in this expression is the transition rate for stimulated emission, because it depends on the number of photons already present in the electromagnetic mode. The stimulated photon is emitted into the same mode. This means that the emission of one additional quantum is stimulated by those already present. This stimulated emission acts like a linear amplifier.

However, this is not all. There is an additional component to the photonic transition rate that is present for all transitions, even when there are no photons present. This is the second term, and it is called appropriately spontaneous emission.

$$\text{Stimulated emission: } W_{2\to1} = n\frac{2\pi q^2 \omega_l y_{2\to1}^2}{V\varepsilon}\sin^2(k_l z)\delta(\Delta E - \hbar\omega) \tag{7.12}$$

and

$$\text{Spontaneous emission: } W_{2\to1} = \frac{2\pi q^2 \omega_l y_{2\to1}^2}{V\varepsilon}\sin^2(k_l z)\delta(\Delta E - \hbar\omega) \tag{7.13}$$

Spontaneous emission always accompanies the transition of a electron from a higher to lower energy state. The energy of the emitted photon is equal to the energy difference between these two states, conserving energy. Maxwell's equations do not account for spontaneous emission because the photon is created from the vacuum or empty state. This is to be expected, because Maxwell's equations describe the continuous fields resulting from many electrons and photons. The transition of an electron between two energy levels accompanied by the creation of a photon can only be described correctly by the quantized expressions of (7.3).

Because spontaneous emission is a quantum event does not mean it is rare. At room temperature, spontaneous emission of photons dominates stimulated emission in most circumstances. The LEDs used for general lighting are bright sources of light beams composed almost entirely of spontaneously emitted photons.

The transition rate for absorption can be determined in the same way. In this case an electron in state 1, with n photons in the field, makes a transition to state 2 accompanied by the absorption of a photon so that the final photon state has $n-1$ photons present. The mode l of the radiation field has frequency ω_l where $\hbar\omega_l = E_2 - E_1 = \Delta E$

$$\mathbf{\Psi}_1 = |1, n\rangle \tag{7.14}$$

$$\mathbf{\Psi}_2 = \langle 2, n-1| \tag{7.15}$$

The transition rate is:

$$W_{2\to 1} = \frac{2\pi q^2 \omega_l}{V\varepsilon} |\langle 2, n-1|ya_l|1, n\rangle|^2 \sin^2(k_l z)\delta(\Delta E - \hbar\omega) \tag{7.16}$$

Using the property of the annihilation operator: $\langle n-1|a|n\rangle = \sqrt{n}$,

$$W_{2\to 1} = n\frac{2\pi q^2 \omega_l y_{2\to 1}^2}{V\varepsilon} \sin^2(k_l z)\delta(\Delta E - \hbar\omega) \tag{7.17}$$

The transition rate for stimulated absorption is equal to the transition rate for stimulated emission. Note that a "spontaneous absorption" term does not occur in the analysis. Such an event would violate energy conservation.

We can directly relate the occupation number n to the intensity of the field as:

$$n = I\frac{n_{\text{refr}} V}{c\hbar\omega} \tag{7.18}$$

and replacing $\sin^2(k_l z)$ by its average value of 1/2, we can rewrite (7.17) as

$$W_{2\to 1} = \frac{1}{\tau_{\text{stim}}} = I\frac{\pi \mathcal{M}_{2\to 1}^2 n_{\text{refr}}}{\varepsilon\hbar c}\delta(\Delta E - \hbar\omega), \tag{7.19}$$

making it explicit that the stimulated emission rate is proportional to the photon population in the lasing mode.

Before proceeding, we should review these key results which create the basis for laser operation.

Laser stands for Light Amplification by Stimulated Emission of Radiation.

By treating the radiation field as a set of modes l, each of which can be thought of as a harmonic oscillator, we have used the creation and annihilation operators to describe the interaction of photons and electrons in a cavity of length L.

The emission of light in this cavity consists of two contributions: Spontaneous emission and Stimulated emission.

The transition rate for stimulated emission is proportional to the number of photons in the cavity. This means that stimulated emission amplifies the photon number in the cavity by moving up the "ladder" of levels of the mode in question. This is an important component of laser action.

If emission occurs in a resonant cavity, for example one with mirrors on each end, then emitted photons will be retro-reflected into the cavity, causing more photons to be emitted. This positive feedback favors emission at the frequency that has the largest dipole transition element.

Spontaneous emission is always present. Spontaneously emitted photons provide the "kindling" that allows stimulated emission to begin.

The transition rate for stimulated absorption is equal to that for stimulated emission.

7.3 Laser Action

A laser consists of two components: a photon amplifier and a positive feedback circuit. The gain spectrum is the range of optical wavelengths (frequencies) over which light emission exceeds absorption. Amplification is generated by stimulated emission which says that the probability for photon emission is proportional to the number of photons already present. Positive feedback is achieved by a pair of reflectors so that light exiting the laser amplifier is returned back to the amplifier medium. In an operational laser, photon amplification creates optical emission gain. Laser action occurs when the gain exceeds the photon absorption rate and photon losses in the resonant cavity.

Laser action is an example of the general principle that an amplifier coupled with positive feedback generates oscillation where most or all of the energy is emitted in a narrow frequency band. Another example is the audio event that occurs when a microphone is placed in front of a loudspeaker. The combination of amplification plus positive feedback produces a piercing tone, the frequency of which occurs at the maximum of the gain spectrum of the amplifier. A third example, (in which you have probably participated) is the rhythmic applause at the end of a popular concert. Starting as random noise, (spontaneous emission) a frequency emerges, and soon almost everyone in the audience is clapping with the same rhythm (stimulated emission). The emergence of this frequency depends on feedback via your ears and the response of your arm muscles which amplify the detected signal.

7.3.1 Analysis of Spontaneous and Stimulated Emission—A Macroscopic Description

Analysis of the quantized electro-magnetic field is a powerful method for understanding the physics of quantum emission and absorption. It is, however, not a helpful method for understanding the design and operation of efficient lasers. This requires a macroscopic analysis based on the dynamics of many photons and electrons simultaneously (Agrawal and Dutta 1986; Bhattacharya 1994).

In Fig. 7.1 we diagram in a very schematic way the three possible transitions that can take place in the absorption or emission of light. We choose a simple two-level system having N_2 states in the upper energy level occupied by electrons and N_1 states in the lower energy level occupied by electrons. In equilibrium, $N_2 < N_1$ and the ratio between the two occupation numbers is given by Boltzmann statistics, that is: $\frac{N_2}{N_1} = e^{-\frac{\Delta E}{k_B T}}$. The probability that an electron can make a transition from the lower energy state to the upper energy state by stimulated absorption (Fig. 7.1) is equal to B_{12}. Because stimulated absorption is the reverse of stimulated emission, the probability for stimulated emission to occur (Fig. 7.1) is given by B_{21}. The probability for spontaneous emission is different and we will call this

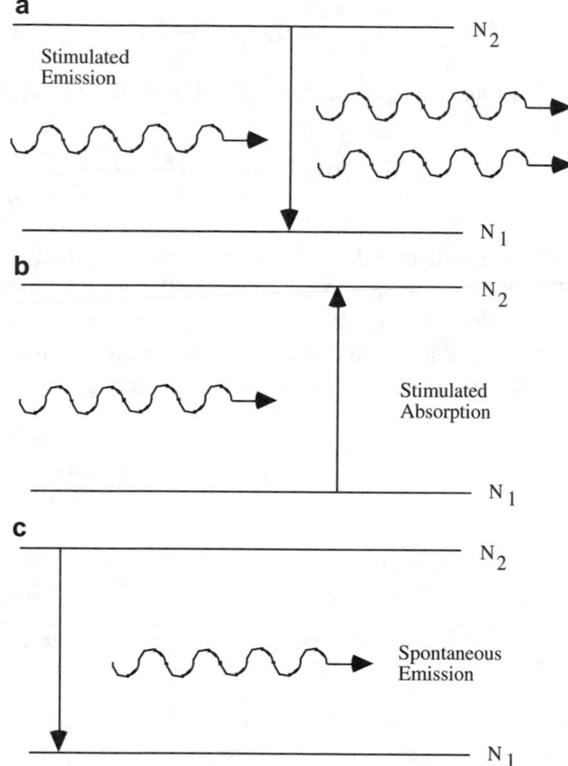

Fig. 7.1 Diagram of the three possible electron-photon interactions. Stimulated emission (**a**) and stimulated absorption (**b**) refer to the fact the probability for absorption or emission depends on how many other photons having the energy difference of the transition are already present. In a spontaneous emission process shown in (**c**), the probability of emission does not depend on the presence of other photons

A_{21}. We would like to compare the number of absorption transitions to the number of emission transitions, in order to calculate the gain.

The number of spontaneous transitions is given by the number of occupied states N_2 multiplied by the probability of a transition: $N_2 A_{21}$. The number of stimulated transitions also depends on $\rho(\omega)$, which is the density of photon states having an energy equal to the transition energy. The number of transitions by stimulated emission is: $N_2 B_{21} \rho(\omega)$. The total number of transitions where a photon is emitted is the sum of these two terms. The number of absorbing transitions depends on the number of occupied states in the lower level N_1 and the number of photons present having the right energy; $\rho(\omega)$: $N_1 B_{12} \rho(\omega)$. Under steady-state conditions, the number of absorbing transitions equals the number of emitting transitions. We can summarize the discussion so far in a set of simple equations:

$$\begin{aligned} \text{Emission rate} &= W_{21} = B_{21}\rho(\hbar\omega) + A_{21} \\ \text{Absorption rate} &= W_{12} = B_{12}\rho(\hbar\omega) \end{aligned} \tag{7.20}$$

and

$$N_1 B_{12}\rho(\hbar\omega) = N_2 B_{21}\rho(\hbar\omega) + N_2 A_{21}$$

This allows us to solve for the photon density at the energy of the transition:

$$\rho(\hbar\omega) = \frac{N_2 A_{21}}{N_1 B_{12} - N_2 B_{21}} = \frac{A_{21}/B_{21}}{\frac{N_1}{N_2}\frac{B_{12}}{B_{21}} - 1} \tag{7.21}$$

Now we will compare this expression for $\rho(\hbar\omega)$ to another one based on the Planck radiation law. We discussed Planck's experiments in Chap. 1. The result of his work was to derive an expression for the energy density of photons. We recall that Planck discovered that the energy density depends on the temperature and on the color, or energy of an individual photon. Planck's radiation law states:

$$\rho(\hbar\omega) = \frac{4\hbar}{\lambda^3}\left[\frac{1}{e^{\frac{\hbar\omega}{k_B T}} - 1}\right] \tag{7.22}$$

In comparing (7.21) and (7.22), we can see some similarities. For example, we know from Boltzmann statistics that: $\frac{N_2}{N_1} = e^{-\frac{\Delta E}{k_B T}}$. Therefore, it follows that: $\frac{N_1}{N_2} = e^{\frac{\Delta E}{k_B T}} = e^{\frac{\hbar\omega}{k_B T}}$. We can see that the two equations are identical when:

$$\begin{aligned} &B_{12} = B_{21} \\ &\text{and} \\ &\frac{A_{21}}{B_{21}} = \frac{4\hbar}{\lambda^3} = \frac{8\pi n^3 \hbar\omega^3}{c^3} \end{aligned} \tag{7.23}$$

The two expressions in (7.23) are called the Einstein relations, in which c is the speed of light, and n is the index of refraction of the medium involved (Casey and Panish 1978).

The ratio of the spontaneous emission rate to the stimulated emission rate is:

$$R = \frac{A_{21}}{\rho(\omega)B_{21}} = e^{\frac{\hbar\omega}{k_B T}} - 1 \tag{7.24}$$

This ratio is normally much greater than unity: that is, the spontaneous emission rate far exceeds the stimulated emission rate. If we consider photons with an energy of 1 eV, ($\lambda = 1240$ nm) then $\frac{\hbar\omega}{k_B T} \approx 40$ at room temperature. R, as a result, is a very big number.

However, in order to have laser action, the reverse must be true: that is the stimulated emission rate must be greater than the spontaneous emission rate. To see how this can happen, read on through the next section.

7.3.2 Optical Gain

Optical gain and optical absorption are closely related. When light is incident on a semiconductor surface, only two things can happen: reflection or transmission. Normally, both occur at the same time. Transmitted photons can subsequently be absorbed. Absorption can take place if the energy of the incident photons is greater than the band gap. Absorption does not occur all in one spot at the surface, but rather progressively as the photons propagate into the semiconductor. At any point inside the semiconductor, the amount of light that gets absorbed is proportional to the total intensity that is present, following the principle of stimulated absorption in (7.17). The constant of proportionality is called the absorption coefficient. This simple model gives an excellent description of reality (Fig. 7.2).

We can write down an equation that describes this situation:

$$I(x + \Delta x) \; - I(x) = \Delta I(x) \tag{7.25}$$

and

$$\Delta I(x) = -\alpha \cdot I(x)$$

in the limit of small Δx,

$$\frac{d}{dx} I(x) = -\alpha I(x);$$
$$\therefore \; I(x) = I_o e^{-\alpha x} \text{ W cm}^{-2}$$

The value of α depends on the material and on the photon energy. For example, in the case of GaAs, α is about 10^4 cm^{-1} for photons having an energy greater than the bandgap ($E_g = 1.43$ eV at room temperature). For photons that are less

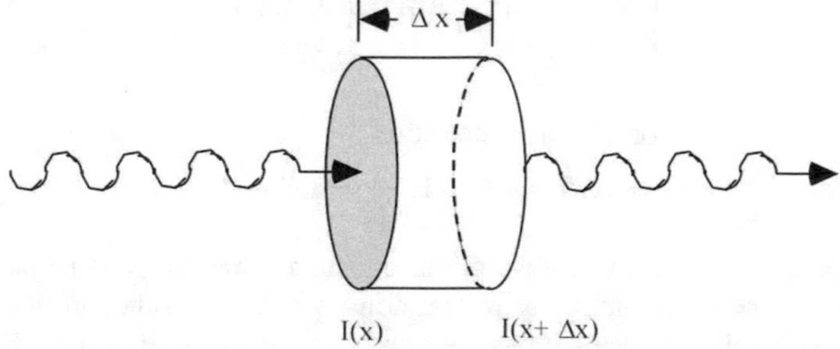

Fig. 7.2 A simple schematic diagram of light passing through a section of material where absorption is taking place. Absorption causes the intensity of light to decrease as a function of distance traveled. The change in the intensity between points x and x + Δx is proportional to the intensity at point x

energetic than the bandgap energy, α is three orders of magnitude smaller. When the absorption coefficient is a positive number, the intensity of the beam decreases as the light propagates through the material. However, suppose that α could be made to be negative, what would be the result?

Example
A beam of light of monochromatic photons of energy 1.5 eV strikes the surface of GaAs at normal incidence. What percentage of the original photon beam penetrates 1 μm beneath the surface? What percentage penetrates 10 μm beneath the surface?

Solution:
There are only two things that can happen to photons incident on an interface. Either they are reflected or transmitted. Some of the transmitted photons are subsequently absorbed. To answer these questions you need to find first of all the percentage of light that is transmitted, and then find out what fraction of those photons are absorbed.

The percentage of light reflected is calculated from Fresnel's equation at normal incidence:

$$R = \frac{(n-1)^2}{(n+1)^2} = \frac{(2.4)^2}{(4.4)^2} = 0.25$$
$$T = 1 - R = 0.75$$

So, 75% of the light is transmitted.

To calculate the intensity:

$$I = I_o \mathrm{e}^{-\alpha x}$$
$$I = I_o \mathrm{e}^{-(10^4\,\mathrm{cm}^{-1})x}$$

$$I = I_o \mathrm{e}^{-1} \text{ for } 1\ \mu\text{m penetration,}$$
$$I = I_o(0.37) = 0.75 \times 0.37 = 0.28$$

For 10 μm penetration,
$$I = 0.75 \times 4.6 \times 10^{-5} = 3.4 \times 10^{-5}$$

To continue our discussion of absorption, consider what happens to the number of photons $\mathcal{N}$, per unit volume, or the photon density, as a function of time. The photon density will decrease with as the number of electron transitions from level 1 to level 2 increases. The density will increase when the number of transitions from state 2 to state 1 increases:

$$\begin{aligned} \frac{d}{dt}\mathcal{N} &= -N_1 \rho(\hbar\omega) B_{12} + N_2 \rho(\hbar\omega) B_{21} \\ &= (N_2 - N_1)\rho(\hbar\omega) B_{21} \end{aligned} \tag{7.26}$$

The photon density is closely related to the energy density: $\rho(\hbar\omega) = \mathcal{N} \cdot \hbar\omega$. Similarly, the intensity is related to the energy density: $I = \rho(\hbar\omega) \cdot \frac{c}{n} = \mathcal{N}\frac{\hbar\omega c}{n}$.

In (7.23) we derived a relationship between intensity and distance. Because of the relationship between the intensity and the photon density, we can write another expression for the gradient.

$$\frac{d}{dx}I(x) = \frac{\hbar\omega c}{n}\frac{d}{dx}\mathcal{N} = \frac{\hbar\omega c}{n}\left[\frac{d}{dt}\mathcal{N} \cdot \frac{dt}{dx}\right] \tag{7.27}$$

For the case of light: $\frac{dx}{dt} = \frac{c}{n}$. Since this is a simple constant, the inverse expression that we would like to substitute in (7.27) is the arithmetic inverse. That is: $\frac{dt}{dx} = \frac{n}{c}$.

Using these results we can determine the condition for generating optical gain.

$$\frac{d}{dt}\mathcal{N} = \frac{1}{hf}\frac{d}{dx}I = \frac{1}{\hbar\omega}I(x) \cdot (-\alpha) = -\frac{\alpha}{\hbar\omega}\rho(\hbar\omega)\frac{c}{n}$$

Using (7.26),

$$\begin{aligned} &\frac{d}{dt}\mathcal{N} = (N_2 - N_1)\rho(\hbar\omega)B_{21} = -\frac{\alpha}{\hbar\omega}\rho(\hbar\omega)\frac{c}{n} \\ &\therefore \alpha = (N_1 - N_2)B_{21}\frac{n\hbar\omega}{c} \end{aligned} \tag{7.28}$$

So α is positive, and absorption occurs when $N_1 > N_2$. On the other hand, α is negative and **amplification occurs when $N_2 > N_1$**. This simple condition is called population inversion. You may notice that although simple, it appears to violate the requirements of Boltzmann statistics. Thus population inversion is not an equilibrium condition, but it can be obtained in steady state (Pearsall 2008). The art of making a laser is understanding how this condition can be achieved in real materials (Fig. 7.3).

When α is negative, the result is gain rather than loss, and instead of using α to describe this condition, we should define a gain coefficient k, where:

$$\begin{aligned} &\text{Gain Coefficient } = k = -\alpha \\ &\therefore\ k = (N_2 - N_1)B_{21}\frac{n\hbar\omega}{c} \end{aligned} \tag{7.29}$$

7.3.3 $N_2 > N_1$: *Population Inversion*

So far we have considered light emission from a system of electrons having two energy levels, E_1 and E_2. Looking at Fig. 7.1 again, you can see that there is one way for electrons to get pumped into the upper level, by stimulated absorption. We have demonstrated that this rate is equal to the stimulated emission rate. However

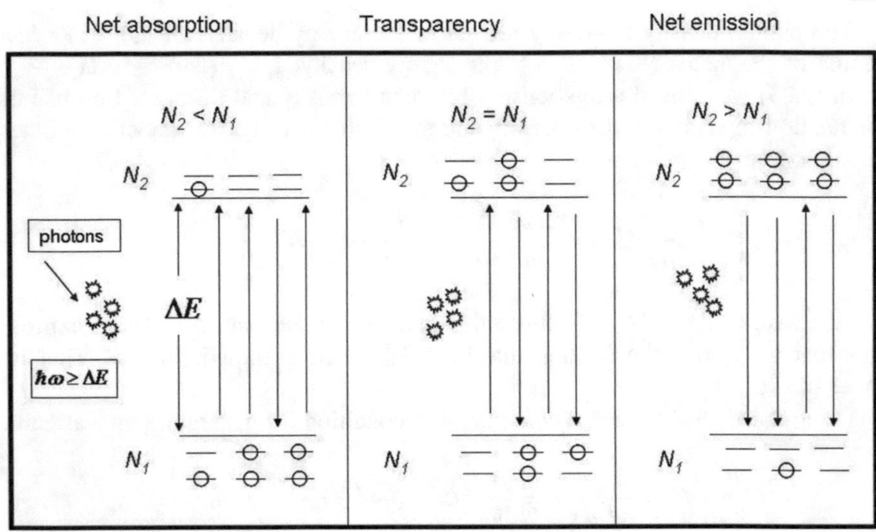

Fig. 7.3 Photon absorption, transparency and emission. The presence of photons in a medium (gaseous, liquid or solid) stimulates transitions of electrons between energy levels. These transitions are accompanied by either photon emission or absorption in order to conserve energy. The net result depends on the difference between the number of occupied states in the two levels. When $N_2 < N_1$, net absorption occurs. When $N_2 > N_1$, net emission occurs. This leads to optical amplification and laser action. The medium is said to be transparent when $N_2 = N_1$

there is a second way for electrons to be de-excited from the upper level, by spontaneous emission. We have shown that this rate is much bigger than the stimulated absorption rate. The result is that there is no way to obtain a population inversion in a 2-level system.

Of course, you might be able to get an inversion if electrons were somehow fed into the upper level by another source—a third level. This turns out to be the road to the solution. In general lasing is easiest to obtain in a four-level arrangement that is diagrammed schematically in Fig. 7.4.

In the beginning of the cycle, all the electrons are in the ground state. If level E_3 is a few $k_B T$ above level E_1, it will be nearly empty by Boltzmann statistics. The cycle starts when a high energy photon with energy $\hbar\omega = E_2 - E_1$ excites an electron from the ground state to the excited state (step 1). The photon is a particle, so all its energy must be absorbed, making a direct transition to E_3 or E_4 impossible. After excitation the electron can be scattered into state E_4 during a collision (step 2). Electrons are more likely to end up in state E_4 than state E_3 or state E_1 because the energy difference is smaller, and so easier to make up by phonon emission. After step 2 there are electrons in state E_4 but not in state E_3. Thus, a population inversion between these two levels now exists. The recombination that follows is an example of optical gain, since emission between these levels far exceeds absorption which is practically 0 (step 3). This transition can be a lasing transition if suitable feedback is provided, which increases the rate

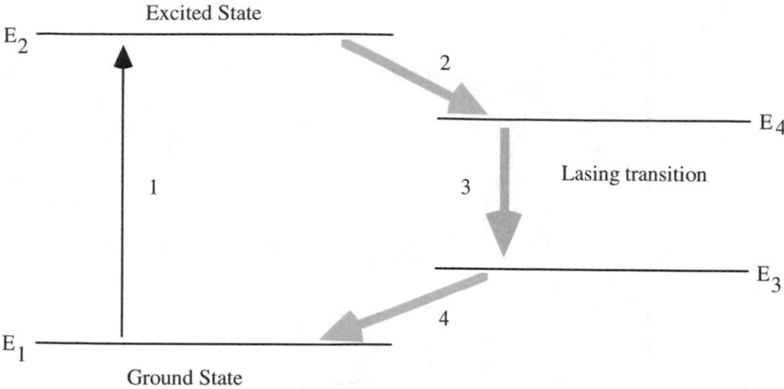

Fig. 7.4 Population inversion can be obtained in a 4-level system, in this case between level 4 and level 3. The excitation and recombination cycle is given in sequence by the numbers in the figure. There are four steps to the complete cycle

of stimulated emission. Finally, electrons that reach level E_3 are recycled to level E_1, leaving state E_3 empty again (step 4). In this example, the number of photons absorbed is still equal to the number of photons emitted. However there is now one set of levels that does most of the absorption, and another set that generates most of the emission.

A semiconductor laser is a good example of a 4-level system, and this can be understood quickly from a simple band structure diagram such as that in Fig. 7.5. Optical stimulation of lasing is relatively easy to demonstrate in a direct gap material, and it proceeds following the cycle outlined above. However, the cycle for obtaining gain by electrical excitation is quite different. In this case the behavior of the p–n junction is used to create a population inversion.

The pumping cycle in Fig. 7.5 is different from the cycle in Fig. 7.4. Initially, level E_3 is fully occupied by electrons. Optical excitation proceeds by the absorption of a photon (1). In order to conserve energy and momentum, the electron that is excited to the conduction band must originate deep in the valence band as shown. Then nearly simultaneously, the excited electron in the conduction band relaxes to state E_4 and the electron in state E_3 relaxes to state E_1 leaving a hole behind. (2) Relaxation takes place by emission of phonons, and is completed in 10^{-12} s. Now there is an electron in state E_4 and a hole in state E_3 creating a population inversion. This situation can persist for about 10^{-9} s. That is three orders of magnitude longer than the relaxation process. Finally, recombination occurs across the gap (3). This transition can be used to make a laser, if suitable feedback is provided.

In a semiconductor material, both spontaneous and stimulated emission proceed by this "4-step" process. No matter what the energy of the optical excitation energy above the bandgap, the energy of the photons emitted during recombination is always close to the bandgap energy. Optical gain only occurs in this range

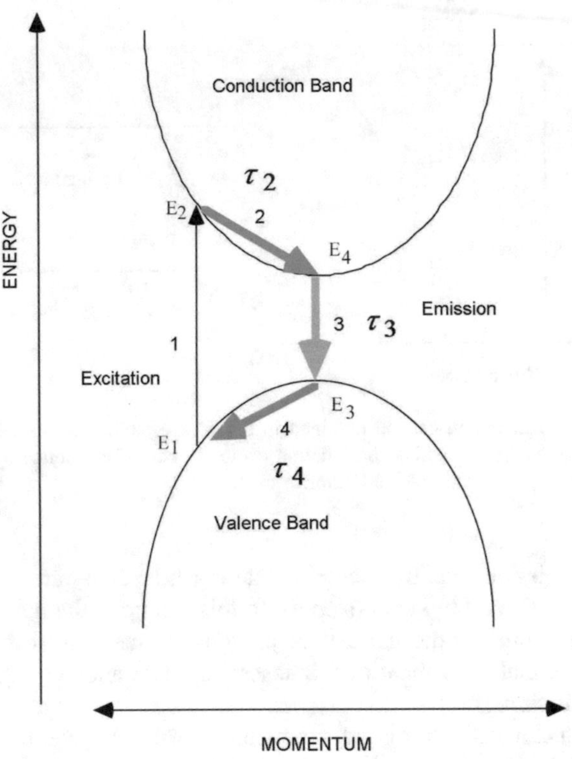

Fig. 7.5 A direct bandgap semiconductor is a good example of a 4-level medium for laser action. There is a significant difference in the pumping scheme, because level E_3 is initially fully occupied by electrons. However, as soon as a photon is absorbed, electrons leave this level to fill the hole in the valence band that is created by the excitation

of energies which we can denote by a distribution function, $g(v)$ called the optical gain spectrum, v being the photon frequency (Fig. 7.6). This is always the situation because the lifetime for optical recombination is orders of magnitude longer than the time for phonon emission. The lower energy portion of the gain curve rises in proportion to the joint density of states. The high energy cutoff is due to self absorption of photon emission. Typical values for the energy width of the gain spectrum lie in the range of 0.01–0.02 eV. In order to take account of this feature we will modify (7.29) for the gain coefficient:

$$k = (N_2 - N_1)P_{21}\frac{n\hbar\omega g(v)}{c} \tag{7.30}$$

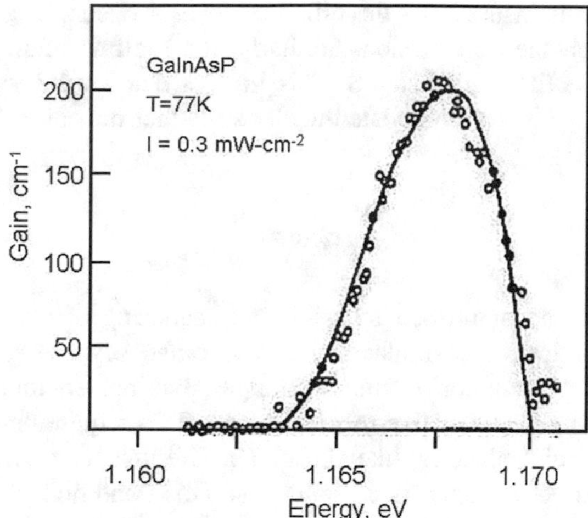

Fig. 7.6 Gain spectrum. The gain spectrum measures the difference between emission and absorption as a function of frequency or energy. The rise in the gain at lower energies is due to the increasing density of states while the fall-off at higher energies is the result of the onset of self-absorption of the emitted photons. The half width of the gain spectrum is approximately equal to kT. The figure shows a comparison between the measured (*dots*) and calculated (*solid line*) gain spectra for GaInAsP, a laser medium which is widely used to make semiconductor diode lasers for optical fiber telecommunications. From Goebel 1982, reproduced by permission

7.4 Semiconductor Diode Lasers

Up to this point, we have considered principles of laser action. We will continue the development in the specific context of semiconductor diode lasers. The semiconductor laser is a prime example of a quantum photonic device.

Today, semiconductor lasers represent 80% of the laser market value, and the percentage is even higher if we consider the number of lasers sold each year. These percentages are on the increase. Other types of lasers, such as fiber lasers or disk lasers, which are widely used for welding and machining of steel, are pumped optically by diode lasers. There is a trend to replace some of these laser systems by direct diode laser processing.

The fabrication of semiconductor lasers integrates material properties, photon resonant cavity design and the design of electron-photon recombination.

7.4.1 Population Inversion in Semiconductor Materials

The most important materials requirement concerns the relative lifetimes of the various steps shown in Fig. 7.5. In order to build up a population inversion in level 4 compared with level 3, the lifetime of step 2 should be shorter than the lifetime of step 3. This insures that the filling of level 4 occurs at a faster rate than it empties by spontaneous emission. Similarly, the lifetime of step 4 should also be shorter that the lifetime of step 3. This insures that level 3 will have enough empty states (holes) to accommodate the electrons that transition from level 4.

$$\begin{aligned} \tau_2 &< \tau_3 \\ \tau_4 &< \tau_3 \end{aligned} \tag{7.31}$$

Steps 2 and 4 are accomplished by interband scattering of electrons by optical phonons. The lifetimes are similar and in the range of $\tau_2 \cong \tau_4 \approx 10^{-12}$ s. The lifetime for optical emission should be longer, but not so long that electrons lose their energy by non-radiative recombination. This requirement is satisfied by direct bandgap semiconductors, like GaAs. The lifetime for radiative recombination across the direct bandgap is $\tau_3 \approx 10^{-9}$ s. This condition is not satisfied by indirect bandgap semiconductors like Si, which explains why laser action by emission across the bandgap has not been measured in this material.

Carrier injection in p–n junction is a convenient and efficient way to introduce excess electrons in the conduction band. This allows us to implement step 1 by electrical pumping. We show in Fig. 7.7 the energy level diagram for a p–n junction, heavily doped so that the Fermi level actually lies in the conduction band on the n-side and in the valence band on the p-side.

Figure 7.7a, b and c show the p-junction in states of increasing forward bias voltage. The physical width of the charge depletion region narrows as the bias voltage increases. When the bias voltage exceeds the built-in voltage, it creates 1-dimensional carrier confinement for both electrons and holes as shown on Fig. 7.7c. In this regime, a state of population inversion can be attained, supporting laser action. Robert Hall of General Electric Laboratories made the first semiconductor p–n diode laser in 1962 using GaAs. While this demonstration validated the physical principle, more invention and engineering is necessary to build a semiconductor laser that can function efficiently at room temperature. The key to the success of semiconductor lasers over all other varieties is that it is possible by materials engineering to modify important parameters that directly impact laser device performance. One such key parameter, shown in Fig. 7.7 is Δw, which is the dimension of the region where population inversion occurs.

To summarize, a laser is an amplifier with positive feedback. We have determined that the condition for amplification to occur is a population inversion, and we have described how this can be achieved in a p–n junction. The other half of the requirement is to create optical feedback.

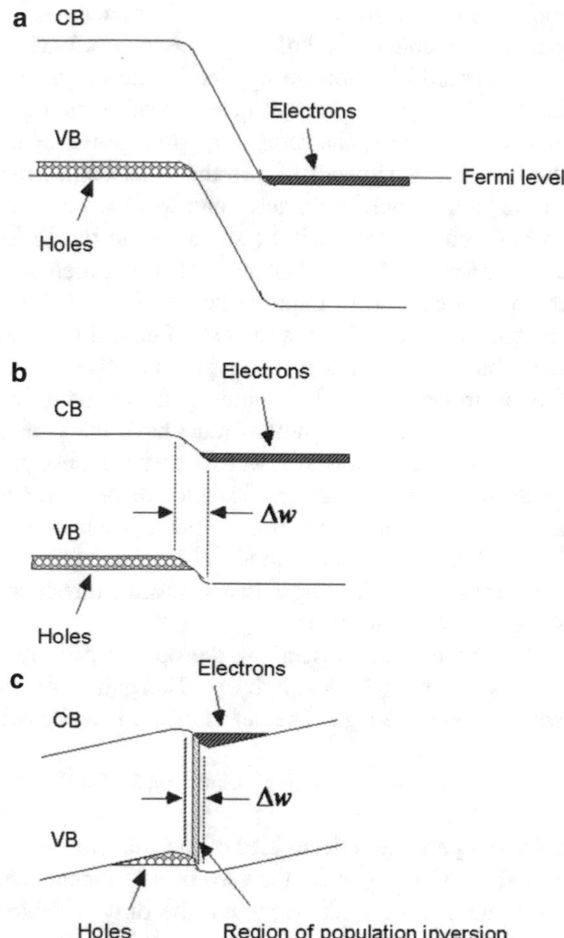

Fig. 7.7 Energy level diagram for a heavily-doped p–n homojunction **a** 0-bias. Fermi level lies in the valence band (VB) on the p-side and the conduction band (CB) on the n-side. **b** Moderate forward bias ($V_{bias} < V_{built\text{-}in}$): The depletion width Δw narrows, but the electron and hole populations do not overlap, and photonic recombination is negligible. **c** Large forward bias ($V_{bias} > V_{built\text{-}in}$): The depletion width reaches a minimum value. For GaAs, this is about 300 nm. The electron and hole populations overlap in space, and photonic recombination takes place

7.4.2 Optical Feedback—Making a Laser

A simple optical resonator can be formed by 2 parallel mirrors. In a semiconductor like GaAs, the mirrors can be obtained by cleaving the GaAs material along crystal planes, which yields atomically-flat, parallel surfaces. The gain coefficient of direct band-gap semiconductors like GaAs is very large compared to that of gas lasers, like He–Ne or solid state laser materials like Nd–YAG. As a result the mirrors at each end of the cavity do not need to be as efficient as those required for other kinds of laser materials. There are two big performance benefits: one is that more power can be extracted from a semiconductor laser at modest input power levels, and the second is that there is a much larger tolerance in the design of the resonator in order to make a working device. There is also a big space savings, too. This is why you can hold a semiconductor laser on the tip of your finger, and you need a table top to hold a gas laser or a YAG laser.

A forward bias voltage applied to the diode creates excess concentrations of electron-hole pairs. Electron-hole recombination generates photons that depart in all directions by spontaneous emission. Some of these photons will travel along the line that is perpendicular to the reflecting surface of the two parallel mirrors. These photons will be reflected and will travel back into the diode along the same path. Of course, there will be some loss in this process. Some photons will be re-absorbed along the way. Others will be scattered out of alignment by defects in the optical path. Most important of all, some will traverse the mirror and be emitted into free space. This is the useful output of the device. Reflected photons will stimulate the emission of other photons. These stimulated photons contribute to the electromagnetic field creating gain. In the development leading to (7.19), we have shown that these photons must have the same energy and the same phase as the stimulating photon. If it were otherwise these photons would interfere destructively with the electromagnetic field. In order for lasing to occur, the gain initiated by a photon during one round trip circuit must be greater than the losses incurred during the same circuit including the fraction of the intensity that is emitted through the reflecting mirrors. Lasing threshold is defined as the point when the gain is equal to the loss.

We can treat any loss along the optical path by an effective absorption coefficient which we will denote by γ. The gain coefficient k behaves just like a loss with the opposite sign. The net laser gain can be expressed:

$$\text{Gain} = R_1 R_2 \mathrm{e}^{(k-\gamma)L} \tag{7.32}$$

R_1 and R_2 are the reflectivities of the mirrors at either end of the gain region, and L is the cavity length. In the case of a semiconductor laser R_1 and R_2 are usually the same. During laser operation, the only variable in this expression is the gain coefficient which depends on the pumping rate, that is, on the forward current in the diode. Everything else remains constant.

The laser threshold is reached when the net gain is unity. This also defines the threshold gain.

$$\text{Threshold Gain} = 1 = R_1 R_2 \mathrm{e}^{(k_{\mathrm{th}}-\gamma)2L}$$

and

$$k_{\mathrm{th}} = \gamma + \frac{1}{2L} \ln\left(\frac{1}{R_1 R_2}\right) \tag{7.33}$$

in the case where R_1 and R_2 are the same,

$$k_{\mathrm{th}} = \gamma + \frac{1}{L} \ln\left(\frac{1}{R}\right)$$

Although spontaneous emission diverts light from lasing modes, reducing the laser efficiency, its presence is absolutely required to make the laser work in the first place. The spontaneous emission "primes the pump" in the beginning by filling all possible radiation modes with photons. Gain and laser action then build up out of the noise in the much smaller number of modes that overlap in energy with the gain spectrum and which are resonant modes of the reflecting cavity (Fig. 7.8).

In general there are a number of modes that lie in the gain spectrum. The exact number can always be calculated from the width of the gain spectrum, the cavity length and the wavelength. Because the gain of semiconductor materials is

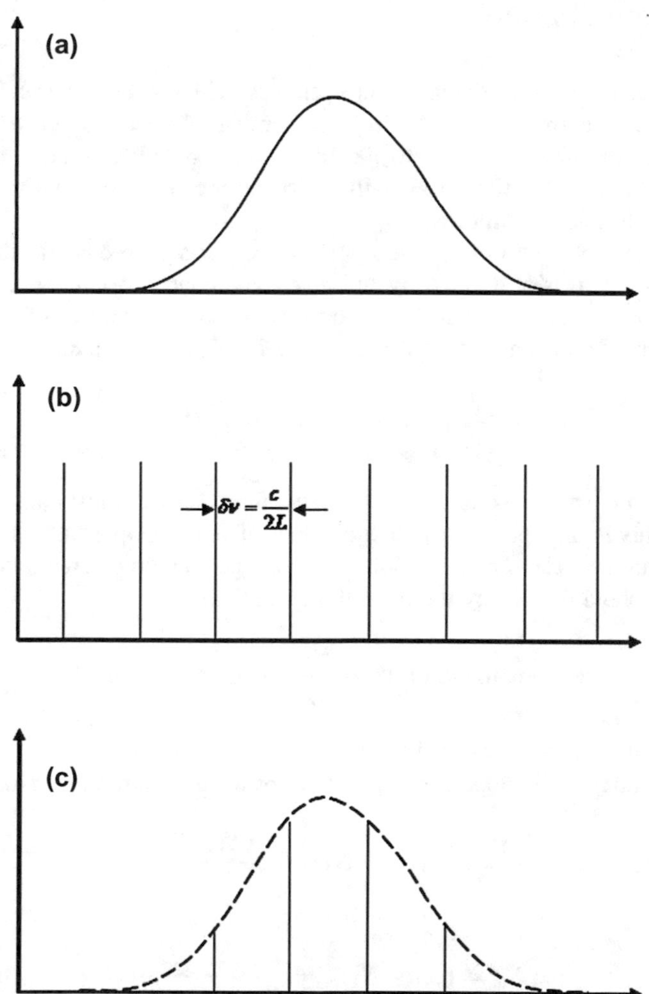

Fig. 7.8 The resonant wavelengths of a laser cavity are those select wavelengths for which a standing wave can be established between the two mirrors at either end of the cavity. These cavity modes and are equally spaced in frequency (**b**). Only the few that happen to occur in the gain spectrum (**a**) of the semiconductor material will participate in the necessary positive photon feedback (**c**)

so large, lasing action often occurs at several modes simultaneously. Single-mode operation can be achieved by making the cavity wavelength selective.

At this point we have assembled all the elements of a laser: gain, population inversion, and a resonant cavity for feedback. It remains only to determine the level of current injection into the diode that is required in order to achieve laser action.

7.4.3 *Laser Threshold*

In a p–n junction laser, population inversion is achieved by electrical pumping. The threshold current is attained when the number of electrons per second being injected into the diode is equal to the threshold population density, taking into account that some of the electrons will be lost to recombination before a suitable population inversion is built up.

The number of electrons injected per second per square cm into the diode is just the current density divided by the electronic charge: $\frac{J}{q}$. If we consider the rate of electrons per second in the recombination region we need to divide this expression by the thickness t ($= \Delta w$ in Fig. 7.7) of the recombination region.

$$\text{Pumping Rate} = \frac{J}{qt} \text{ (electrons s}^{-1}\text{ cm}^{-3}\text{)} \tag{7.34}$$

What goes in must come out, and so the recombination rate must equal the pumping rate. This is the optoelectronic equivalent of the principle that absorption be equal to emission. The recombination rate is the population inversion necessary to achieve threshold divided by the recombination time:

$$\text{Recombination rate} = \frac{N_{\text{th}}}{\tau_r} \text{(electrons s}^{-1}\text{ cm}^{-3}\text{)} \tag{7.35}$$

where τ_r is the minority carrier lifetime.

We have already developed an expression for the population inversion in (7.28).

$$k_{\text{th}} = (N_2 - N_1)_{\text{th}} B_{21} \frac{nh\nu g(\nu)}{c} \tag{7.36}$$

Therefore,

$$N_{\text{th}} = (N_2 - N_1)_{\text{th}} = \frac{k_{\text{th}}}{B_{21}} \cdot \frac{c}{nh\nu g(\nu)}$$

where B_{21} is the stimulated emission coefficient. In (7.23) we related B_{21} to the spontaneous recombination rate. This is a useful relationship to know because you can measure this rate directly.

$$A_{21} = \frac{8\pi n^3 h\nu^3}{c^3} \cdot B_{21}$$
$$B_{21} = A_{21} \frac{c^3}{8\pi n^3 h\nu^3} \tag{7.37}$$

Substituting (7.37) into (7.36),

$$N_{th} = \frac{k_{th}}{A_{21}} \cdot \frac{8\pi n^2 \nu^2}{c^2 g(\nu)} \tag{7.38}$$

The spontaneous emission rate A_{21} is determined by the inverse of the spontaneous emission lifetime τ_{21}. This lifetime can be measured by exciting the laser material with a light pulse from an external laser emitting photons with energy above the band gap of the semiconductor. The semiconductor will emit photoluminescence that dies out with the spontaneous emission lifetime. Typical values for τ_{21} are several nanoseconds.

The threshold current density can be expressed by combining (7.34), (7.35), and (7.38),

$$J_{th} = \frac{qtN_{th}}{\tau_r} = qt\left(\frac{\tau_{21}}{\tau_r}\right)\frac{k_{th} n^2 \nu^2}{c^2 g(\nu)} \text{ A cm}^{-2} \tag{7.39}$$

You would like to have a lower threshold current. There are some variables in this expression that are under the control of the laser designer. The thickness of the recombination region can be reduced physically. This was first done by making a heterostructure, and has been developed into the currently-used quantum well design, where the recombination is restricted to a potential well of thickness comparable to the de Broglie wavelength; that is, about 10 nm.

In a semiconductor laser having a bandstructure similar to that shown in Fig. 7.5, the gain function can be adequately represented by a Gaussian distribution. The value of the distribution at its maximum value can be expressed in terms of its full-width at half-maximum (usually abbreviated *FWHM*).

$$g\left(\nu_{max}\right) = \frac{2}{\pi \Delta\nu} \quad \text{(Gaussian gain distribution)} \tag{7.40}$$

The exact form of the gain distribution function is almost never known. It can be adequately approximated by:

$$g\left(\nu_{max}\right) \approx \frac{1}{\Delta\nu} \tag{7.41}$$

The fundamental nature of stimulated emission dictates that the laser will want to emit light whose frequency lies as close as possible to the peak of the gain distribution. We can use this approximation to make a practical estimate of the threshold current density.

$$J_{th} = \frac{qtN_{th}}{\tau_r} = qt\left(\frac{\tau_{21}}{\tau_r}\right)\frac{k_{th}n^2 v_{max}^2 \Delta v}{c^2} \text{ A cm}^{-2}$$

or (7.42)

$$J_{th} = \frac{qtN_{th}}{\tau_r} = qt\left(\frac{\tau_{21}}{\tau_r}\right)\frac{k_{th}n^2 \Delta v}{\lambda_{max}^2} \text{ A cm}^{-2}$$

All the variables in this expression are easily accessible. The ratio $\left(\frac{\tau_r}{\tau_{21}}\right)$ is referred to as the internal quantum efficiency, and is typically taken to be unity. The width of the recombination region, t, is controlled during fabrication. In a heterostructure laser this is typically about 10^{-5} cm, and in a quantum well laser, about one order of magnitude less $\sim 10^{-6}$ cm. The width of the gain spectrum in energy is about kT or 0.02 eV at room temperature.

7.5 Laser Engineering

7.5.1 Heterostructure Lasers

The p–n junction structure of Fig. 7.7 can be improved to make a better laser in two important ways. One is to reduce the thickness of the recombination region ($= t$) in (7.42), and the other is to impose a structure that confines both the emitted light and the electron-hole pairs to the same region in space. Confinement is the key concept, in emission, the presence of a photon stimulates emission of a second photon. The photonic field must overlap with the population inversion of electrons. A semiconductor heterostructure can be used to confine simultaneously both photons and electron-hole pairs to the same region of space. The simplest realization of such a semiconductor heterostructure was designed and demonstrated by I. Hayashi, M.B. Panish and P. Foy. It enabled room-temperature CW operation of a semiconductor laser in 1969 for the first time, making the semiconductor laser a practical reality, 7 years after first being demonstrated by R.N. Hall. Zhores Alferov and Herbert Kroemer shared part of the 2000 Nobel Prize in Physics for conceiving the use of a semiconductor heterostructure to transform the GaAs laser diode from a laboratory experiment to a useful device (Fig. 7.9).

7.5.2 Optical Confinement—Waveguiding

We can use Snell's law for refraction to determine the critical angle for total internal reflection in a heterostructure of $Al_xGa_{1-x}As/GaAs$. This effect creates optical waveguiding in the plane perpendicular to the direction of propagation that is defined by the resonator mirrors. As diagrammed schematically in Fig. 7.10, a

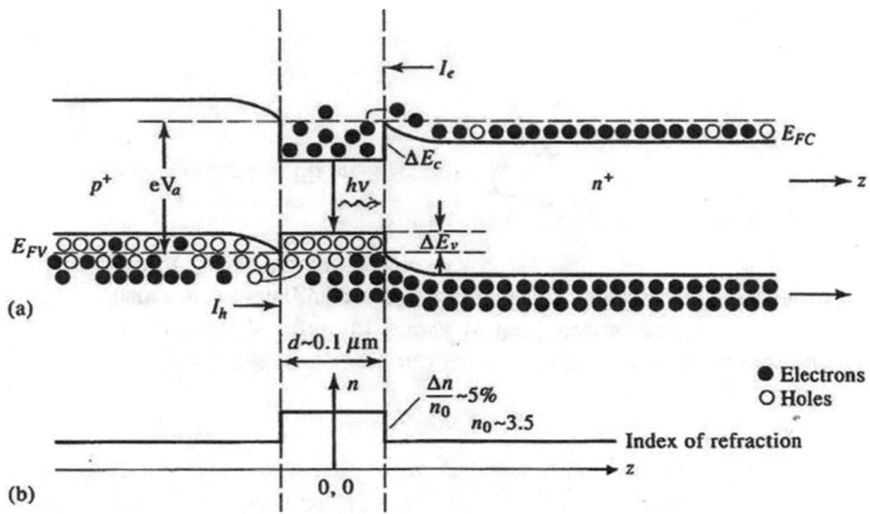

Fig. 7.9 Energy-level diagram for an $Al_xGa_{1-x}As$-GaAs heterostructure diode laser. **a** The energy band edges of the p–n junction in strong forward bias. Electrons are trapped in the potential well of the conduction band and holes are trapped in the potential well of the valence band achieving significant spatial overlap of the population inversion. The width of the recombination region is fixed physically by the materials growth process, and can now be significantly narrower than the 300 nm lower limit of the homostructure diode. Note the position of the quasi Fermi level for electrons in the conduction band (E_{FC}) and the quasi Fermi level for holes in the valence band, (E_{FV}). **b** The index of refraction of GaAs is higher than the index of refraction of $Al_xGa_{1-x}As$. Thus the GaAs recombination region acts like the core of an optical waveguide. Much of the recombination radiation will be confined by total internal reflection, so that the density of photons in the GaAs recombination region is greatly increased. This increases stimulated emission of radiation. Reproduced from Amnon Yariv, *Optical Electronics in Modern Communications*, 1996, by permission of Oxford University Press, USA

ray incident on the interface between two materials at an angle of θ_1 relative to a line perpendicular to the interface is refracted at an angle θ_2 in the second material. If the index of refraction in region 1 is greater than that in region 2, then it follows from Snell's law that $\theta_2 > \theta_1$. The angle θ_1 for which $\theta_2 = \frac{\pi}{2}$ radians is defined as the critical angle: θ_c. The critical angle for total internal reflection for the $Al_{0.30}Ga_{0.70}As/GaAs$ waveguide is:

$$\theta_c = \sin^{-1}\left(\frac{n_2}{n_1}\right) = \sin^{-1}\left(\frac{3.4}{3.6}\right) = 71^\circ$$

Spontaneous emission that lies within the "escape cone" of 142° is radiated out of the heterostructure. The direction of laser emission, is defined by the axis of the resonators. This is strongly guided by the heterostructure which serves to reinforce the ratio of stimulated to spontaneous emission in the output beam (Coldren et al. 2012).

The simple heterostructure concept shown in Fig. 7.9 can be further improved. First of all, confinement of the optical mode as well as confinement of the electron-

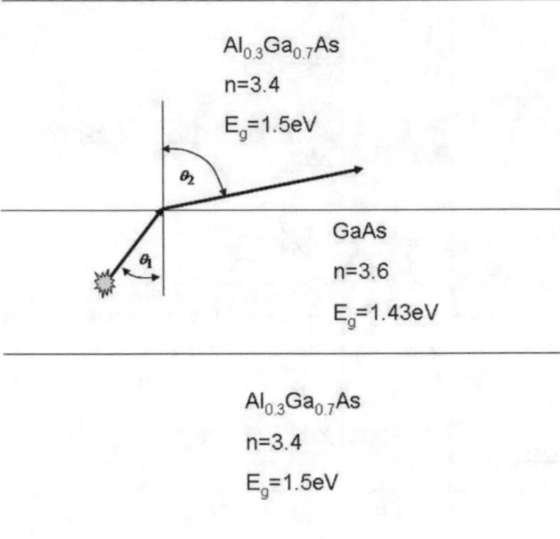

Fig. 7.10 Spontaneous emission occurs in all directions with equal probability. Given the indices of refraction of GaAs and $Al_{0.30}Ga_{0.70}As$, the "escape cone" according to Snell's law is 142°. A photon emitted within this cone will leave the laser structure. Otherwise, the photons are trapped by total internal reflection and are waveguided by the laser structure. Resonators at the ends of the laser cavity determine the modes that are highly reflected, and which will have the highest photon density leading to the strongest amplification by stimulated emission

hole population can be achieved not just in 1-dimension, as shown in Fig. 7.8, but in 2-dimensions perpendicular to the direction of propagation. This is implemented by the buried heterostructure laser. Second, the confinement dimension of electrons and holes can be optimized independently of the confinement of the optical mode. This is called the separate-confinement heterostructure, or SCH. This concept leads naturally to the quantum well laser. We will treat both of these structures.

7.5.3 *Buried Heterostructure Lasers*

The buried heterostructure laser is an extension of the simple heterostructure laser. In this design the active region is surrounded on all sides by higher bandgap and lower index AlGaAs (Fig. 7.11).

In the usual buried heterostructure design, the vertical dimension of the active region is about 200 nm, and is defined during epitaxial growth. The lateral dimension of the active region is typically 15 times larger and is defined by optical lithography (Fig. 7.12).

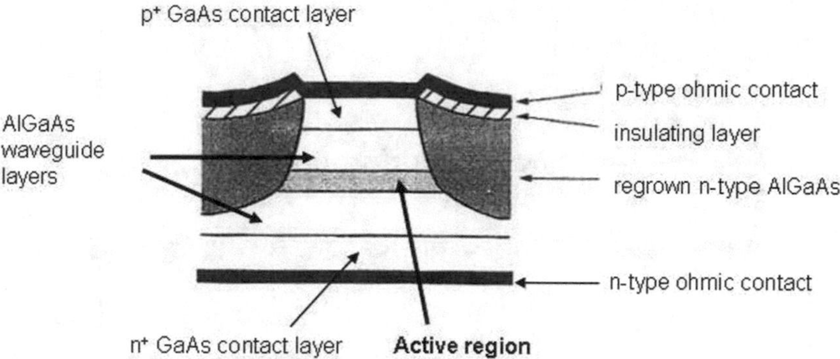

Fig. 7.11 Cross section of a buried heterostructure laser. Waveguide confinement and feedback assure that laser emission occurs in the direction perpendicular to the plane of the cross-section

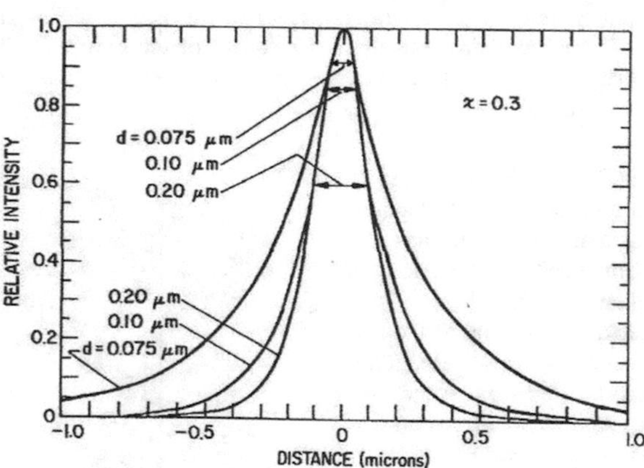

Fig. 7.12 The near-field intensity of laser emission has an angular divergence that depends on the inverse thickness of the active region. This is a diffraction effect similar to that which occurs when a photon passes through a slit. Reproduced from H.C. Casey, Jr. and M.B. Panish, Heterostructure Lasers, Part A, 1978, by permission of Elsevier Press

The optical mode intensity is shared between the active region where it contributes to stimulated emission, and as an evanescent wave in the confinement regions (Yariv 1996). Since AlGaAs has a larger bandgap than GaAs, laser radiation can stimulate additional recombination only in the GaAs active region. As the thickness of the active region is decreased, the intensity of the photonic radiation in the recombination region also decreases. The sharing of the optical mode volume between the active region and the waveguiding region is characterized by the confinement factor Γ. The confinement factor is easily calculated from Maxwell's equations, given the index of refraction profile (Casey and Panish 1978). The result obtained for the $Al_{0.3}\,Ga_{0.7}As/GaAs$ laser is shown in Fig. 7.13.

The confinement factor modifies the expression for the threshold current density:

$$J_{\mathrm{th}} = \frac{q\left(\frac{t}{\Gamma}\right)N_{\mathrm{th}}}{\tau_r} = q\left(\frac{t}{\Gamma}\right)\left(\frac{\tau_{21}}{\tau_r}\right)\frac{k_{\mathrm{th}}n^2\Delta\nu}{\lambda_{\mathrm{max}}^2}\ \mathrm{A\ cm^{-2}} \tag{7.43}$$

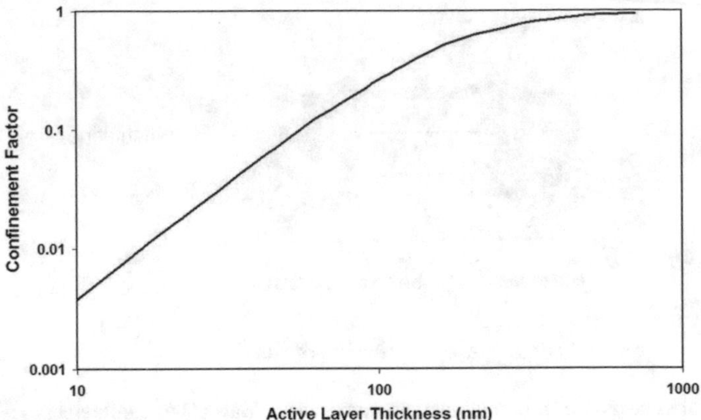

Fig. 7.13 The optical confinement factor is the fraction of the mode volume that is confined within the geometric boundary of the active region of the laser. The confinement factor increases approximately as the square of the active region thickness and saturates at unity when the active region thickness is greater than the wavelength of the laser emission in the active region

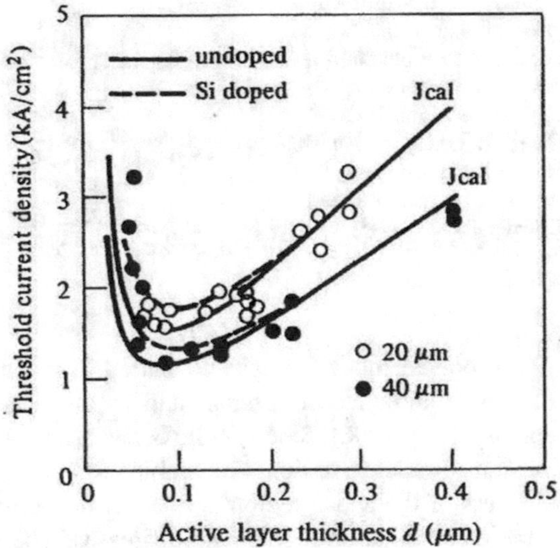

Fig. 7.14 The threshold current density of a double heterostructure laser has a distinct minimum value when the active layer thickness is equal to the laser emission wavelength. Reproduced from Amnon Yariv, *Optical Electronics in Modern Communications*, 1996, by permission of Oxford University Press, USA

The confinement factor approaches unity when the active region thickness is greater than the wavelength of the laser emission in the GaAs medium: $t > \frac{885\ \text{nm}}{3.6} > 250$ nm.

As the active layer thickness decreases below this limit, the confinement factor decreases exponentially. Thus the ratio $\frac{t}{\Gamma}$ increases, resulting in a dramatic rise in the threshold current density due to the reduction of the photon density in the electron-hole recombination region (Fig. 7.14).

7.5.4 Separate Confinement of Photons and Electrons

In the double heterostructure laser design of the previous section, photons and electrons are confined by the same physical structure. Separate confinement of photons and electrons is a straightforward way to optimize optical confinement and carrier confinement in order to reduce the threshold current density (Chuang 2010).

The optical waveguide should be adjusted to give a confinement factor close to unity. We have just shown that the optimum waveguide dimension is approximately equal to the photon wavelength in the guiding medium. For a GaAs/AlGaAs laser this thickness is about 100–200 nm.

Equation (7.43) shows that the reduction of the active region thickness reduces the threshold current. At the same time, this reduces the overlap of the active region with the optical mode. However it does not change the optical confinement factor. If the thickness of the active region and the waveguide thickness are reduced together, this gives rise to the dramatic increase in threshold current demonstrated in Fig. 7.14. If the thickness of the active region is decreased while the waveguide thickness stays constant, then the threshold current is increased, to be sure, but only by the ratio of the thickness of the optical waveguide region divided by the thickness of the recombination region.

In this case, we can define an effective confinement factor Γ^{sch}

$$\Gamma^{sch} = \Gamma \frac{t}{d} \tag{7.44}$$

Equation (7.44) shows that this optimization by separate confinement produces no major change in the threshold current because improvement by lowering the thickness of the active region is balanced by a decrease in the overlap of the active region with the optical mode. However, when the active region thickness reaches the quantum confinement regime, the physics of photon generation changes in a dramatic way.

The dimension of the active region can be reduced to the quantum regime. This means that recombination takes place in a 2-dimensional quantum well. The change from 3-d to 2-d is the key to understanding why the quantum well laser design can reduce significantly the threshold current density.

The number of occupied states in the conduction band k-space is the volume enclosed by the Fermi surface. In a direct gap semiconductor with parabolic bands like GaAs, this is a sphere of radius k. The volume is

$$\mathcal{N}(k) = 2x\frac{4}{3}\pi k^3 = \frac{8}{3}\pi k^3 \tag{7.45}$$

The energy dependence on k in a direct bandgap semiconducting material like GaAs is parabolic near the conduction band edge where photon emission occurs. That is:

$$E(k) - E_c = \frac{\hbar^2 k^2}{2m_c} \tag{7.46}$$

Solving for k

$$k = \left(\frac{2m_c}{\hbar^2} (E(k) - E_c) \right)^{\frac{1}{2}} \tag{7.47}$$

And the number of states as a function of energy is expressed;

$$\mathcal{N}(E) = \left(\frac{8\pi}{3} \right) \left(\frac{2m_c}{\hbar^2} \right)^{\frac{3}{2}} (E(k) - E_c)^{\frac{3}{2}} \tag{7.48}$$

The density of states per unit energy in k-space is therefore:

$$\frac{d\mathcal{N}(E)}{dE} = 4\pi \left(\frac{2m}{\hbar^2} \right)^{\frac{3}{2}} (E - E_c)^{\frac{1}{2}} \tag{7.49}$$

Changing from k to real space we compare the volume in k-space to that of real space. For simplicity and without any loss of generality, we consider the simple cubic crystal structure. The volume in k-space is:

$$V_{k\text{-space}} = \frac{2\pi}{a_x} \cdot \frac{2\pi}{a_y} \cdot \frac{2\pi}{a_z} = \frac{8\pi^3}{V_{\text{real-space}}} \tag{7.50}$$

To convert to real-space dimensions, we divide (7.47) by (7.48) which relates real space volume to volume in k-space. The result is the energy density of states per unit volume in real space.

$$\rho_{3-d}(E) = \frac{1}{2\pi^2} \left(\frac{2m}{\hbar^2} \right)^{\frac{3}{2}} \left((E(k) - E_c)^{\frac{1}{2}} \right) \tag{7.51}$$

The gain factor $\gamma(E)$ is expressed as:

$$\begin{aligned} \gamma(E_0) &= \frac{\lambda_0^2}{8\pi^2\hbar^2 n^2 \tau} (2m_r)^{\frac{3}{2}} (E_0 - E_c)^{\frac{1}{2}} (f_c - f_h) \\ &= \frac{\lambda_0^2 \hbar}{4n^2 \tau} \rho_{3-d}(E_0)(f_c - f_h) \end{aligned} \tag{7.52}$$

Equation (7.49) tells us that the density of states is near zero at the conduction band edge and rises slowly as a function of photon energy. Thus, laser emission cannot occur at the band edge, and additional band filling, that is additional pumping current density, is needed to increase the gain function to the point where laser action can occur. These features are summarized in Fig. 7.15.

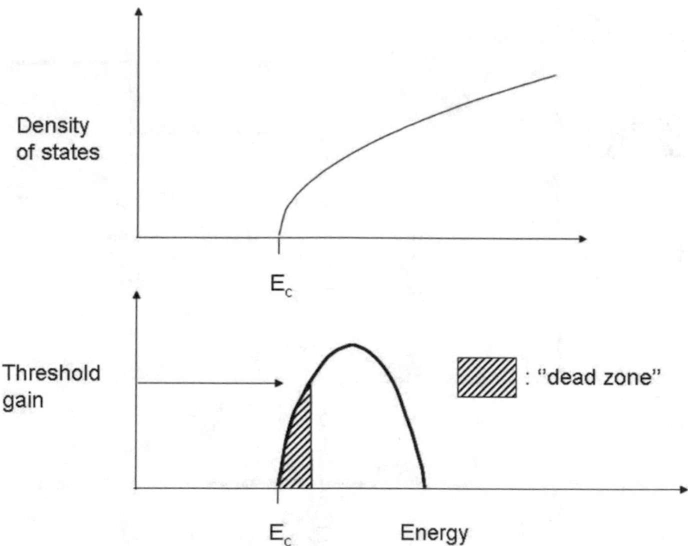

Fig. 7.15 As current is injected in a 3-d semiconductor laser, the gain function $g(v)$ increases proportionally to the density of states. Threshold gain is reached when the device starts lasing. This means that there is a "dead-zone" where current is injected, photons are emitted, but the round-trip gain is insufficient to overcome losses

In the case of a 2-dimensional environment, such as that of an electron in a quantum well, the same approach yields a different result. There are 2 states for each value of k, but the total number of k-states is given by a the area of a circle having
radius $= k$.

In two dimensions, the procedure is the same, but k-space is now a 2-dimensional entity

$$A_{k\text{-space}} = \frac{2\pi}{a_x} \cdot \frac{2\pi}{a_y} = \frac{4\pi^2}{A_{\text{real-space}}} \tag{7.53}$$

The number of occupied states is the area in k-space:

$$\mathcal{N}(E) = 2 \cdot \pi k^2 = 2\pi \left(\frac{2m}{\hbar^2}\right)(E - E_c) \tag{7.54}$$

and the density of states is:

$$\rho_{2-d}(E) = \frac{4\pi m}{4\pi^2\hbar^2} = \left(\frac{m}{\pi\hbar^2}\right) \tag{7.55}$$

That is, the density of states is a constant, independent of energy for electrons occupying the $n = 1$ quantum level. For a considerable range of energy, the density of states is constant, which means that the gain factor is also constant and independent of the drive current of the laser. This situation is advantageous for

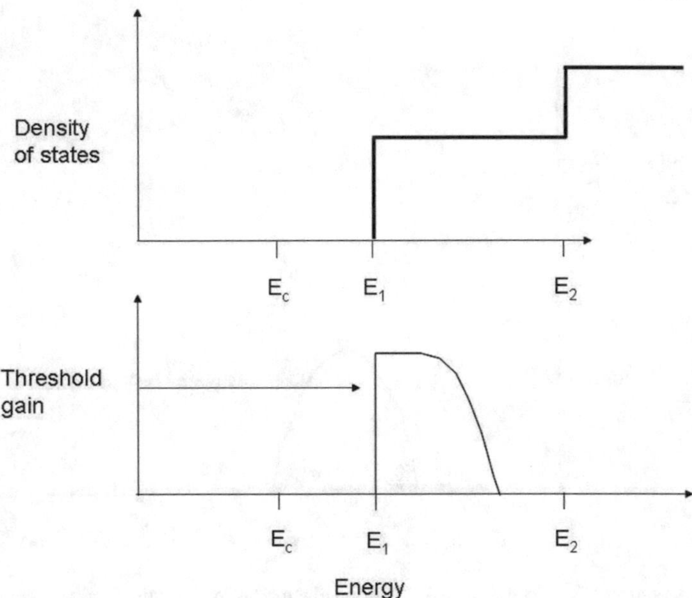

Fig. 7.16 In a quantum well structure, laser action can take place for photon energy greater than E_1, the ground state energy of the quantum well. Current injected into the structure populates the lowest energy states at this level where the density of states is already high. As a result threshold gain can be reached as soon as there is current present. Because the density of states in a 2-d structure "turns on" in a step-function manner, there is no "dead-zone" as in the case of a 3-d laser. This results in a significant reduction in threshold current required to initiate laser action

direct current intensity modulation bringing both stability and improvement in modulation bandwidth. When the energy of the electron permits occupation of the $n = 1$ and the $n = 2$ quantum levels, then the density of states doubles, being the sum of the densities of the two levels. In this new regime, the gain factor also doubles at this step, and then remains constant over a considerable range of electron-hole energies.

Above the quantized energy level of the quantum-confined state, the density of states rises like a step-function to a constant value. In a quantum well environment, laser action can first occur at the energy of this step in the density of states, and this feature leads directly to a significant reduction in threshold current density. In Fig. 7.16, we can see how the "dead-zone" is eliminated.

Equation (7.55) is of course an areal density of states. We can transform this expression into a volumetric density of states so that it can be compared with (7.49). To do this we divide by the thickness of the quantum well, L_z.

$$\rho_{2-d}, V(E) = \left(\frac{m}{\pi \hbar^2 L_z}\right) \tag{7.56}$$

The 2–d density of states is constant everywhere in the quantum well. Thus when we integrate the volumetric density of states over the quantum well width, we recover (7.55).

7.5.5 *The Quantum Well Laser*

The separate confinement quantum-well laser takes advantage of the dramatically lower threshold current provided by the 2-d nature of electrons confined in a quantum well. Typical well dimensions that provide this confinement are ~10 nm for a $GaAs/Al_{0.30}Ga_{0.70}As$ heterostructure. Quantum well dimensions larger than this will not assure the 2-d behavior of the density of states. Quantum well dimensions smaller than this result in the $n = 1$ energy level lying so near in energy to the top of the well that electrons are no longer well-confined. (so to speak!)

There are two main approaches to remedy this result. One is to grade the index in the photon waveguide so that the optical field intensity is maximized in the quantum well region. This structure is named **gr**aded-**in**dex **s**eparate **c**onfinement **h**eterostructure, or GRINSCH (Fig. 7.17).

The second approach is to insert several quantum wells across the width of the optical waveguide. This is called a multi-quantum well laser.

Laser engineering using these approaches has resulted in quantum well lasers that have a threshold current an order of magnitude lower than that which obtains for a double heterostructure design.

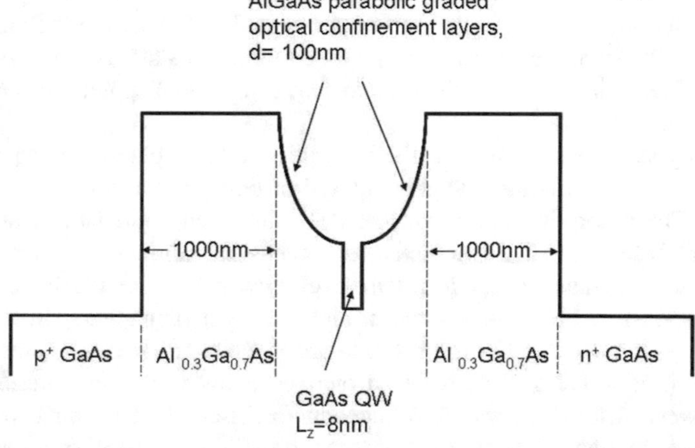

Fig. 7.17 GRINSCH design. The graded-index design creates a self-focussing waveguide that enhances the optical field strength at the center of the wave guide where the quantum well(s) active region is located. This boosts stimulated emission. In this structure, the effective confinement factor depends on the simple ratio of the active region thickness to the width of the optical mode

7.6 Summary

A laser is an amplifier with positive feedback. Amplification is generated by stimulated emission of photons, and positive feedback is achieved using mirrors.

Quantum photonics concerns the exchange of energy electrons with the excitations of the electromagnetic field. The physics of spontaneous and stimulated emission of photons is directly seen though quantization of the electromagnetic fields. This physics is the foundation of laser action. An electron can be promoted from a lower to higher energy state provided that a photon is annihilated, having the corresponding energy difference. This is stimulated absorption. An electron can transition from a higher to lower energy state by emitting a photon having the corresponding energy difference. This transition can be stimulated by the presence of another photon, or the transition can occur spontaneously in the absence of a photon.

The transition rate for stimulated emission depends on the number of photons already present in the electro-magnetic mode. The stimulated photon is emitted into the same mode. This stimulated emission acts like a linear amplifier that duplicates all the properties that define the mode of the stimulating photon: frequency, polarization and phase. The matrix element for stimulated emission is the same as that for stimulated absorption.

Spontaneous emission can only be understood as a quantum reaction with the vacuum. The emitted photon emerges from the vacuum state with an energy that corresponds to the energy difference between the initial and final states of the electron. Photons created by spontaneous emission act as seeds for subsequent stimulated optical transitions. Under conditions of thermal equilibrium, the spontaneous emission rate greatly exceeds the rate of stimulated emission. On the other hand, population inversion, which is a highly non-equilibrium state, is a necessary condition to raise the stimulated emission rate above the rate of spontaneous emission. This leads to laser action.

In steady-state laser operation, the absorption rate of photons is equal to the emission rate. The emission rate is composed of both spontaneous and stimulated emission. The absorption rate is due only to stimulated absorption. Optical gain will occur if the stimulated emission rate exceeds the stimulated absorption rate. Such a situation cannot occur in a two level system because the same two levels are responsible both for absorption and emission of photons. In a three or four level system, the equilibrium between absorption and emission is maintained, but the absorption takes place between one set of levels and the emission takes place between a different set. A semiconductor laser is an example of a four-level system. Putting a forward current through the diode causes recombination to occur, generating photons. Some of these photons will be emitted into the resonant modes of the cavity created by the mirrors. Only these photons will be amplified by stimulated emission. Their properties are the same as those of the photons that stimulated the emission. As the current is increased, these amplified modes will account for a greater percentage of the total recombination. Threshold is reached when the amplification per round trip in the cavity exceeds the absorption and scattering losses for the same round trip.

Semiconductor lasers were first demonstrated by engineering stimulated emission generated by electrons that make a direct transition between occupied states in the conduction band and empty states in the valence band. This was soon extended to direct transitions between any two states where a population inversion could be engineered (Unterrainer et al. 1990). An outstanding example of this approach is the quantum cascade laser, which is discussed in the following chapter.

A critical parameter for laser performance is the threshold current density. The development of the double-heterostructure laser concept by Alférov and Kroemer is based on confinement of the optical mode and the recombination of electrons and holes to the same region of space. This idea paved the way to room-temperature lasers and was awarded the year-2000 Nobel prize. Application of quantum confinement to two dimensions has dramatically improved this design by separate optimization of the optical waveguiding region and the electron-hole recombination region. This results in a steep reduction in the threshold current and a gain factor that is independent of operating drive current.

7.7 Exercises

7.1 Laser action can occur when the stimulated emission rate exceeds the spontaneous emission rate (see 7.39). What would happen if you reduced the spontaneous emission rate to zero? Would you have a threshold-less laser? Explain your answer.

7.2 Calculate the ratio of the spontaneous to stimulated emission rates at thermal equilibrium at $T = 300$ K, for an electron transition of 0.2 eV (infrared) 4 eV (near UV).

7.3 Estimate the threshold current density in A cm^{-2} of a GaAs-based laser with the following properties:

Emission wavelength = 850 nm
Linewidth of the gain spectrum = 1.5×10^{13} Hz
Internal losses = 30 cm^{-1}
Index of refraction = 3.5
Cavity length = 400 μm
Thickness of the recombination region = 200 nm.

7.4 Compare the threshold current density of a GaN laser to that of a GaAs laser. What can be changed?

References

G.P. Agrawal, N.K. Dutta, *Long Wavelength Semiconductor Lasers* (Van Nostrand Reinhold, New York, 1986). ISBN-13: 978-0442209957

P.K. Bhattacharya, *Semiconductor Optoelectronic Devices* (Prentice-Hall, Englewood Cliffs, 1994). ISBN 0-13-805748-6

H.C. Casey Jr., M.B. Panish, *Heterostructure Lasers* (Academic Press, Orlando, 1978). ISBN 0-12-163101-X

S.L. Chuang, *Physics of Optoelectronic Devices*, 2nd edn. (Wiley, New York, 2010). ISBN 978-0-0470-23919-5

L.A. Coldren, S.W. Corzine, M.L. Mashanovitch, *Diode Lasers and Photonic Integrated Circuits*, 2nd edn. (Wiley, Hoboken, 2012). ISBN-13: 978-0470484128

E.O. Goebel, *GaInAsP Alloy Semiconductors*, ed by T.P. Pearsall (Wiley, Chichester, 1982). ISBN 0-471-10119-2

R.N. Hall, G.E. Fenner, J.D. Kingsley, T.J. Soltys, R.O. Carlson, Phys. Rev. Lett. **9**, 366–369 (1962), https://journals.aps.org/prl/abstract/10.1103/PhysRevLett.9.366

T.P. Pearsall, *Photonics Essentials*, 2nd edn. (New York McGraw-Hill, 2008). ISBN: 978-0-07-162935-5

K. Unterrainer, C. Kremser, E. Gornik, C.R. Pidgeon, Y.L. Ivanov, E.E. Haller, Tunable cyclotron-resonance laser in germanium. Phys. Rev. Lett. **64**, 2277–2280 (1990), https://journals.aps.org/prl/abstract/10.1103/PhysRevLett.64.2277

A. Yariv, *Optical Electronics in Modern Communications* (Oxford University Press, USA, 1996). ISBN 0-19-510626-1

Chapter 8
Quantum Cascade Lasers

Abstract The quantum cascade laser represents an important accomplishment in quantum photonics, combining quantum excitation of the electromagnetic field, electron transport and controllable quantum mechanical tunneling all working cooperatively in a single device. In cascade lasers, the carriers are "recycled" from one stage to the next, so that the evacuation region of stage 1 is connected to the injection region of stage 2. A single carrier traveling through the structure may emit a photon at each stage. The successful operation of the quantum cascade laser is a balancing act between population inversion maintained by resonant phonon evacuation of carriers from the lower level and non-radiative depletion of the upper level population caused by optical phonon scattering. The interband cascade laser eliminates this major difficulty of the quantum cascade laser, by using a conduction band to valence band optical transition to generate the laser emission. As a result, the threshold power density in the interband cascade laser is lower by an order of magnitude compared to the quantum cascade laser.

8.1 Introduction

The quantum cascade laser, like its cousin, the interband cascade laser, unites purely quantum behavior like tunneling with photonics and electronics, creating a kind of concerto for quantum photonics. The quantum cascade laser (QCL) and the interband cascade laser (ICL) are commercially-available components, and they demonstrate that active quantum effects like tunneling can be used in a controllable way at room temperature with macroscopic effects.

These devices are based on classic laser principles: a gain region implemented by population inversion, inside an optical resonator that creates positive feedback for optical frequencies within the gain spectrum. In addition there is a carrier injection region and a collector evacuation region. Laser action requires that the optical recombination lifetime be longer than the pumping and evacuation lifetimes, enabling population inversion.

In cascade lasers, the carriers are "recycled" from one stage to the next, so that the evacuation region of stage 1 is connected to the injection region of stage

© Springer Nature Switzerland AG 2020

T. P. Pearsall, *Quantum Photonics*, Graduate Texts in Physics,
https://doi.org/10.1007/978-3-030-47325-9_8

2. A single carrier traveling through the structure may emit a photon at each stage. Thus the same carrier may emit several photons as it traverses the device structure. Unlike a p-n junction laser, the quantum efficiency of a cascade is in general much greater than unity. As always energy is conserved. A better figure of merit is the "wall plug" efficiency which is the ratio of the emitted optical energy to the electrical energy consumed. The wall plug efficiency is always less than unity.

The cascade laser represents an important accomplishment in quantum photonics, combining quantum excitation of the electromagnetic field, electron transport and controllable quantum mechanical tunneling all working cooperatively in a single device.

There are two varieties of this device: *interband* cascade lasers (ICL) where photons and laser action are created by electron-hole interband recombination. Laser action in the ICL follows the analysis presented in Chap. 7. The cascade structure enables the construction in series of multiple optical gain sections, but does not play an important role in the active regions of the device. In the intersubband cascade laser (QCL), photons and laser action are created by transitions between quantum well states in the conduction band. The intersubband quantum cascade laser is a unipolar device. In this chapter we will consider primarily the intersubband quantum cascade laser.

In Fig. 8.1a we show a schematic diagram of potential energy versus distance for a series of quantum wells. In Fig. 8.1b we show how this potential distribution changes under an applied electric field **E**. Note that for a specific value of the applied electric field, the n_1 level of the jth well can be made degenerate with respect to the n_2 level of the $j + 1$th quantum well. This situation enables resonant tunneling to occur between the jth and $j + 1$th wells.

The quantum cascade laser uses voltage-tunable tunneling to transfer electrons from the ground state quantum level in a quantum well to a higher quantum level in the adjacent quantum well. This electron can then relax to a lower quantum energy level by emitting a photon. And, it can then be injected by tunneling into the next stage, allowing the above procedure to repeat for as many such stages as make up the structure.

Shortly after the announcement by Hayashi et al. (1970) of a semiconductor laser operating continuously at room temperature, Kazarinov and Suris in 1971 proposed an optical amplifier (laser) based on photon-assisted tunneling between levels in a multi-quantum well structure. The core of their idea was to exploit electron tunneling from a level in one quantum well into a virtual state in the forbidden region of an adjacent quantum well, driven by an applied bias voltage. In the absence of an external photon flux, the tunneling probability is near zero. They postulated that in the presence of photons having an energy ε equal to the difference between the energy of the virtual state and of that the next lowest quantum well state would enhance the tunneling probability and stimulate the emission of an additional photon by laser action. In Fig. 8.2, we show a schematic diagram of this proposal.

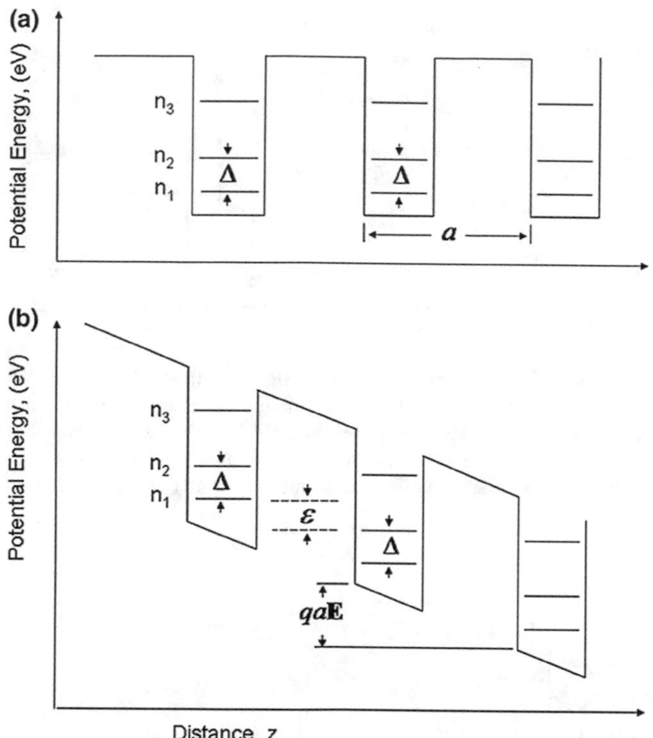

Fig. 8.1 Schematic diagram of a superlattice potential in the direction perpendicular to the plane of the layers forming the superlattice. **a** Potential distribution as a function of distance. The periodicity of the superlattice is a. The energy separation between the subbands n_1 and n_2 in the same quantum well is Δ. **b** In the presence of an electric field **E**, the bands are tilted by the applied voltage $qa\mathbf{E}$. The difference in energy between the subband n_1 and n_2 in the adjacent quantum well is given by ε

While a device based on the proposal of Kazarinov and Suris has yet to be demonstrated, it does contain important elements of working quantum cascade lasers.

In Fig. 8.3 we show a typical structure for a quantum cascade laser (Sirtori et al. 1996). As shown in the figure we can separate the structure into 3 regions: carrier injection, photon emission, and carrier evacuation. We will consider each of these regions separately. The prime requirement for a laser is the generation of stimulated emission. Stimulated emission requires population inversion between the energy levels where photons are generated in the photon emission region.

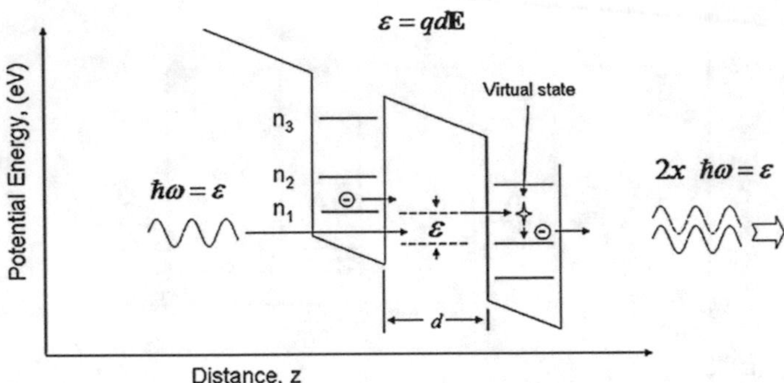

Fig. 8.2 In the proposal of Kazarinov and Suris of (1971), an electron would tunnel from a stable quantum well state into a virtual state in the adjoining well. In the presence of a photon the energy of which is equal to ε, the electron transitions to a final allowed state in the adjoining well by stimulated emission of a second photon of energy ε. The energy ε is determined by the applied voltage, and so the proposed photon amplifier would be tunable in energy

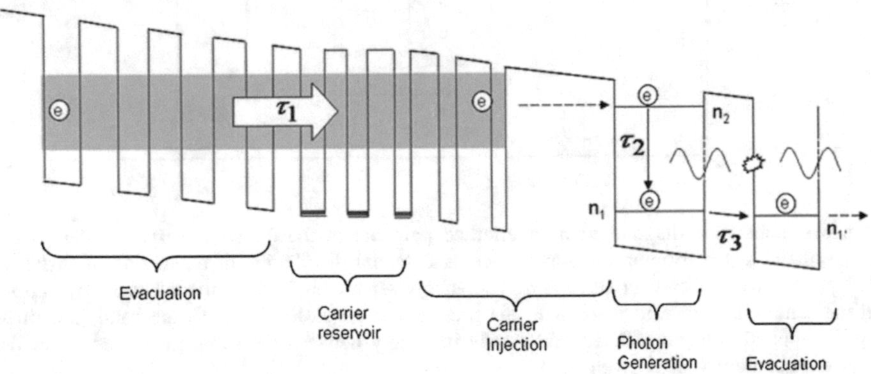

Fig. 8.3 Diagram of a QCL showing the injector region, emission and evacuation regions. In this structure we show emission of a photon via an intersubband transition within the same quantum well. This is one among a wide variety of structures in which QCL operation has been demonstrated

8.2 Requirements for an Intersubband Laser

Intersubband optical transitions were discovered and characterized by Gornik and Tsui (1976). A laser based on intersubband optical transitions is above all a laser, and the basic laser requirements still apply:

- Population inversion:
  Laser action requires a population inversion between two levels. In Chap. 7 we proved that this requires at least 3 energy levels for electron excitation, photon emission and electron evacuation. The lifetimes of the transitions between these states must assure a population inversion.

We will label τ_1 as the lifetime for carriers to resupply the upper level for emission. The lifetime for photon emission will be labelled τ_2 and the scattering time to evacuate electrons from the lower level will be labelled τ_3. A basic requirement for laser action is that the lifetime for photon emission be long enough to ensure that the upper level is populated and that the lower level is depopulated, (refer to (7.31)) The lifetime for photon emission in the photon generation region must also be less than the phonon scattering time, in order to minimize dissipation of the electron energy in the upper state by non-radiative inelastic scattering (Faist et al. 1994).

$$\begin{aligned} \tau_1 &< \tau_2 \\ \tau_3 &< \tau_2 \end{aligned} \tag{8.1}$$

and

$$\tau_2 < \tau_{\text{phonon}}$$

- Resonator
  Emitted photons must be re-circulated in the active region in order to stimulate further emission. The resonator can be formed by the usual cleaving process to provide two parallel reflecting faces, or by a distributed feed back grating, the same as the case for a conventional semiconductor laser.
- Gain is greater than round-trip losses
  Laser action occurs when photon gain exceeds the round-trip losses, typically due to waveguide imperfections and optical phonon scattering.

In the following sections we will see how the design of a working quantum cascade laser satisfies the criteria for laser action.

8.3 Injector Region

The injector region consists of a series of quantum wells containing a reservoir of electrons, formed by heavy doping ($N_D \geq 10^{12}\,\text{cm}^{-2}$). The reservoir is connected to the active region and to the evacuation region by a series of quantum wells. The structure shown in Fig. 8.3 is based on a working design, consisting of 9 quantum wells with dimensions between 1.7 and 3.5 nm. (The complete specifications of this structure are given by Faist (2013) Appendix A) In the structure under consideration, the total width of this region is about 30 nm.

We recall from Chap. 3 that sequential quantum wells have resonant levels that form energy minibands with finite width rather than sharp levels, as shown in Fig. 8.4. Such minibands facilitate transport by tunneling for two reasons. The first is that the range of voltage for which efficient transfer occurs is broader, which favors stable transport of electrons. The second reason is related to the wave-function of the electron.

In the calculations of free-electron tunneling in Chap. 3, we assumed that the electron wavefunction has a single, well-defined k-vector. This assumption is needed to calculate accurately the energy levels in quantum wells (Harrison 2006). However, the wavefunction of real electrons is localised in both space and time, and it is expressed as a wave packet: a superposition of free-electron wavefunctions:

$$\Psi(k,r) = \sum_k a_k e^{ik\cdot r} \tag{8.2}$$

When an electron represented by such a wavepacket is incident on a sequence of quantum wells, it will be transmitted or reflected depending on the relationship of each of its component wavevectors with the transmission coefficient of the quantum well structure. It is obvious that a quantum well structure with a miniband structure will transmit a wavepacket more completely than that of a single resonant quantum well (refer to Fig. 3.17) (Fig. 8.4).

When the bias voltage aligns the levels so that electrons in the reservoir region can tunnel into the photon region, electrons are driven into the photon generation region by carrier diffusion, the direct result of the carrier concentration gradient between the two regions. When a carrier disappears due to a photon generation transition, it is resupplied by diffusion. This time is estimated from the diffusion relation:

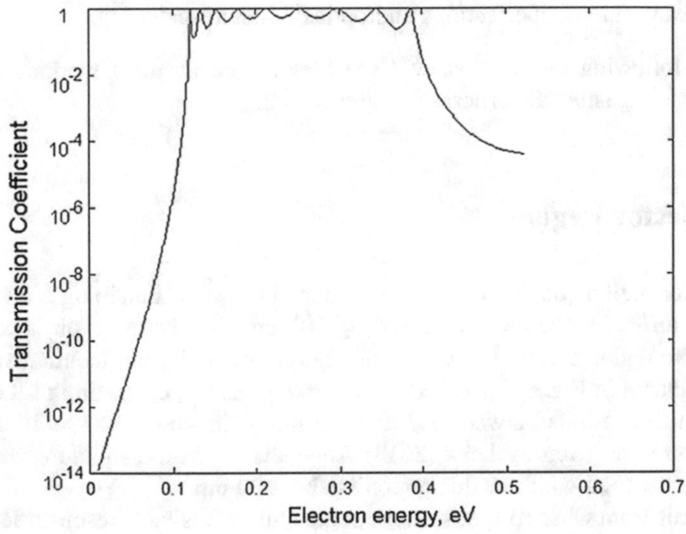

Fig. 8.4 The transmission coefficient for a sequence of 10 quantum wells having well width = barrier width = 2 nm. The well material is $Ga_{0.47}In_{0.53}As$ and the barrier material is $Al_{0.48}In_{0.52}As$. The energy level structure is a miniband having a width of ~0.3 eV

$$\tau_1 = \frac{L^2}{2D} = \frac{qL^2}{2\mu k_B T} \tag{8.3}$$

(the factor 2 is appropriate for 1-dimensional diffusion).

The diffusion time depends on the square of the distance, and for rather small distances appropriate for quantum wells, this time can be very short. For example, using parameters appropriate for electrons in GaAs:

$$\mu_e \approx 5000 \text{ cm}^2 \text{ V}^1 \text{ s}^{-1}$$
$$k_B T = 0.026 \text{ eV}$$
$$L = 50 \text{ nm}$$

We can estimate:

$$\tau_1 \cong 10^{-13} \text{ s} \tag{8.4}$$

8.4 Electron Evacuation from the Lower Level

Before considering the photon generation region, we will treat the evacuation of electrons from the n = 1 level of the photon generation region. In order to ensure population inversion, this level must be empty of carriers relative to the population of the n = 2 level. Electron-phonon scattering is an efficient mechanism to transfer electrons out of this lower level, provided that there is a suitable final state for them to scatter to. This is provided by a second quantum well adjacent to the photon generation region.

Phonons are mechanical vibrations of the crystal lattice. In a semiconductor like GaAs with two atoms per unit cell, there are two basic phonon types: acoustic and optical modes. Acoustic phonons correspond to large scale compressive or shear waves. Adjacent atoms are displaced in the same direction, and there is little relative movement between nearest neighbors.

Optical phonon excitation describes vibrations where nearest neighbors are displaced opposite to each other and their relative motion is considerable. Optical phonons have polarization. Transverse optical phonons have displacement amplitude that is perpendicular to the direction of propagation. Longitudinal optical phonons have displacement in the direction of propagation. The two ions making up the unit cell of III–V semiconductors carry different net charge, and the relative motion of these ions creates an electric displacement field that interacts strongly with electrons. On the other hand, acoustic phonon excitation does not create as significant a displacement field, and their interaction with electrons is less significant.

Electron evacuation is achieved by inelastic (non-reversible) scattering of the electron from its state following photon emission to another band lying lower in energy. Scattering by optical phonons is an efficient way to achieve this objective because the phonon scattering rate is an order of magnitude larger than the photon transition rate. Thus an electron that makes an intersubband optical transition can be scattered out of the lower energy subband at a rate that far exceeds the arrival rate of additional electrons, preventing population build-up in this state. Scattering by optical phonons is accompanied by energy loss. The optical phonon energy is transferred to the crystal lattice as heat. Under most conditions, this energy is large enough to ensure that the scattered electron remains in the lower energy band, but small enough, relative to the intersubband photon energy, to have only a moderate effect on device efficiency.

Electron-phonon scattering is mediated by the interaction between the electron and the piezo-electric potential created by local lattice deformation. The Hamiltonian for this interaction is written (Ziman 1967):

$$\mathcal{H}_{e\text{-ph}} = -\int \text{Disp}(r) \cdot \text{Polar}(r) d\text{r} \tag{8.5}$$

Equation (8.5) makes it clear that only phonons having a polarization aligned with the electron displacement vector will contribute to scattering. These are the longitudinal optical phonons. The transverse phonons are polarized perpendicular to the displacement. This feature: a single polarization, along with the constant energy of the optical phonon as a function of wavevector greatly simplifies treatment of optical phonon scattering.

An optical phonon can be approximated as an excitation having a finite constant energy for a wide range of wavevectors from $\mathbf{Q} = 0$ to the maximum defined by the unit cell boundaries.

This means that highly efficient electron-phonon scattering out of the $n_{2'}$ level into the $n_{1'}$ level can take place with a constant energy shift and conservation of momentum ($\mathbf{k}_f - \mathbf{k}_i = \mathbf{Q}_{\text{OP}}$, where $\mathbf{k}$ is the wavevector of the electron). To appreciate this, we recall that a quantum well is a 3-dimensional object. Although there is strong confinement in the z-direction, the electron can move freely in the 2-dimensional space orthogonal to the direction of strong confinement. In Fig. 8.5 we show a schematic diagram of the energy-momentum relationship in the x-y plane. An electron at the lowest energy state i.e., $k_x = 0$, $k_y = 0$ in the upper band can change its state in one of three ways.

1. Intraband phonon scattering: The electron scatters to another state in the upper subband by absorbing an optical phonon and gaining the optical phonon energy $= \hbar\omega_{\text{OP}}$.
2. Direct intersubband optical transitions: The electron can make a direct optical transition to the lower subband emitting a photon with energy $= \Delta$.
3. Interband phonon scattering: The electron scatters to another state in the lower subband by emitting an optical phonon and giving up the optical phonon energy $= \hbar\omega_{\text{OP}}$.

It is known from measurements that the optical phonon scattering time ($\sim 10^{-10}$ s) is an order of magnitude shorter than the photon recombination time (10^{-9} s). So, phonon scattering dominates.

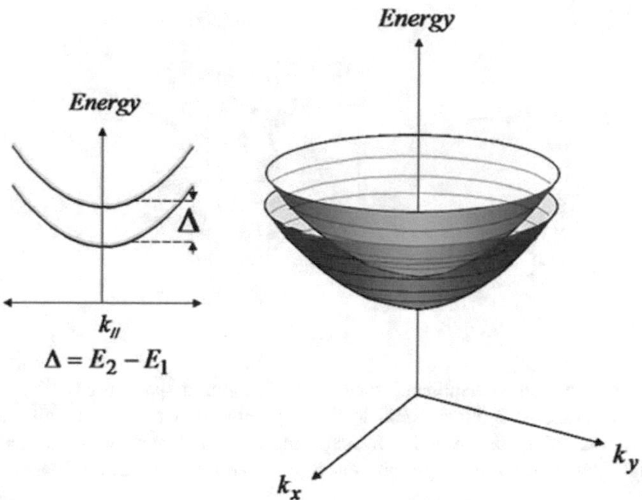

Fig. 8.5 If the energy spacing Δ between the $n_{2'}$ and $n_{1'}$ level is at least as large as the optical phonon energy, then highly efficient electron-phonon scattering can be harnessed to evacuate electrons from the $n_{2'}$ to the $n_{1'}$ level

The relative probability of phonon absorption to phonon emission can be estimated using Boltzmann statistics

$$\frac{N_{\text{absorption}}}{N_{\text{emission}}} = \mathrm{e}^{-\frac{2\hbar\omega_{\text{OP}}}{kT}} \tag{8.6}$$

For example, in the case of GaAs, $\hbar\omega_{\text{OP}} = 0.029$ eV. At room temperature, we estimate intersubband phonon scattering to be about 20 times more likely than intrasubband scattering.

In Fig. 8.5 it is also apparent that an electron can scatter from the upper subband to the lower subband by phonon emission only if $\Delta > \hbar\omega_{\text{OP}}$. This puts a critical constraint on the design of the quantum well dimension.

$$E_{n_{2'}} - E_{n_{1'}} > \hbar\omega_{\text{OP}} \tag{8.7}$$

The phase space for the final state is described by a circle of states lying on the energy surface of the lower subband. The magnitude of the momentum transfer $\hbar\mathbf{Q}$ is given by (see Fig. 8.6a):

$$\hbar|Q| = \hbar\sqrt{k_i^2 + k_f^2 - 2k_i k_f \cos\theta} \tag{8.8}$$

For example, the optical phonons in GaAs have an energy of about 0.029 eV. An electron in the lowest energy state in the n_2 quantum level can transition to

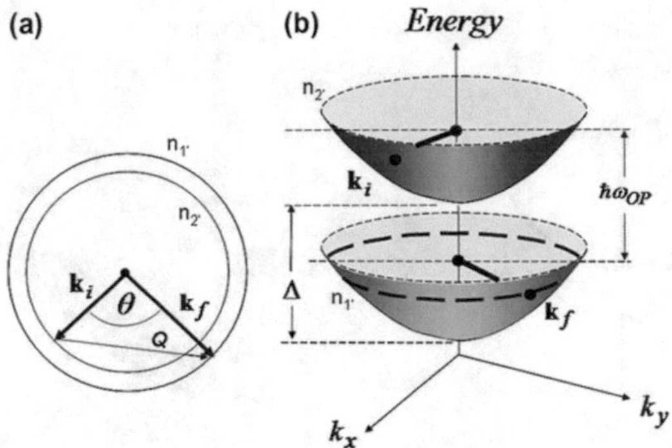

Fig. 8.6 Electron phonon intersubband scattering. The phase-space for optical phonon emission $= \hbar\omega_{\mathrm{OP}}$ consists of a 2-dimensional locus of points (shown schematically as a *circle*) in the plane parallel to the quantum wells. **a** Scattering angle between the initial and final states. **b** Relationship between the optical phonon energy $= \hbar\omega_{\mathrm{OP}}$ and the intersubband energy difference $= \Delta$

any one of a number of states in the n_1 level, lying lower in energy by 0.029 eV (Fig. 8.6b).

Electron-phonon scattering is the result of the electric field displacement vector of the electron interacting with the polarization of the phonon vibrational mode.

Phonon excitations in a crystal can be described just like photons as an ensemble of harmonic oscillators. The Hamiltonian can therefore be written using creation and annihilation operators as:

$$\mathcal{H} = \frac{1}{2}\sum_q \hbar\omega_q \left(a_q^* a_q + a_q a_q^*\right) \tag{8.9}$$

The phonon energy spectrum can be written down immediately as

$$E = \sum_q \hbar\omega_q \left(n_q + \frac{1}{2}\right) \tag{8.10}$$

Electron-phonon scattering is mediated by the interaction between the electron and the piezo-electric potential created by local lattice deformation.

The matrix element for the interaction can be written in the usual way:

$$\mathcal{M} = \int_V \Psi_{k'}^* \Phi(r) \Psi_k dr \tag{8.11}$$

Using normal coordinates and raising lowering operators, this can be written

$$\mathcal{M} = \langle n_Q| \int_V \mathrm{e}^{-ik'r} A(r-l)\mathrm{e}^{-iQl}\mathrm{e}^{ikr} dr |n_Q + 1\rangle \tag{8.12}$$

where $A(r-l)$ gives the displacement from equilibrium, l being the equilibrium coordinate, and e^{iQl} is the wave associated with the phonon.

$$\mathcal{H}_{e-\text{ph}} = 2\pi i \left(\frac{q^2\hbar\omega}{2\pi\varepsilon_P V}\right)^{\frac{1}{2}} \sum_Q \frac{1}{Q}\left(a_Q^* e^{-iQ\cdot\mathbf{r}} - a_Q e^{iQ\cdot\mathbf{r}}\right) \tag{8.13}$$

$$\mathcal{M} = \hbar\left(\frac{n_Q+1}{2NVm^*\hbar\omega}\right)^{\frac{1}{2}} \delta_{k'-Q} I(k,k'), \tag{8.14}$$

within a reciprocal lattice vector. $I(k,k')$ is a form factor:

$$I(k,k') = -i\mathbf{p}_{q,p}\cdot(\mathbf{k}-\mathbf{k}')\int_r \Psi_{k'}^*\nabla U\Psi_k dr \tag{8.15}$$

and is zero for transverse polarizations. So only longitudinal phonons, where the polarization is aligned with displacement, interact strongly with electrons.

The electron-phonon scattering rate can be expressed using Fermi's golden rule:

$$\frac{1}{\tau_{ij}} = \frac{2\pi}{\hbar}\sum_{k_j} |\langle\Psi_j|\mathcal{H}_{\text{scatt}}|\Psi_i\rangle|^2 \delta(E(k_i)-E(k_j)-\Delta E) \tag{8.16}$$

$$\frac{1}{\tau_{ij}} = \frac{m^*q^2\hbar^2\omega_{LO}^2}{2\hbar^4\varepsilon_P}\sum_j d\theta\frac{I^{ij}(Q)}{Q} = \frac{m^*q^2(E_{\text{OP}}^2)}{2\hbar^4\varepsilon_P}\int_0^{2\pi}\frac{I^{ij}(Q)}{Q}d\theta \tag{8.17}$$

where,

$$\frac{1}{\varepsilon_P} = \frac{1}{\varepsilon_\infty} - \frac{1}{\varepsilon_{\text{static}}}$$

Now we can see that resonant scattering (that is, $Q \to 0$) is the key (Fig. 8.7).

To favor population inversion and stimulated emission, we want to maximize the scattering rate out of the upper subband to the lower subband, and this will occur as $Q \to 0$; that is, when the energy separation between the two bands is equal to the optical phonon energy.

In the quantum cascade laser, the condition for resonant phonon scattering is the key to efficient evacuation of electrons from the $n = 1$ level. If this condition is implemented in the same quantum well where photon generation occurs, it will put an unwanted restriction on the energy of the upper level. Thus the final state for electron-phonon scattering should be located in a quantum well spatially adjacent to that in which the photon generation is achieved, as shown in Fig. 8.3 (Beck et al. 2002). An alternative is to place the upper level in a separate quantum well "upstream" from the well containing the lower energy state for photon gener-

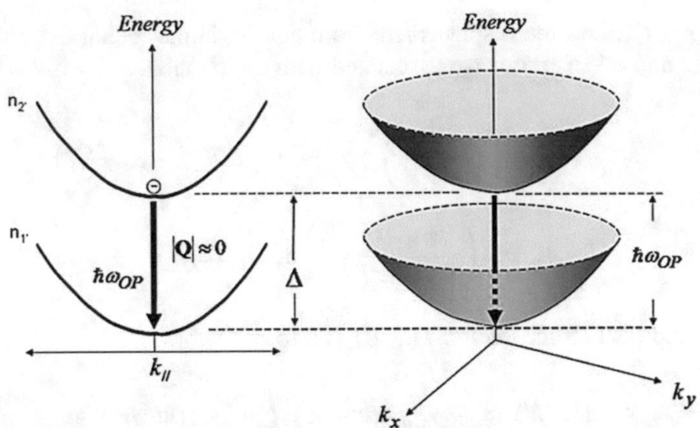

Fig. 8.7 Resonant electron-phonon intersubband scattering occurs when the momentum transfer between the initial and final state $|\mathbf{Q}| = |\mathbf{k}_j - \mathbf{k}_f| \approx 0$. In this regime, the scattering rate is significantly increased over the bulk value of $\sim 10^{10}\ \mathrm{s}^{-1}$

ation. Then resonant phonon scattering to evacuate this state can be incorporated in the same well as the lower energy state. Successful operation of QCL has been demonstrated using both structures.

The energy alignment of the $n = 1$ band in well 1 with the $n' = 1$ level in well 2 is diagrammed in Fig. 8.8. The energy difference depends directly on the applied electric field and the physical separation between the wells. As the applied electric field is increased, the energy difference between level $n = 1$ in well 1 and $n' = 1$ in well 2 increases in proportion. When this energy difference is equal to the LO phonon energy, (= 0.029 eV for GaAs and 0.034 eV for $Ga_{0.47}In_{0.53}As$) we can expect that any carriers in the $n = 1$ level will be transferred efficiently to the $n' = 1$ level in well 2. This situation creates, as a result, a population inversion between the $n = 2$ and the $n = 1$ levels in well 1.

Example: Estimate the electric field needed to align the levels $n = 1$ and $n' = 1$ for the structure shown in Fig. 8.3.

In this structure composed of $Al_{0.48}In_{0.52}As$ and $Ga_{0.47}In_{0.53}As$, the distance between the center of well 1 and well 2 is 8 nm. In well 1, the $n = 1$ level occurs at 0.10 eV above the bottom of the well. In well 2, the $n' = 1$ level occurs at 0.166 eV above the bottom of the well. At 0-applied field, the energy difference $\Delta E = 0.1 - 0.166 = -0.066$ eV. By applying an external bias such that the potential difference between the two wells is 0.10 eV, we see that the $n = 1$ level now lies higher than the $n' = 1$ level by 0.034 eV, which is the energy of the LO phonon.

The applied field $\varepsilon = \frac{0.10}{q8x10^{-9}} = 1.2 \times 10^5\ \mathrm{V\ cm}^{-1}$. We note that this field is close to the electric field that will cause breakdown by direct electron tunneling from the valence to conduction band in $Ga_{0.47}In_{0.53}As \sim 2.5 \times 10^5\ \mathrm{V\ cm}^{-1}$.

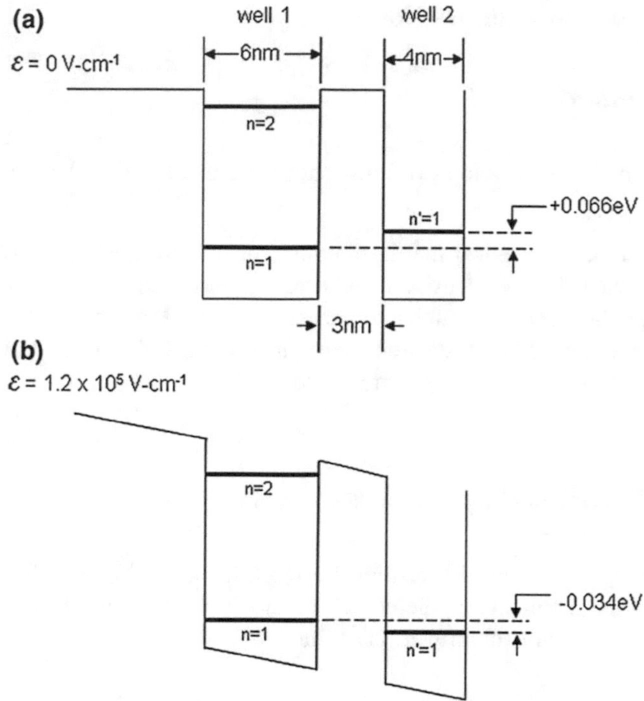

Fig. 8.8 Alignment of quantum levels in the photon generation region. **a** At 0 applied bias, quantum confinement causes the $n' = 1$ level to lie 0.066 eV higher than the $n = 1$ level in well 1. **b** In the presence of an applied electric field of 1.2×10^5 V cm^{-1}, the $n' = 1$ level now lies below the $n = 1$ level by 0.034 eV. This design enables independent choice of the energy separation between the states $n = 2$ and $n = 1$ states, thus fixing the photon emission energy. It also enables tuning the energy difference between the $n = 1$ and the $n' = 1$ states to achieve resonant phonon evacuation of the $n = 1$ level

The applied voltage per section would be $\mathrm{V} = 1.2 \times 10^7\ \mathrm{V\ m^{-1}} * 5 \times 10^{-8}\ \mathrm{m} = 0.6$ V per section.

In the particular structure of the example above, we estimate that a bias voltage of 0.6 V per quantum cascade section is needed to create the conditions for resonant LO phonon scattering of electrons out of the $n = 1$ state in well 1, and creating a population inversion enabling laser action. Increasing the bias beyond this point will not improve the population inversion situation. In addition, because carrier supply is driven by diffusion, rather than by the external electric field, there is no benefit to be gained by increasing the bias voltage once conditions for lasing have been reached. This represents a significant difference in behavior that distinguishes the QCL from a conventional interband diode laser.

8.5 Photon Emission Region

8.5.1 Introduction

In this section we will develop two important features of intersubband optical transitions.

1. The quantum mechanical matrix element for optical transitions between subbands created by 1-dimensional quantum confinement is non-zero when the difference between their subbands indices is an odd integer.
2. The magnitude of the transition element is comparable to that for band-to-band recombination in p-n junction lasers.

8.5.2 Intersubband Optical Transitions

The probability of an optical transition between the states Ψ_2 and Ψ_1 is given by the square of the dipole matrix element. The dipole matrix element μ between two electronic states in a quantum well is defined as:

$$\mathcal{M}_{2\rightarrow 1} \equiv \int_{-L/2}^{L/2} \langle \Psi_1 | \mathcal{H}_{\text{dipole}} | \Psi_2 \rangle dz = \int_{-L/2}^{L/2} \Psi_1 q z \Psi_2 dz, \tag{8.18}$$

where L is the width of the well.

The oscillator strength f of a dipole transition gives the relative strengths of the optical transitions for absorption or emission between two states.

$$f_{21} = \frac{2m_e}{q^2\hbar^2}(E_2 - E_1)(\mathcal{M}_{2\rightarrow 1})^2 \tag{8.19}$$

The transition rate is determined using Fermi's golden rule:

$$\frac{1}{\tau_{ij}} = \frac{2\pi}{\hbar} \sum_{k_j} |\langle \Psi_j | \mathcal{H}_{\mathbf{dipole}} | \Psi_i \rangle|^2 \delta(E_j - E_i) \tag{8.20}$$

The energy of an electron in a quantum well state is expressed:

$$E_n(k) = \frac{\hbar^2}{2m^*}\left(n^2 k_q^2 + k_t^2\right), \tag{8.21}$$

where $k_q = \frac{\pi}{L}$ and $k_t =$ the transverse wavevector. The energy difference between two transitions is:

$$E_m - E_n = \frac{\pi^2 \hbar^2}{2L^2 m^*}\left(m^2 - n^2\right) \tag{8.22}$$

In Chap. 3 we determined the wavefunctions for the electronic states in an infinitely deep quantum well as follows:

$$\Psi(x) = \left\{\sqrt{\frac{2}{L}} \cos\left(\frac{n\pi}{L} z\right), n = 1, 3, 5, \ldots. \right. \tag{3.13a}$$

$$\Psi(x) = \left\{\sqrt{\frac{2}{L}} \sin\left(\frac{n\pi}{L} z\right), n = 2, 4, 6, \ldots. \right. \tag{3.13b}$$

The dipole matrix element between the states $n = 1$ and $n = 2$ is:

$$\mathcal{M}_{2\to 1} = \frac{2q}{L} \int_{-L/2}^{L/2} z \cos\left(\frac{\pi}{L} z\right) \sin\left(\frac{2\pi}{L} z\right) dz = \frac{16L}{9\pi^2} q \tag{8.23}$$

This integral can be solved in a straightforward if somewhat tedious manner, but it is more interesting to view the calculation of the dipole moment graphically. First we will consider the transition between the $n = 2$ quantum level and the $n = 1$ quantum level. The dipole matrix element has been calculated using (8.10) and is non-zero. In Fig. 8.9, we show why this must be the case.

The solution of Schrödinger's equation gives the electron wavefunction for each level. It is apparent from inspection of (3.13a) and (3.13b), that each solution is orthogonal to all the others over the range $-\frac{L}{2} \leq z \leq \frac{L}{2}$. In Fig. 8.9, for example, it can be seen that the $n = 2$ wavefunction is antisymmetric about the center of the well; whereas the $n = 1$ wavefunction is symmetric. Their product results in a function ϕ_{21} that is as much negative as positive, (i.e. antisymmetric) so that the integral of this function is zero when taken over the range of z. However, the calculation of the dipole matrix element involves multiplying the function ϕ_{21} by qz. The result is a function that is positive across the entire range of z. Thus the dipole matrix element is non-zero.

We can compare this result with the dipole matrix element for the transition between the levels $n = 3$ and $n = 1$. The elements of this calculation are diagrammed in Fig. 8.10.

The result of the calculation of the dipole matrix element for the transition $n_3 \to n_1$ is clearly equal to zero. That is, the transition is forbidden. The selection rule governing optical transitions in centro-symmetric quantum wells is the following:

> Optical transitions between confined states n and m in quantum wells are allowed if and only if their quantum numbers differ by an odd integer.

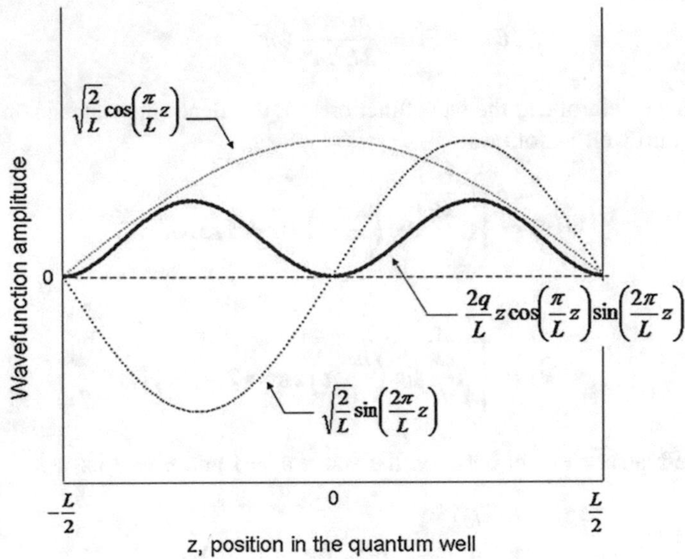

Fig. 8.9 Diagram of the functional elements in the calculation of the dipole matrix element between the $n = 2$ and $n = 1$ levels in a quantum well. The *bold curve* shows the function that is integrated to give the dipole matrix element

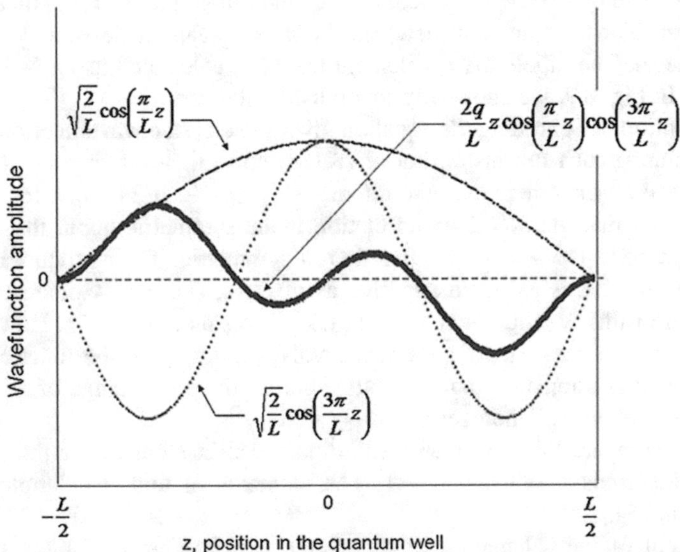

Fig. 8.10 Diagram of the elements of the dipole matrix element between the $n = 3$ level and the $n = 1$ level of a quantum well. The dipole matrix element function, shown in heavier print, is antisymmetric with respect to the centre of the well and the value of its integral over the domain is zero

We also note the important result that the polarization of the emitted or absorbed photons is parallel to the z-axis.

The oscillator strength of the transition between the state $n = 2$ and $n = 1$ can be evaluated from (8.19)

$$f_{21} = \frac{2m_e}{q^2\hbar^2}(E_2 - E_1)\mathcal{M}^2_{2\to 1} = \frac{m_e}{m^*}\frac{256}{27\pi^2} = 0.96\frac{m_e}{m^*} \tag{8.24}$$

which was first derived by West and Eglash (1985). The optical transition rate is proportional to the oscillator strength. Since the oscillator strength for a band-to-band transition in GaAs is also close to unity, it follows that the transition rate for spontaneous photon emission between the $n = 2$ and $n = 1$ subbands will be similar to that for a bulk band-to-band transition. In the case of GaAs this rate is approximately $10^8\ s^{-1}$ (see 't Hooft et al. 1987).

Photon generation is diagrammed schematically in Fig. 8.11. Photon generation is composed of spontaneous emission (τ_{21}) and stimulated emission (τ_{stim}), and thus

$$\frac{1}{\tau_2} = \frac{1}{\tau_{21}} + \frac{1}{\tau_{stim}}. \tag{8.25}$$

Reviewing the conditions for population inversion, we have:

$$\begin{aligned} \tau_1 &< \tau_2 \\ \tau_3 &< \tau_2 \end{aligned} \tag{8.26}$$

For the structure under consideration, we have shown that

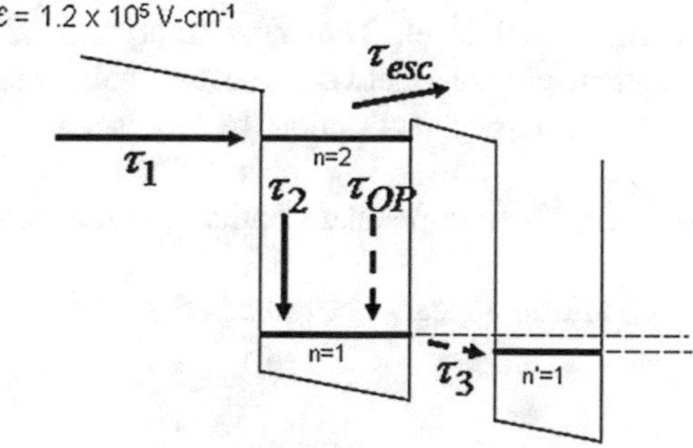

Fig. 8.11 Photon generation region of a quantum cascade laser. Electrons are supplied the $n = 2$ subband by diffusion. Electrons in the $n = 2$ band can give up their energy by transitions to the $n = 1$ subband, either by emitting a photon (spontaneous or stimulated emission) or a series of phonons (*dashed line*). Electrons in the $n = 1$ level can transition by resonant phonon scattering to the $n' = 1$ subband in the adjacent quantum well

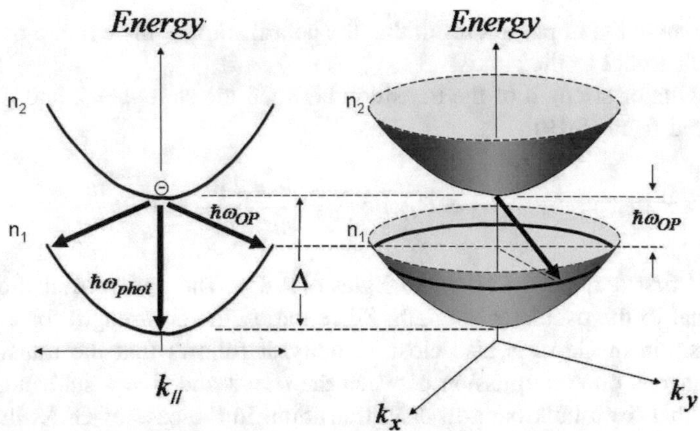

Fig. 8.12 The two main pathways for energy relaxation from the $n = 2$ to the $n = 1$ subband are photon generation and sequential optical phonon scattering. In a transition where a photon is emitted, the energy relaxation takes place in one step. Optical phonon scattering can take the electron from the $n = 2$ to $n = 1$ subband, and sequential intraband optical phonon scattering in the $n = 1$ subband takes place until the electron reaches the energy minimum at $\mathbf{k} = (0 \;\; 0 \;\; 0)$

$$\tau_1 \cong 10^{-13} \text{ s}$$
$$\tau_2 \cong 10^{-8} \text{ s}$$
$$\tau_3 \cong 5 \times 10^{-11} \text{ s (resonant optical phonon scattering)}$$

So the inequalities in (8.26) are satisfied.

There is an additional condition which is that the optical transition between $n = 2$ and $n = 1$ levels take place at a faster rate than other competing processes that can depopulate the $n = 2$ level. The most important of these is intersubband scattering by optical phonons between the two subbands. This means that τ_2 should be less than the non-resonant optical phonon scattering time (Fig. 8.12).

$$\tau_2 < \tau_{\text{OP}} \cong 10^{-9} \text{ s (non-resonant optical phonon scattering)}$$

We can see that this condition appears not to be satisfied.

8.5.3 Spontaneous and Stimulated Emission

The spontaneous emission rate $A_{21} = \frac{1}{\tau_{21}}$ has been developed in Chap. 7. The ratio of spontaneous to stimulated emission near equilibrium is given in (7.24).

$$R = \frac{A_{21}}{\rho(\omega) B_{21}} = e^{\frac{\hbar\omega}{k_B T}} - 1$$

For the quantum cascade structure under consideration, $\hbar\omega = 0.3$ eV, and at room temperature, the spontaneous emission rate is about 10^5 greater than the stimulated emission rate. While the spontaneous emission rate remains constant, the stimulated emission rate is variable as demonstrated in (6.18):

$$W_{2\to 1} = \frac{1}{\tau_{stim}} = I\frac{\pi \mathcal{M}_{2\to 1}^2 n_{refr}}{\varepsilon \hbar c}\delta(\Delta E - \hbar\omega) \tag{7.19}$$

where I is the intensity of the lasing mode.

As soon as current is present in the structure, photons are generated leading to stimulated emission, which increases the stimulated emission rate $= \frac{1}{\tau_{stim}}$.

With no external electric field applied, the ratio of the number of carriers in the level $n = 2$ to that in level $n = 1$ is given by the Boltzmann relation;

$\frac{n_2}{n_1} = e^{-\frac{\Delta E}{kT}} \cong 10^{-5}$for an energy difference of 0.3 eV at room temperature.

By applying an electric field, this situation is dramatically changed. As soon as the applied bias is sufficient to empty carriers from the $n = 1$ level by resonant phonon scattering to level $n' = 1$, population inversion occurs. This non-equilibrium situation enables stimulated photon emission into the lasing mode. As the photon population in the lasing mode increases, so does the stimulated emission rate, which quickly becomes greater than the optical phonon intersubband non-resonant scattering rate. That is:

$$\tau_{stim} < \tau_{OP} \cong 10^{-9} \text{ s (non-resonant)} \tag{8.27}$$

Now all conditions for laser action are satisfied.

8.6 Current-Voltage and Light-Current Relationship

The quantum cascade intrasubband device operates in a quite different way from the interband laser discussed in Chap. 7. In this device, population inversion is achieved by structural design that creates the condition for efficient evacuation of electrons from the lowest lying state in the photon generation region. In the interband laser, population inversion is obtained by injection of minority carriers into a recombination region.

The quantum cascade intrasubband device performance is a function of the applied electric field which plays a critical role in aligning the quantum well levels in the structure. Once this electric field is reached, electrons can move efficiently in the structure, and each electron can, in principle, generate one photon in the photon generation region of every stage of the structure. In this situation, the current needed to create the threshold photon flux (where optical gain exceeds round-trip losses) is reduced by the number of cascade stages in the structure. Although this increases the operating voltage, the cascade feature enables laser operation at a lower current and reduced power dissipation.

In the interband laser, electric field plays no role in determining laser operation. With the *p-n* junction in forward bias, the electric field in the laser device remains well below the threshold for band-to-band breakdown. The laser light output is proportional to the injected current, and ohmic heating limits the output optical power.

Once lasing is achieved in a quantum cascade laser, increasing the electric field further will cause the energy levels to be misaligned, and optical output power will saturate and eventually decline. In the example studied above, the electric field needed to optimally align the energy levels is already close to the breakdown electric field where electrons in the valence band are ionized directly into the conduction band of the host material.

8.6.1 Rate Equations

Treating one period of the cascade laser structure,

$$\frac{d}{dt}n_2 = \frac{J}{q} - n_2\left(\frac{1}{\tau_{\mathrm{TOT}}}\right) - \mathrm{S}g_m(n_2 - n_1) \tag{8.28}$$

where $\frac{1}{\tau_{\mathrm{TOT}}} = \frac{1}{\tau_{21}} + \frac{1}{\tau_{\mathrm{OP}}} + \frac{1}{\tau_{\mathrm{esc}}} + \cdots$ includes spontaneous emission, plus all the scattering channels from state $n = 2$, and $\mathrm{S}g_m(n_2 - n_1)$ represents electron depopulation from state $n = 2$ by stimulated emission.

$$\frac{d}{dt}n_1 = n_2\left(\frac{1}{\tau_{21}} + \frac{1}{\tau_{\mathrm{OP}}}\right) + \mathrm{S}g_m(n_2 - n_1) - \frac{n_1 - n_1^{\mathrm{thermal}}}{\tau_3} \tag{8.29}$$

$$\frac{d}{dt}S = \frac{c}{n_{\mathrm{refr}}}\left\{[g_m(n_2 - n_1) - \alpha_{\mathrm{rt-loss}}]S + \beta\frac{n_2}{\tau_{21}}\right\} \tag{8.30}$$

where β is the fraction of the spontaneous emitted photons in the lasing mode. The quantity $\alpha_{\mathrm{rt-loss}}$ represents all the round-trip optical losses.

In steady state, the time derivatives are all 0.

Re-arranging (8.28):

$$n_2 = \tau_{\mathrm{TOT}}\left[\frac{J}{q} + \mathrm{S}g_m(n_2 - n_1)\right] = \tau_{\mathrm{TOT}}\left[\frac{J}{q} + \mathrm{S}g_m(\Delta n)\right] \tag{8.31}$$

Solving (8.29) for the quantity Δn:

$$n_1 = \left(\frac{\tau_3}{\tau_2}\right)\left(\frac{J\tau_{\mathrm{TOT}}}{q} + \mathrm{S}g_m\Delta n\tau_{\mathrm{TOT}}\right) + \mathrm{S}g_m\Delta n\tau_3 + n_1^{\mathrm{therm}} \tag{8.32}$$

$$n_2 - n_1 = \Delta n = \left(\frac{J\tau_{\mathrm{TOT}}}{q} + \mathrm{S}g_m\Delta n\tau_{\mathrm{TOT}}\right)\left(1 - \frac{\tau_3}{\tau_2}\right) - \mathrm{S}g_m\Delta n\tau_3 - n_1^{\mathrm{therm}}$$

$$\Delta n = \frac{\frac{J\tau_{\text{TOT}}}{q}\left(1 - \frac{\tau_3}{\tau_2}\right) - n_1^{\text{therm}}}{1 - \text{S}g_m\left[\tau_{\text{TOT}}\left(1 - \frac{\tau_3}{\tau_2}\right) - \tau_3\right]} \tag{8.33}$$

In the sub-threshold regime, where $S \approx 0$, (8.33) can be simplified to

$$\Delta n = \frac{J\tau_{\text{TOT}}}{q}\left(1 - \frac{\tau_3}{\tau_2}\right) - n_1^{\text{therm}} \tag{8.34}$$

This is population inversion "by design". It is evident that the evacuation time from the $n = 2$ to the $n' = 1$ level must be much shorter than the combined effect of optical phonon scattering and spontaneous emission in order to achieve population inversion.

We can examine the condition for threshold using (8.30).

$$(g_m\Delta n - \alpha_{\text{rt-loss}})S + \beta\frac{n_2}{\tau_{21}} = 0 \tag{8.35}$$

$$\left[g_m\left(\frac{J\tau_{\text{TOT}}}{q}\left(1 - \frac{\tau_3}{\tau_2}\right) - n_1^{\text{therm}}\right) - \alpha_{\text{rt-loss}}\right]S + \beta\frac{n_2}{\tau_{21}} = 0 \tag{8.36}$$

$$J_{\text{threshold}} = \left(\frac{q}{g_m\tau_{\text{TOT}}\left(1 - \frac{\tau_3}{\tau_2}\right)}\right)\left(\alpha_{\text{rt-loss}} - \frac{\beta n_2}{\tau_{21}S} + g_m n_1^{\text{therm}}\right) \tag{8.37}$$

For a well-designed quantum cascade structure, (8.37) can be simplified to:

$$J_{\text{threshold}} = \left(\frac{q}{\tau_{\text{TOT}}}\right)\left(\frac{\alpha_{\text{rt-loss}}}{g_m} - \frac{\beta n_2}{g_m\tau_{21}S} + n_1^{\text{therm}}\right) \tag{8.38}$$

8.7 Interband Cascade Lasers

The interband cascade laser (ICL) has a structure based on an assembly of quantum wells. Unlike the quantum cascade laser, the ICL is a bipolar device. Both electrons and holes are required for charge transport and photon emission. Laser action is achieved by recombination of conduction-band electrons with valence-band holes. In the conventional laser diode that we discussed in Chap. 7, electrons and holes are injected into the recombination region using a p-n junction. In the ICL, there is no p-n junction. In the ICL, holes are created by direct tunneling of electrons from the valence band to conduction band, similar to the situation of the breakdown by tunneling of p-n junctions that was discussed in Chap. 3. However, instead of using an electric field to align the energy valence band and conduction band, the ICL exploits the unusual energy band alignment of the constituent materials: GaSb and InAs that are used to make the quantum-well structure (Fig. 8.13).

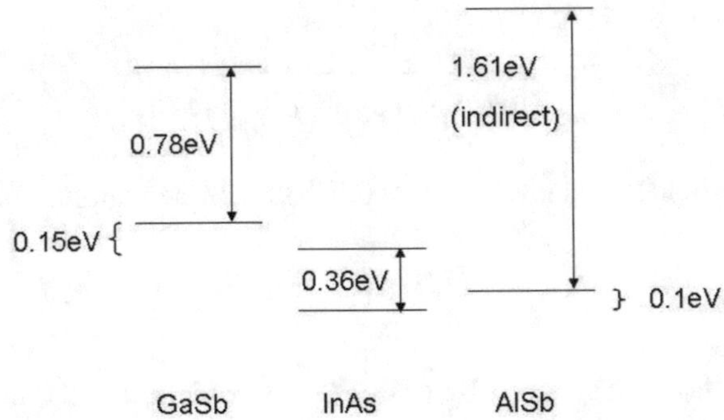

Fig. 8.13 GaSb, InAs and AlSb have the same crystal structure and nearly the same lattice parameter. In addition, they span a wide range of band gap energy. Thus they make an interesting system of materials for devices based on quantum confined energy states. However, their energy band alignment is quite unusual. GaSb/AlSb forms a type-I band alignment with the conduction and valence bands of GaSb lying inside those of AlSb, similar to GaAs/AlAs or GaInAs/InP. The pair AlSb/InAs shows a straddling alignment. However, a unique situation arises with the alignment between GaSb and InAs. The valence band energy of electrons in GaSb lies above the conduction band of InAs. This enables the direct tunneling of valence-band electrons from GaSb to the conduction band of InAs. This is a key component of the Interband Cascade Laser

Efficient laser action depends on the level of stimulated emission. In the case of the QCL, the degree of stimulated emission is determined by a competition between the rates of intersubband optical transitions and non-radiative losses through phonon emission and tunneling escape from the active region (8.25). In the ICL design, these particular non-radiative losses are dramatically reduced by causing the optical recombination to occur between the lowest-lying energy state in the conduction band and the highest-lying state in the valence band, just as in the case of a conventional quantum-well laser. Phonon scattering between intersubband states is no longer a concern. This results in a recombination lifetime for the ICL which is similar to that for a conventional quantum-well laser, and about 2 orders of magnitude larger than that for QCL devices. This longer recombination time reduces the electrical pumping rate needed to reach population inversion. This in turn lowers the laser threshold and increases the available output power.

8.7.1 ICL Active Region Design

GaSb ($E_g = 0.78$ eV) and InAs ($E_g = 0.36$ eV) are direct bandgap zinc-blende semiconductors which have nearly the same lattice parameter (Table 8.1). Extended heterostructures of these material can be grown without introducing strain and strain-related defects. Both carrier confinement and optical confinement are assured by barriers of AlSb that surround the region of electron-hole recombination. Optical emission occurs between the conduction band minimum of the quantum-confined state in InAs and the valence band maximum of the quantum-confined state in GaSb (Fig. 8.14).

Table 8.1 Material parameters for semiconductors used in fabrication of interband cascade lasers (ICL)

Material	Temp (K)	E_{gap} (eV)	a_0, A lattice constant	VB offset (eV)	m_e	m_{hh}	m_{ih}
GaSb	295	0.78	6.095	0	0.041	0.4	0.05
InAs	295	0.36	6.058	−0.51	0.023	0.41	0.026
AlSb	295	1.61 indirect	6.14	−0.41	0.12	0.98	

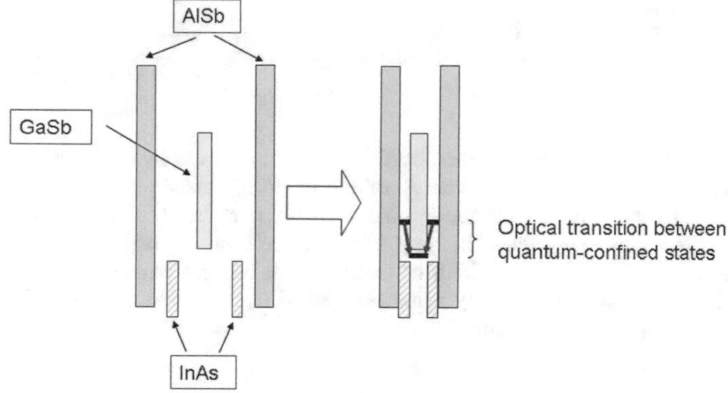

Fig. 8.14 Active-region structure of the GaSb/InAs ICL. Quantum confinement raises the bottom of the conduction-band in InAs above the valence-band maximum in GaSb (These are indicated by *horizontal black line* segments.) This creates an optical transition that is direct in **k**-space, but indirect in real space. Both energy and momentum are conserved. The optical matrix element depends on the spatial overlap of the wavefunctions in InAs and GaSb

8.7.2 Hole Injection by Tunneling

Optical recombination and laser action require the presence of holes in the valence band of GaSb and electrons in the conduction band of InAs, leading to population inversion. Consider first the introduction of holes in the valence band of GaSb. The hole injection structure is shown in Fig. 8.15.

Design involves the determination of the widths of the layers of GaSb, InAs and AlSb that are required to give the desired progression of quantum confined energy states in the GaSb valence band and the InAs conduction band. On application of an external electric field, these levels can be made resonant. Tunneling of electrons takes place from the valence band of GaSb (where the density of occupied initial states is high) to the conduction band of InAs (where the density of occupied states is lower). This creates a hole population in the valence band of GaSb and an electron population in the conduction band of InAs. In an extended structure, the applied electric field will separate electrons and holes, leading to a population inversion condition in the active region (Fig. 8.16).

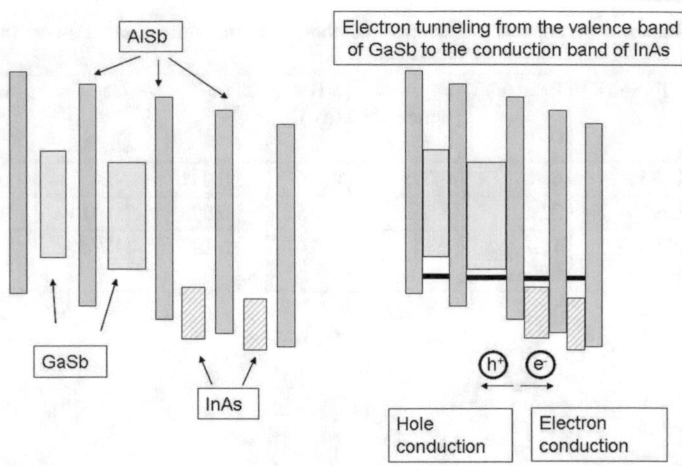

Fig. 8.15 Hole injection into the valence band of GaSb is achieved by resonant tunneling. The *horizontal black line* indications the energy of the confined valence band in p-type material and the bottom of the confined conduction band in *n*-type material. The conduction band of bulk InAs lies lower in energy than the top of the valence band in GaSb. In the quantum-well structure the valence band state of GaSb of GaSb can be brought within near resonance with the conduction band state in InAs. By using an applied electric field these two states can be made resonant implementing tunneling

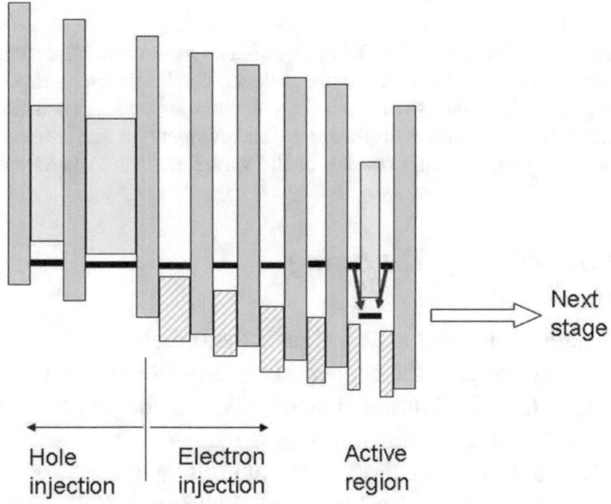

Fig. 8.16 One complete period of an interband cascade laser structure. GaSb regions are shown in *yellow*, with the height of the *yellow* region corresponding to the forbidden band separating the valence and conduction bands. Similarly, the *solid blue* regions correspond to AlSb, and the shaded region to InAs materials. The horizontal *black line* indicates the energy of the confined valence band in p-type material and the bottom of the confined conduction band in *n*-type material. Optical recombination leading to laser action occurs between electrons confined to InAs and confined hole states in GaSb (Lee et al. 1999)

8.8 Summary

The quantum cascade laser is an excellent example of innovative device design using quantum photonics. Quantum cascade laser devices have been continuously developed following the first demonstration in 1994. Twenty years later, these lasers can be produced in commercial volumes and are increasingly used in mid-IR sensor applications (Razeghi et al. 2015).

The quantum cascade laser is a unipolar device based on electron transport through a series of engineered quantum wells. Photon radiation occurs when electrons make a quantum transition between two subband energy levels. The successful operation of the quantum cascade laser is a balancing act between population inversion maintained by resonant phonon evacuation of carriers from the lower level and non-radiative depletion of the upper level population caused by optical phonon scattering. The electron non-radiative relaxation between subbands with energy separation higher than the optical phonon energy is much shorter (>1 p s), than the radiative recombination time (>1 ns), resulting in a very low radiative efficiency ($<10^{-3}$) compared to interband diode lasers.

A resolution to this difficulty was proposed in 1995 by Yang. His design retained the multiple-section laser structure by using tunneling between quantum wells and introduced an optical gain region based on interband recombination. This design, known as the interband cascade laser eliminates a major difficulty of the quantum cascade laser, which is the depletion of population inversion via non-radiative, inter subband optical phonon scattering (Fig. 8.17).

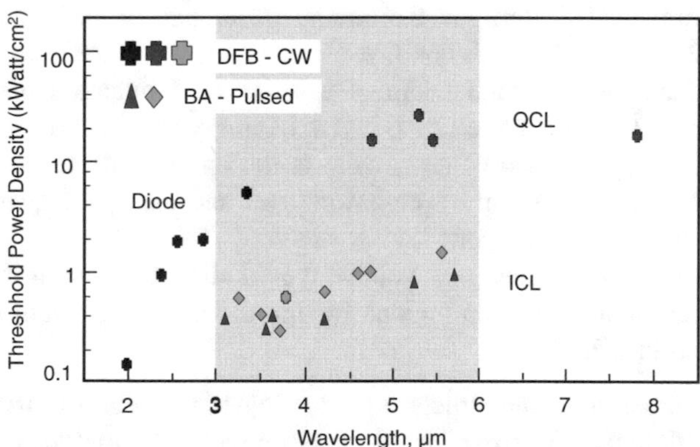

Fig. 8.17 Threshold power density (expressed in kW cm^{-1}) versus operation wavelength for some mid-IR semiconductor lasers (Nähle 2013). The interband cascade laser (ICL) requires a significantly reduced threshold power density compared to that for the quantum cascade laser. This difference reflects the elimination of non-radiative losses due to undesired optical photon scattering. Diode lasers are color-coded in *black* (results from University of Würzburg, Germany and Nanoplus, Germany). Quantum-cascade lasers are *red* (results from Northwestern University, USA and Alpes Lasers SA, Switzerland). Interband-cascade lasers show significantly lower threshold power densities in the 3–6 μm range (*green* results from Naval Research Laboratory, USA; *blue* results from University of Würzburg, Germany). (Reprinted from the May 2013 edition of *Laser Focus World* with permission from L. Nähle, L. Hildebrandt, M. Kamp, and S. Höfling, "Interband Cascade Lasers: ICLs open opportunities for mid-IR sensing" Laser Focus World, May 2013, Copyright 2016 by PennWell)

The interband cascade laser, unlike the quantum cascade laser, is a bipolar device, containing both n-type and p-type regions, determined by the unusual type-II valence band alignment between InAs and GaSb, rather than by doping (Kim et al. 2008).

Because of lower losses from non-radiative recombination the threshold power density in ICL is lower by an order of magnitude compared to QCL (Lee 1999). In remote sensing operations, where power consumption is an important parameter, the ICL has a superior performance relative to the QCL in the wavelength region of 3–6 μm.

8.9 Exercises

8.1 Evaluate the intersubband optical dipole matrix element between 2 levels i and j, and show, following West and Eglash, that the general expression is:

$$\mathcal{M}_{ij} = qL_w \frac{8}{\pi^2} \frac{ij}{(i^2 - j^2)^2}$$

8.2 Quantum cascade lasers have been successfully designed using a wide variety of structures. This direction-transition device, emitting at 4.6 μm was designed by Carlo Sirtori, and co-workers (1996) (Fig. 8.18).

(a) Using the data given, calculate the position of the quantized levels in each quantum well. Determine the energy difference between the $n = 1$ and $n = 2$ levels in the emission layer.
(b) Calculate the miniband width of a series of 5 quantum wells having a width of 1.8 nm confined by 6 layers having width of 2.0 nm.
(c) At zero applied bias voltage, what is the energy difference between the $n = 1$ quantum well in the emission layer and the $n = 1$ quantum well in the photon transfer layer.
(d) What range of applied bias voltage would align the quantum levels in the evacuation and injection regions (within the miniband width that you calculated in 8.2b)?

8.3 Design a direct tunneling structure for the injection section of the ICL shown in Fig. 8.16: that is, 5 barriers and 4 quantum wells. Assume that the materials used are AlSb for the barrier and InAs for the well regions.

Assume that the operating electric field is $E = 10^4$ V cm^{-1}, and that the emission energy is 0.37 eV.

(a) Determine appropriate thicknesses for each of the barriers and wells. (Hint treat each of the quantum wells independently to determine the lowest energy bound state in each of the 4 quantum wells. Then apply the elec-

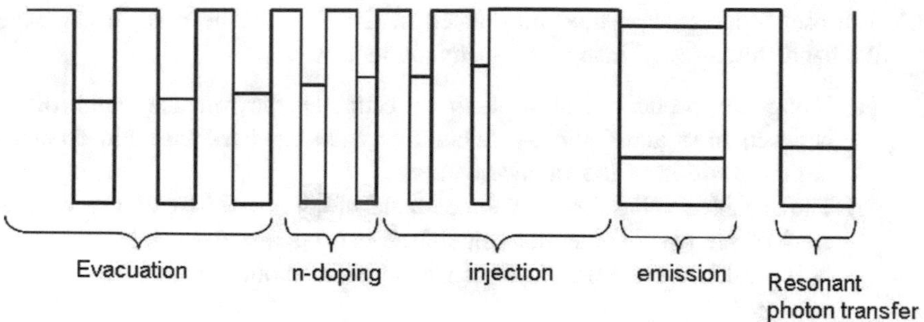

Fig. 8.18 Quantum cascade laser structure. Epitaxial layers are formed using $Ga_{0.47}In_{0.53}As$ and $Al_{0.52}In_{0.48}As$. Layer dimensions are given in Table 8.2

Table 8.2 Layer geometry for a direct transition quantum cascade laser operating at 4.6 μm

	$Al_{0.52}In_{0.48}As$, m* = 0.08 (nm)	$Ga_{0.47}In_{0.53}As$, m* = 0.043 (nm)
Evacuation	2.7	
Evacuation		2.2
Evacuation	1.2	
Evacuation		2.1
Evacuation	2.0	
Evacuation		1.8
N doping	1.8	
N doping		1.7
N doping	2.0	
N doping		1.6
Injection	2.2	
Injection		1.6
Injection	2.4	
Injection		1.4
Injection barrier	6.8	
Emission		4.8
Evacuation barrier	2.8	
Resonant phonon transfer		3.9

tric field and adjust the well parameters so the ground-state energies are aligned energetically at this value of the electric field. Show that this sequence will also give the desired emission energy.

(b) Estimate the energy width of the lowest energy tunneling miniband for the integrated 5-barrier/4-well structure? (Hint: use the average of the well widths determined in 8.3a and compute the transmission coefficient for the complete structure, ignoring the electric field).

8.4 The pseudopotential method introduced in Chap. 4 can be used to calculate the band structures of many III-V semiconductors.

(a) Using the pseudopotential method, estimate the valence band offset between InAs and GaSb by calculating each bandstructure and comparing the position of the valence bands.

(b) The accepted value for the valence band offset is −0.51 eV. The parameters of the pseudopotential calculation can be varied to yield this value of the valence band offset. Which parameters should be changed, and by how much?

References

M. Beck, D. Hofstetter, T. Aellen, J. Faist, U. Oesterle, M. Ilegems, E. Gini, H. Melchior, Continuous wave operation of a mid-infrared semiconductor laser at room temperature. Science **295**, 301–305 (2002). http://science.sciencemag.org/content/295/5553/301.long

J. Faist, *Quantum Cascade Lasers* (Oxford University Press, Oxford, 2013)

J. Faist, F. Capasso, D.L. Sivco, C. Sirtori, A.L. Hutchinson, A.Y. Cho, Quantum cascade laser. Science **264**, 553–556 (1994). https://science.sciencemag.org/content/264/5158/553

E. Gornik, D. Tsui, Voltage-tunable far-infrared emission from Si inversion layers. Phys. Rev. Lett. **37**, 1425–1428 (1976). https://journals.aps.org/prl/abstract/10.1103/PhysRevLett.37.1425

P. Harrison, *Quantum Wells, Wires and Dots*, 3rd edn. (Wiley Interscience, Chichester, 2006)

I. Hayashi, M.B. Panish, P.W. Foy, S. Sumski, Junction lasers which operate continuously at room-temperature. Appl. Phys. Lett. **17**, 109 (1970). https://aip.scitation.org/doi/abs/10.1063/1.1653326

R.F. Kazarinov, R.A. Suris, Possibility of the amplification of electromagnetic waves in a semiconductor with a superlattice. Sov. Phys. Semicond. **5**, 707–709 (1971)

M. Kim, C.L. Canedy, W.W. Bewley, C.S. Kim, J.R. Lindle, J. Abell, I. Vurgaftman, J.R. Meyer, Interband cascade laser emitting at $\lambda = 3.75$ μm in continuous wave above room temperature. Appl. Phys. Lett. **92**, 191110 (2008). Bibcode:2008ApPhL..92s1110 K. https://doi.org/10.1063/1.2930685. https://aip.scitation.org/doi/abs/10.1063/1.2930685

H. Lee, L.J. Olafsen, R.J. Menna, W.W. Bewley, R.U. Martinelli, I. Vurgaftman, D.Z. Garbuzov, C.L. Felix, M. Maiorov, J.R. Meyer, J.C. Connolly, A.R. Sugg, G.H. Olsen, Room-temperature type-II W quantum well diode laser with broadened waveguide emitting at $\lambda =$ 3.30 μm. Electron. Lett. **35**, 1743–1745 (1999). https://ieeexplore.ieee.org/abstract/document/811158

L. Nähle, L. Hildebrandt, M. Kamp, S. Höfling, Interband cascade lasers: ICLs open opportunities for mid-IR sensing. Laser Focus World, **49**, no 5, 70–73 (2013). https://www.laserfocusworld.com/lasers-sources/article/16557000/interband-cascade-lasersicls-open-opportunities-for-midir-sensing

M. Razeghi, Q.Y. Lu, N. Bandyopadhyay, W. Zhou, D. Heydari, Y. Bai, S. Slivken, Quantum cascade lasers: from tool to product. Opt. Expr. **23**, 8462–8475 (2015). http://cqd.eecs.northwestern.edu/pubs/journals.php?type=authors&search=Razeghi

C. Sirtori, J. Faist, F. Capasso, D.L. Sivco, A.L. Hutchinson, A.Y. Cho, Long wavelngth vertical transition quantum cascade lasers operating CW at 110 K. Superlattices Microstruct. **19**, 367–363 (1996) https://www.sciencedirect.com/science/article/abs/pii/S0749603696900397

G.W. 't Hooft, W.A.J.A. van der Poel, L.W. Molenkamp, C.T. Foxon, Giant oscillator strength of free excitons in GaAs. Phys. Rev. B **35**, 8281 (1987). https://journals.aps.org/prb/abstract/10.1103/PhysRevB.35.8281

L.C. West, S.J. Eglash, First observation of an extremely large-dipole infrared transition within the conduction band of a GaAs quantum well. Appl. Phys. Lett. **46**, 1156–1158 (1985). https://aip.scitation.org/doi/abs/10.1063/1.95742

R.Q. Yang, Infrared laser based on intersubband transitions in quantum wells. Superlattices Microstruct. **17**, 77–83 (1995). https://www.sciencedirect.com/science/article/abs/pii/S0749603685710178

J. Ziman, *Electrons and Phonons* (Oxford University Press, London, 1967)

Chapter 9
Non-linear Optics: Second-Harmonic Generation and Parametric Oscillation

Abstract The electric field of a beam of photons propagating through a medium produces polarization by spatially deforming the charge distribution in the outer shell of electrons. The relationship between the polarization and the electric field is the susceptibility. The polarization of the charge distribution is opposed by a restoring force of each atom. The first order susceptibility represents the linear response of the medium to the electric field. When the medium lacks a center of inversion symmetry, it follows that the relationship between the electric field and the polarization is no longer linear. The non-linear optical response of materials is a field of study that is rich with phenomena and applications. Second harmonic generation converts two photons having frequency $= \omega$ into one photon having frequency $= 2\omega$. Parametric amplification involves beams of two different frequencies at the input: a stronger pump beam at ω_3 and a weaker signal beam at ω_1. A special case of parametric amplification occurs when the signal amplitude at ω_1 is zero. Analysis by Maxwell's equations shows that no amplification occurs at the output. However experiment shows that measureable spontaneous fluorescence occurs. This fluorescence is called spontaneous parametric down conversion. Its existence demonstrates the limitation of Maxwell's equations to the classical domain and beams of light composed of many photons. Spontaneous parametric down conversion is widely used to prepare single photons and pairs of entangled photons.

9.1 Introduction

An applied electric field polarizes the medium in which it exists. The relation between the two is written

$$\mathbf{P} = \varepsilon_0 \chi \mathbf{E} \tag{9.1}$$

where χ is the susceptibility, and ε_0 is the permittivity of free space. The susceptibility is in general a complex quantity and accounts for both dielectric (loss-free) and conductive (lossy) behavior. In this chapter we will consider only dielectric media, so that χ is a real parameter. In the case of an anisotropic material, $\chi \to \chi_{ij}$ and relation (9.1) becomes:

© Springer Nature Switzerland AG 2020

T. P. Pearsall, *Quantum Photonics*, Graduate Texts in Physics,
https://doi.org/10.1007/978-3-030-47325-9_9

$$\mathbf{P}_i = \varepsilon_0 \sum_j \chi_{ij} \mathbf{E}_j \tag{9.2}$$

The electric field of a beam of photons propagating through a medium produces polarization by spatially deforming the charge distribution in the outer shell of electrons. The polarization is synchronous with the electric field and in the case of an ideal dielectric, loss-less as well, so that the polarization is in phase with the displacement.

$$\mathbf{P}(t) = Nq\mathbf{x}(t) \tag{9.3}$$

9.2 Non-linear Response of Optical Materials

The wavelength of an optical wave propagating through a solid dielectric material is about 10^3 times longer than the interatomic spacing. While the deformation of the charge distribution takes places at optical frequencies ($\sim 10^{15}$ Hz) it is distributed over a significant distance compared to the electronic cloud that surrounds an individual atom or crystalline unit cell.

The polarization of the charge distribution is opposed by a restoring force of each atom, leading to a situation similar to that of the harmonic oscillator. In a one-dimensional model we can express the potential:

$$V(x) = \frac{m}{2}ax^2 + \frac{m}{3}bx^3 + \cdots \tag{9.4}$$

In a material with a center of inversion symmetry, $V(x) = V(-x)$, and thus $b \equiv 0$.

$$V(x) = \frac{m}{2}ax^2 + \cdots \tag{9.5}$$

We recognize the first term in (9.5) as the potential for a harmonic oscillator. We recognize that $a = \omega_0^2$, and the restoring force on the electron is:

$$F(x) = qE(x) = -\frac{d}{dx}V(x) = -m\omega_0^2 x \tag{9.6}$$

$$x(t) = -\frac{q}{m\omega_0^2}E(t)$$

However, if the material lacks a center of inversion symmetry then the potential can take the form of (9.4). In this case restoring force on the electron is:

$$F(x) = -\frac{d}{dx}V(x) = -m\omega_0^2 x - mbx^2 \tag{9.7}$$

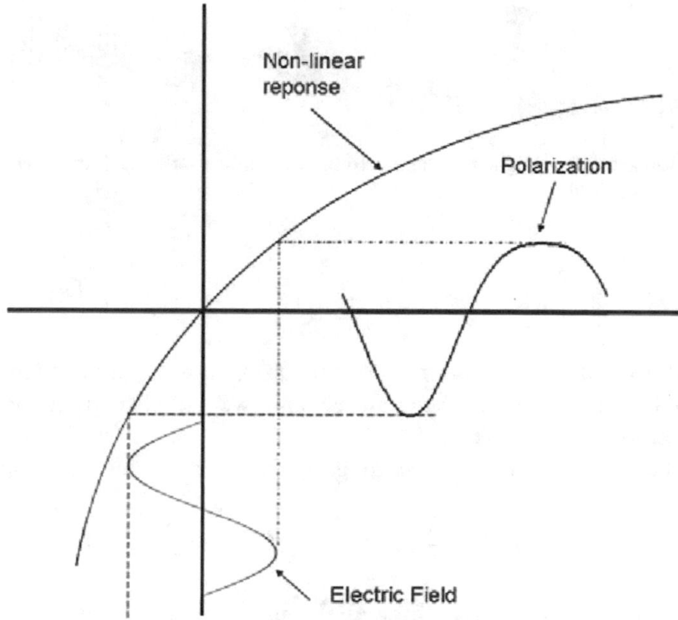

Fig. 9.1 The polarization is the response of a medium to the presence of an electric field. When a photon passes through a material lacking inversion symmetry, the polarization becomes a non-linear function of the electric field. The sinusoidal oscillation of the electric field is distorted

It can be seen immediately from (9.7) that the relationship between the electric field and the polarization is no longer linear, and that for a given magnitude of electric field, the polarization will be less when the amplitude of the field is positive than when the amplitude of the field is negative. This situation is diagrammed schematically in Fig. 9.1. It can be seen that the sinusoidal oscillation of the electrical field is transformed by the non-linear potential into a polarization that no longer has a pure sinusoidal form (Yariv 1975).

The driving force comes from the electric field of the incident photon

$$F(t) = \frac{qE}{2}\left(e^{i\omega t} + e^{-i\omega t}\right) \tag{9.8}$$

and the inertial force is

$$F(t) = m\left(\frac{d^2}{dt^2}x(t) + \gamma\frac{d}{dt}x(t)\right) \tag{9.9}$$

The second term in (9.9) represents possible losses.

These three forces act together following the schematic diagram in Fig. 9.2.

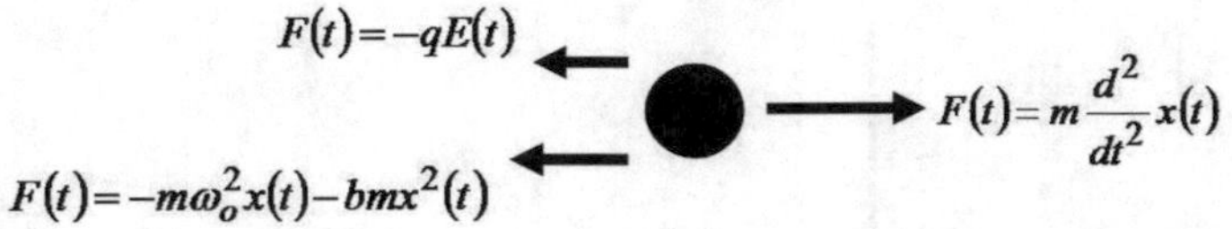

$F(t) = -qE(t)$

$F(t) = m\frac{d^2}{dt^2}x(t)$

$F(t) = -m\omega_o^2 x(t) - bmx^2(t)$

Fig. 9.2 Schematic force diagram on an electronic state in the outer shell of a material due to the photon electric field

$$\frac{d^2}{dt^2}x(t) + \gamma\frac{d}{dt}x(t) + \omega_0^2 x(t) + bx^2(t) = -\frac{qE}{2m}\left(e^{i\omega t} + e^{-i\omega t}\right) \tag{9.10}$$

To proceed, we will assume that propagation takes place in an ideal dielectric, so that $\gamma = 0$. For greater clarity in the presentation we will also ignore the complex conjugate terms (for example $e^{-i\omega t}$).

We will analyse (9.10) by assuming a trial solution for the displacement function $x(t)$.

$$x(t) = \frac{1}{2}\left(k_1 e^{i\omega t} + k_2 e^{2i\omega t} + cc\right) \tag{9.11}$$

This substitution transforms the differential equation into an algebraic expression. Substituting this trial solution into (9.10) gives

$$\begin{aligned} &-\frac{1}{2}\left(k_1\omega^2 e^{i\omega t} + 4k_2\omega^2 e^{2i\omega t}\right) + \frac{\omega_0^2}{2}k_1 e^{i\omega t} + \frac{\omega_0^2}{2}k_2 e^{2i\omega t} \\ &+ \frac{b}{4}\left(k_1^2 e^{2i\omega t} + 2k_1k_2 e^{3i\omega t} + k_2^2 e^{4i\omega t}\right) = -\frac{qE}{2m}\left(e^{i\omega t}\right) \end{aligned} \tag{9.12}$$

This equation is valid for all t, and this requires that the coefficients of the terms involving $e^{i\omega t}$ be equal on both sides of the equation, and the same for the coefficients of $e^{2i\omega t}$.

Consider first the terms involving $e^{i\omega t}$:

$$-\frac{1}{2}\left(k_1\omega^2 e^{i\omega t}\right) + \frac{\omega_0^2}{2}k_1 e^{i\omega t} = -\frac{qE}{2m}\left(e^{i\omega t}\right) \tag{9.13}$$

Solving for k_1:

$$k_1 = -\frac{qE}{m}\left(\frac{1}{\left[\omega_0^2 - \omega^2\right]}\right) \tag{9.14}$$

The polarization at ω is written as:

$$P(\omega) = -\frac{Nq}{2}x(t) = -\frac{Nq}{2}\left(k_1 e^{i\omega t} + cc\right) \equiv \frac{\varepsilon_0}{2}\left[\chi(\omega)Ee^{i\omega t} + cc\right] \tag{9.15}$$

where we have introduced the linear susceptibility $\chi(\omega)$.

Solving for $\chi(\omega)$,

$$\chi(\omega) \equiv \chi_L^{\omega} = -\frac{Nq}{\varepsilon_0 E} k_1 = -\frac{Nq}{\varepsilon_0 E}\left(-\frac{qE}{m}\frac{1}{\left[\omega_0^2 - \omega^2\right]}\right) = \frac{Nq^2}{\varepsilon_0 m}\left(\frac{1}{\omega_0^2 - \omega^2}\right) \tag{9.16}$$

It follows that:

$$\chi_L^{(2\omega)} = \frac{Nq^2}{\varepsilon_0 m}\left(\frac{1}{\omega_0^2 - 4\omega^2}\right)$$

Next consider the terms in $e^{2i\omega t}$:

$$-2k_2\omega^2 + \frac{\omega_0^2}{2}k_2 + \frac{b}{4}k_1^2 = 0$$

$$k_2 = \frac{k_1^2}{2}\frac{b}{4\omega^2 - \omega_0^2} = \frac{q^2E^2}{2m^2}\frac{b}{\left(\omega_0^2 - \omega^2\right)^2\left(\omega_0^2 - 4\omega^2\right)} \tag{9.17}$$

The non-linear susceptibility $d_{\mathrm{NL}}^{(2\omega)}$ is defined from the expression for the polarization at 2ω (Bloembergen 1996; Shen 2003).

$$P^{(2\omega)} = \frac{Nq}{2}\left(k_2 e^{2i\omega t} + cc\right) \equiv \frac{1}{2}\left[d_{\mathrm{NL}}^{(2\omega)} E^2 e^{i\omega t} + cc\right] \tag{9.18}$$

$$d_{\mathrm{NL}}^{(2\omega)} = \frac{-bNq^2}{\left(\omega_0^2 - \omega^2\right)^2\left(\omega_0^2 - 4\omega^2\right)} = \frac{bm\varepsilon_0^2\left(\chi_L^{\omega}\right)^2\chi_L^{(2\omega)}}{2N^2q^3} \tag{9.19}$$

In a 3-dimensional system, the non-linear susceptibility is not a constant, but rather a 3rd order tensor. In general, the definition is

$$P_k^{\omega_3=\omega_1+\omega_2} = d_{ijk}^{\omega_3} E_j^{\omega_1} E_k^{\omega_2} \tag{9.20}$$

The tensor $d_{ijk}^{\omega_3}$ has 27 components. There is, however, no difference between the results when the indices j and k are interchanged. That is,

$$d_{ijk} = d_{ikj} \tag{9.21}$$

This reduces the number of independent components to 18. When the propagation medium is a crystal, symmetry simplifies the situation even further, so that in practice for interesting non-linear optical crystalline materials there are only 1 or 2 independent elements (Boyd 2008). Thus, the subscripts kj can be replaced by a single number.

$$\begin{aligned} xx &\rightarrow 1 \\ yy &\rightarrow 2 \\ zz &\rightarrow 3 \\ yz = zy &\rightarrow 4 \\ xz = zx &\rightarrow 5 \\ xy = yx &\rightarrow 6 \end{aligned} \tag{9.22}$$

The electric fields are combined into a 1 by 6 vector

$$\begin{pmatrix} E_x^2 \\ E_y^2 \\ E_z^2 \\ 2E_zE_y \\ 2E_zE_y \\ 2E_xE_y \end{pmatrix} \tag{9.23}$$

The polarization is calculated in the usual way:

$$\begin{pmatrix} P_x \\ P_y \\ P_z \end{pmatrix} = \begin{pmatrix} d_{11}\; d_{12}\; d_{13}\; d_{14}\; d_{15}\; d_{16} \\ d_{21}\; d_{22}\; d_{23}\; d_{24}\; d_{25}\; d_{26} \\ d_{31}\; d_{32}\; d_{33}\; d_{34}\; d_{35}\; d_{36} \end{pmatrix} \begin{pmatrix} E_x^2 \\ E_y^2 \\ E_z^2 \\ 2E_zE_y \\ 2E_zE_y \\ 2E_xE_y \end{pmatrix} \tag{9.24}$$

Example KDP—potassium dihydrogen phosphate belongs to symmetry class $\overline{4}2m$ (tetragonal a 4-fold axis, two 2-fold axes and a mirror plane). The non-linear susceptibility tensor for this class is:

$$d_{ijk}^{\omega_3} = \begin{pmatrix} 0\;0\;0\; d_{14} & 0 & 0 \\ 0\;0\;0\; 0 & d_{14} & 0 \\ 0\;0\;0\; 0 & 0 & d_{36} \end{pmatrix} \tag{9.25}$$

For KDP (Gamdan Optics 2016),

$$\begin{aligned} d_{14} &= 0.35\, pm - \mathrm{V}^{-1} \\ d_{36} &= 0.44\, pm - \mathrm{V}^{-1} \end{aligned}$$

GaAs is a cubic zinc-blende structure with symmetry class $\overline{4}3m$. The non-linear susceptibility tensor for this crystal class is:

$$d_{ijk}^{\omega_3} = \begin{pmatrix} 0\,0\,0\,d_{14} & 0 & 0 \\ 0\,0\,0 & 0 & d_{14} & 0 \\ 0\,0\,0 & 0 & 0 & d_{14} \end{pmatrix} \tag{9.26}$$

For GaAs, $d_{14} = 94\ pm - \mathrm{V}^{-1}$ (Skauli 2002).

β-Barium Borate (BBO) $\mathrm{Ba(BO_2)_2}$ trigonal class 3m

$$d_{ijk}^{\omega_3} = \begin{pmatrix} 0 & 0 & 0 & 0 & d_{15} & -d_{22} \\ -d_{22} & d_{22} & 0 & d_{15} & 0 & 0 \\ d_{31} & d_{31} & d_{33} & 0 & 0 & 0 \end{pmatrix} \tag{9.27}$$

For BBO (Coherent 2016),

$$d_{22} = 2.22\ pm - \mathrm{V}^{-1}$$
$$d_{31} = 0.16\ pm - \mathrm{V}^{-1}$$
$$d_{33} \approx d_{15} < d_{31}$$

A non-linear optical response can occur in any material that lacks a center of inversion symmetry. Typically, non-linear effects are measured in crystals, and attention is paid to the relationship between the direction of photon propagation and the axes of symmetry of the crystalline structure (Wong 2002). However, a non-linear optical signal can occur in other circumstances. An important example concerns interfaces (Heinz 1991). In most cases, the interface between 2 different materials does not have a center of inversion symmetry: for example, the interface between two liquids. When it is also the case that neither material forming the interface has a non-zero non-linear susceptibility, then non-linear effects can be used as a sensitive probe of the interface itself. This opportunity arises in the study of biological molecular domains that are supported in aqueous solution. As a result, non-linear optical spectroscopy is an important analytical tool in biochemical research.

9.3 Electromagnetic Wave Propagation in Non-linear Materials

Summarizing the results of Sect. 9.2, the polarization can be expressed in the general case as:

$$P_i = \varepsilon_0 \sum_j \chi_{ij}^{(1)} E_j + \varepsilon_0 \sum_j \sum_k d_{ijk}^{(2)} E_j E_k + \cdots \tag{9.28}$$

We will consider the propagation in the z-direction of the photon with a well-defined linear polarization along a single Cartesian axis. This assumption simplifies the analysis and corresponds to most experimental situations. Since we will be working with only one relevant component of the susceptibility tensor, we will leave out the dimensional subscripts for clarity in the presentation.

Consider now, the co-linear propagation of two monochromatic photon beams, of frequency ω_1 and ω_2. The electric field of the total electric field is just the arithmetic sum of the two components:

$$\mathbf{E}_T(t) = \mathbf{E}_1(t) + \mathbf{E}_2(t)$$

where $\mathbf{E}_1(t) = E_1 e^{i\omega_1 t} + E_1^* e^{-i\omega_1 t} = E_1 e^{i\omega_1 t} + cc$ and $\mathbf{E}_2(t) = E_2 e^{i\omega_2 t} + cc$.

The 2nd-order non-linear polarization is:

$$P^{(2)}(t) = \frac{\varepsilon_0 d^{(2)}}{2}\left(E_1 e^{i\omega_1 t} + cc + E_2 e^{i\omega_2 t} + cc\right)^2 \tag{9.29}$$

Without making the full expansion of (9.29), it can be seen that the second order polarization will contain terms with different frequency dependencies, notably:

$$\begin{aligned} &\text{Second harmonic generation: } E_1^2 e^{2i\omega t} \\ &\text{Frequency sum: } |E_1||E_2| e^{i(\omega_1+\omega_2)t} \\ &\text{Frequency difference : } |E_1||E_2| e^{i(\omega_1-\omega_2)t} \\ &\text{And dc polarization } = \text{ optical rectification : } E_1^2 + E_2^2 \end{aligned} \tag{9.30}$$

For example, in the case of second harmonic generation, $\omega_1 = \omega_2$, and two terms result, one with frequency $= 2\omega_1$, and a dc polarization. This is diagrammed schematically in Fig. 9.3.

We use Maxwell's equations to describe the propagation of a photon wave in a non-linear medium in the z-direction:

$$\frac{d^2}{dz^2}\mathbf{E}(\mathbf{r},t) = \mu_0\sigma\frac{\partial}{\partial t}\mathbf{E}(\mathbf{r},t) + \mu_0\varepsilon_0\frac{\partial^2}{\partial t^2}\left[\mathbf{E}(\mathbf{r},t) + \frac{\mathbf{P}_L(\mathbf{r},t) + \mathbf{P}_{\mathrm{NL}}(\mathbf{r},t)}{\varepsilon_0}\right] \tag{9.31}$$

where we assume the absence of free charges (i.e. $\nabla \cdot \mathbf{E}(\mathbf{r},t) = 0$).

Without any loss of generality, we will presume that a planar wave propagates in the z-direction, so that $\frac{\partial}{\partial x} = \frac{\partial}{\partial y} = 0$.

Next consider 3 waves with distinct frequencies and wave-vectors propagating simultaneously in the non-linear medium:

$$\begin{aligned} E_i(z,t) &= \frac{1}{2}\left[E_{1i}(z)e^{i(\omega_1 t - k_1 z)} + cc\right] \\ E_j(z,t) &= \frac{1}{2}\left[E_{2j}(z)e^{i(\omega_2 t - k_2 z)} + cc\right] \\ E_k(z,t) &= \frac{1}{2}\left[E_{3k}(z)e^{i(\omega_3 t - k_3 z)} + cc\right] \end{aligned} \tag{9.32}$$

and $|k_l|^2 = n_l^2 \frac{\omega_l^2}{c^2}$.

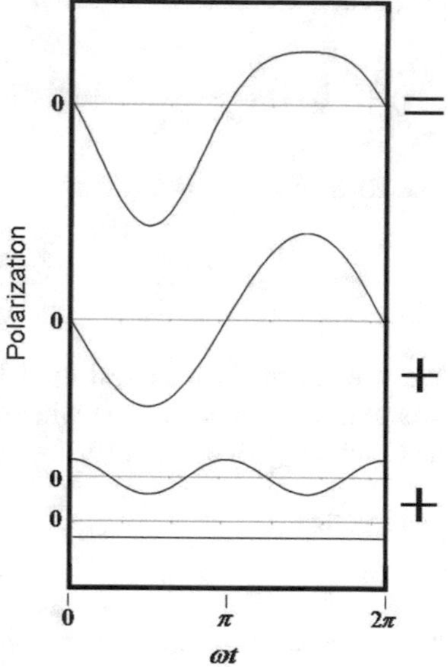

Fig. 9.3 As a result of non-linear restoring forces, the optical polarization is a sum of the incident photon frequency, its second harmonic and a dc polarization. The dc electric field that results from the dc polarization is the manifestation of optical rectification

In the case of propagation in a linear dielectric material, the polarization and the electric field are related by the linear susceptibility:

$$P_L^{\omega_1}(z,t) = \varepsilon_0 \chi_L E_i(z,t) = \frac{\varepsilon_0 \chi_L}{2}\left[E_{1i}(z)e^{i(\omega_1 t - k_1 z)} + cc\right] \tag{9.33}$$

While the non-linear polarization is a product that mixes the 3 wave components. For example:

$$P_{\mathrm{NL}}^{\omega_1} = \frac{d_{ijk}}{2}\left[E_{3k}E_{2j}^* e^{i[(\omega_3-\omega_2)t-(k_3-k_2)z]} + cc\right] \tag{9.34}$$

Next we consider wave propagation in a non-linear medium. We will take the wave to be linearly polarized. This simplifies the presentation of the algebra and corresponds to the usual experimental conditions. Taking the first wave (at ω_1) of (9.32), we expand the LHS of (9.31):

$$\frac{d^2}{dz^2}E_i(z,t) = \frac{1}{2}\left[-k_l^2 E_{1i}(z)e^{i(\omega_1 t - k_1 z)} - 2ik_1 e^{i(\omega_1 t - k_1 z)}\frac{d}{dz}E_{1i}(z) + e^{i(\omega_1 t - k_1 z)}\frac{d^2}{dz^2}E_{1i}(z) + cc\right] \tag{9.35}$$

We presume that the amplitude of the electric field varies slowly over the interval corresponding to the wavelength of the wave, i.e.:

$$|k_l^2 E_{1i}(z)| \gg \left|k_1 \frac{d}{dz}E_{1i}(z)\right| \gg \left|\frac{d^2}{dz^2}E_{1i}(z)\right| \tag{9.36}$$

Neglecting the 2nd spatial derivative of $E_{1i}(z)$,

$$\frac{d^2}{dz^2}E_i(z,t) = -\frac{1}{2}\left[k_1^2 E_{1i}(z)e^{i(\omega_1 t - k_1 z)} + 2ik_1 e^{i(\omega_1 t - k_1 z)}\frac{d}{dz}E_{1i}(z) + cc\right] \quad (9.37)$$

The RHS of (9.31) is written:

$$\text{RHS} = \left[\frac{i\omega_1\mu_0\sigma}{2} - \frac{\omega_1^2\mu_0\varepsilon_0}{2}\right]E_{1i}(z)e^{i(\omega_1 t - k_1 z)} + \mu_0\frac{\partial^2}{\partial t^2}P_{\text{NL}} \quad (9.38)$$

We will restrict our interest to dielectric materials which can be considered lossless in the frequency region where they are transparent; that is, $\sigma \approx 0$.

Combining (9.37) and (9.38),

$$-ik_1 e^{i(\omega_1 t - k_1 z)}\frac{d}{dz}E_{1i}(z) + cc = \mu_0\frac{\partial^2}{\partial t^2}P_{\text{NL}} \quad (9.39)$$

where we have used the relationship: $k_1^2 E_{1i} = \omega_1^2\mu_0\varepsilon_0 E_{1i}$ to simplify the expression.

$$\mu_0\frac{\partial^2}{\partial t^2}P_{\text{NL}} = -\frac{\mu_0}{2}(\omega_3 - \omega_2)^2 d_{ijk}E_{3k}E_{2j}^* e^{i[(\omega_3-\omega_2)t-(k_3-k_2)z]} + cc \quad (9.40)$$

$$ik_1 e^{i(\omega_1 t - k_1 z)}\frac{d}{dz}E_{1i}(z) = \frac{\mu_0}{2}(\omega_3 - \omega_2)^2 d_{ijk}E_{3k}E_{2j}^* e^{i[(\omega_3-\omega_2)t-(k_3-k_2)z]} + cc \quad (9.41)$$

We require that photon energy be conserved; that is $\hbar\omega_1 = \hbar\omega_3 - \hbar\omega_2$. Recall that $\omega_1^2 = \frac{k_1^2}{\mu_0\varepsilon}$.

$$\frac{d}{dz}E_{1i}(z) = -\frac{i\omega_1}{2}\sqrt{\frac{\mu_0}{\varepsilon}}d_{ijk}E_{3k}E_{2j}^* e^{-i[(k_3-k_2-k_1)z]} + cc \quad (9.42)$$

Likewise,

$$\frac{d}{dz}E_{2j}(z) = \frac{i\omega_2}{2}\sqrt{\frac{\mu_0}{\varepsilon}}d_{ijk}E_{3k}^* E_{1i}e^{-i[(k_2+k_1-k_3)z]} + cc \quad (9.43)$$

$$\frac{d}{dz}E_{3k}(z) = -\frac{i\omega_3}{2}\sqrt{\frac{\mu_0}{\varepsilon}}d_{ijk}E_{1i}E_{2j}e^{-i[(k_1+k_2-k_3)z]} + cc \quad (9.44)$$

The physical explanation behind 2nd harmonic generation is based on the non-linear response of valence electrons to polarization displacement imposed by the electro-magnetic field of a photon flux. This description is accurate, but gives an incomplete accounting of non-linear effects, because it does not include creation of photons from vacuum states. Equations (9.42)–(9.44) tell us that the non-linear optical tensor causes the electro-magnetic field of one ray to couple to fields of the two other rays.

This mixing gives rise to other effects in addition to 2nd harmonic generation: frequency upconversion, frequency down conversion, parametric amplification, and spontaneous parametric down conversion, to name some important effects that can be harnessed using 2nd order non-linearities.

9.3.1 Application to Second Harmonic Generation

Equations (9.42)–(9.44) can be applied to the generation of frequency sum or difference in non linear materials. An important particular case is second harmonic generation. For this situation, $\omega_1 = \omega_2$, and $\omega_3 = 2\omega_1$. Second harmonic generation is thus an interaction between 2 rays of photons, diagrammed schematically in Fig. 9.4.

$$\frac{d}{dz}E_{3k}(z) = -i\omega_1\sqrt{\frac{\mu_0}{\varepsilon}}d_{ijk}E_{1i}E_{2j}e^{-i[(k_1+k_2-k_3)z]} + cc \tag{9.45}$$

where we have substituted $\omega_1 = \frac{\omega_3}{2}$.

Integrating across the sample from 0 to L

$$E_{3k}(L) - 0 = \omega_1\sqrt{\frac{\mu_0}{\varepsilon}}d_{ijk}E_{1i}E_{2j}\frac{e^{-i[(k_1+k_2-k_3)L]} - 1}{k_1 + k_2 - k_3} + cc \tag{9.46}$$

We recall from (5.43) that $\mathbf{k}_1 + \mathbf{k}_2 - \mathbf{k}_3 = \Delta\mathbf{k}$ is the vector sum of the momenta of the photons involved in 2nd harmonic generation. In the semi-classical approach being developed here, power can be transferred efficiently from the 1st to the 2nd harmonic only is the waves propagate with the same phase velocity and remain in phase with each other, i.e., that $\Delta\mathbf{k} = 0$. We refer to $\Delta\mathbf{k}$ as the phase mismatch.

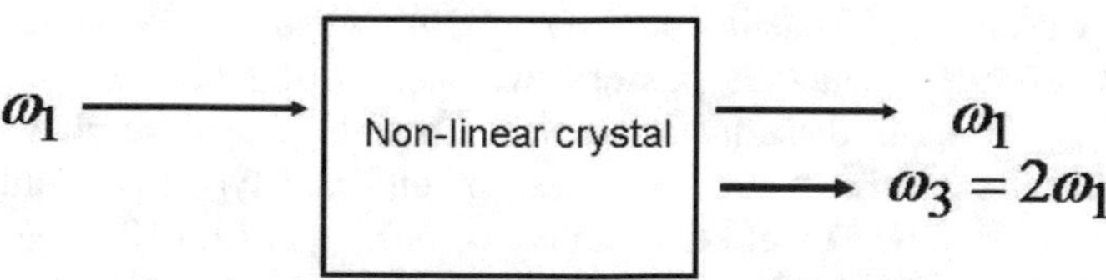

Fig. 9.4 A schematic diagram of second harmonic generation in a nonlinear optical material. A photon beam entering the material with frequency $= \omega_3$ polarizes the valence electron distribution. Because of the absence of inversion symmetry, which characterizes all non-linear materials, oscillation of the electron distribution produces a second harmonic photonic signal $\omega_3 = 2\omega_1$ and a dc polarization of the material

$$E_{3k}(L) - 0 = \omega_1 \sqrt{\frac{\mu_0}{\varepsilon}} d_{ijk} E_{1i} E_{2j} L \frac{e^{-i\Delta kL}}{\Delta kL} + cc \tag{9.47}$$

The intensity of the 2nd harmonic signal is

$$|E_{3k} E_{3k}^*| = \omega_1^2 \frac{\mu_0}{\varepsilon} (d^{ijk})^2 |E_{1i} E_{2j}|^2 L^2 \frac{\sin^2\left(\frac{\Delta kL}{2}\right)}{\left(\frac{\Delta kL}{2}\right)^2} \tag{9.48}$$

In the general case, when $\Delta\mathbf{k} \neq 0$, (9.48) is a periodic function of the path length of propagation in the non-linear material. It oscillates between 0, where all the optical power is present as the first harmonic, and some maximum where power is shared between first harmonic and second harmonic emission. The maximum value depends inversely on the degree of phase mismatch. This result is diagrammed schematically in Fig. 9.5.

It follows from (9.48), that useful second-harmonic emission can be obtained if the non-linear material could be cut to a precise thickness that depends on the phase-mismatch. This is not at all practical. However, if we can create phase-matching between the first and second harmonic waves, so that $\Delta\mathbf{k} = 0$, the situation changes dramatically;

$$\lim_{\Delta\mathbf{k}\to 0} \left[\frac{\sin^2\left(\frac{\Delta kL}{2}\right)}{\left(\frac{\Delta kL}{2}\right)^2} \right] = 1, \quad \text{and thus: } \left|E_{3k} E_{3k}^*\right| = \omega_1^2 \frac{\mu_0}{\varepsilon} \left|d^{ijk} E_{1i} E_{2j}\right|^2 L^2 \tag{9.49}$$

Under phase-matched conditions, the optical power in the second harmonic increases with the square of the propagation distance. At some point, our initial assumption of non-depletion of the first harmonic pump beam will no longer apply, and the conversion rate will diminish. This does not change the important result that nearly 100% of the pump power can in principle be converted to the 2nd harmonic using phase-matched conditions.

Phase-matching can be achieved by using the natural birefringence of non-linear optical materials. The phase-matched condition can be reduced simple to the requirement that the index of refraction of the 1st harmonic be equal to that of the 2nd harmonic wave. In the usual case, the index of refraction is a decreasing function of wavelength.[1] Thus it would be not possible to achieve phase-matching. However, birefringent materials possess two indices of refraction: $n_{\text{ordinary}} = n_0$ and $n_{\text{extraordinary}} = n_e$. In these materials, it is possible to achieve phase-matching when the 1st and 2nd harmonic waves are of different types: one ordinary and the other extraordinary. The effective index of refraction ($n_e(\theta)$) experienced by the extraordinary wave in a birefringent crystal depends on the angle between it

[1]In normally dispersive materials, the dependence of index on wavelength can be calculated from the Sellmeier Equations. To take the example of BBO: $n_0^2 = 2.7359 + \frac{0.01878}{\lambda^2 - 0.01822} - 0.01354\lambda^2$ and $n_e^2 = 2.3753 + \frac{0.01224}{\lambda^2 - 0.01667} - 0.01516\lambda^2$, where λ is measured in μm.

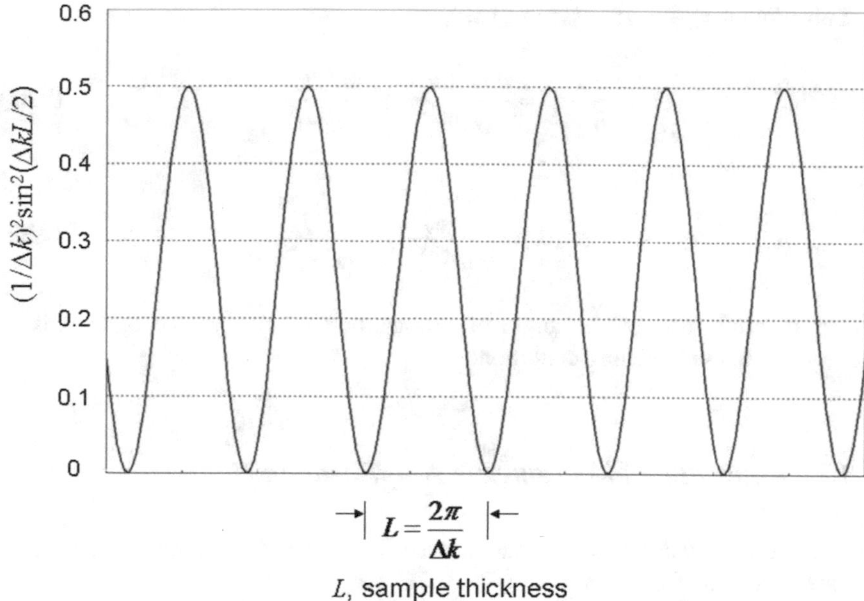

Fig. 9.5 When the first and second harmonic waves are not phase-matched, the output power is shared between the first and second harmonic waves. The amount of second harmonic emission varies periodically with propagation length in the non-linear material

propagation direction and the optic axis of the crystal. A simple geometric formula relates $n_e(\theta)$ to n_e and n_0:

$$\frac{1}{n_e^2(\theta)} = \frac{\cos^2(\theta)}{n_0^2} + \frac{\sin^2(\theta)}{n_e^2} \tag{9.50}$$

If $n_e^{2\omega} < n_0^{\omega}$, then there will be an angle θ_m, the phase-matching angle, where $n_e^{2\omega}(\theta_m) = n_0^{\omega}$. If the 1st harmonic is launched along the direction θ_m as an ordinary ray, (that is, polarized perpendicular to the plane formed by the optic axis and the direction of propagation) the 2nd harmonic component will be generated having the same index refraction and direction of propagation, but as an extraordinary ray (that is, polarized in the plane formed by the optic axis and the direction of propagation, see Fig. 5.3). Thus, the phase-matched 2nd harmonic ray is collinear with the 1st harmonic. We call this type-I phase-matching. Because this non-linear interaction produces only one additional ray, type-I phase-matching is the only possible solution, due to conservation of momentum. However, as we shall see shortly, frequency up conversion and frequency down conversion produce 2 additional rays. This enables the generation of each optical ray at an angle to the pumping beam provided that overall photon momentum is conserved in 3 dimensions. This is called type-II phase matching. It has the advantage that the 3 optical waves are separated spatially, because they propagate in different directions.

Substituting n_0^ω for $n_e^{2\omega}(\theta_m)$ in (9.50),

$$\frac{1}{\left(n_0^\omega\right)^2} = \frac{\cos^2(\theta_m)}{\left(n_0^{2\omega}\right)^2} + \frac{\sin^2(\theta_m)}{\left(n_e^{2\omega}\right)^2} \tag{9.51}$$

and

$$\sin^2(\theta_m) = \frac{\left(n_0^\omega\right)^{-2} - \left(n_0^{2\omega}\right)^{-2}}{\left(n_e^{2\omega}\right)^{-2} - \left(n_0^{2\omega}\right)^{-2}} \tag{9.52}$$

allows determination of the angle of propagation relative to the optic axis to achieve the phase-matching condition.

9.3.2 Application to Parametric Amplification

Parametric amplification is the general case of non-linear interaction. Three rays are involved where $\omega_3 = \omega_1 + \omega_2$ (Fig. 9.6).

Starting with (9.42)–(9.44), we make a change of variable that simplifies the expressions:

$$F_i = \sqrt{\frac{n_i}{\omega_i}} E_i, \quad \text{for } i = 1, 2, 3 \tag{9.53}$$

$|F_i|^2$ is thus proportional to the photon flux.

The equations for the electric fields are transformed to:

$$\frac{d}{dz} F_{1i}(z) = -\frac{i}{2} \kappa d_{ijk} F_{3k} F_{2j}^* e^{-i\Delta kz} + cc \tag{9.54}$$

$$\frac{d}{dz} F_{2j}(z) = \frac{i}{2} \kappa d_{ijk} F_{3k}^* F_{1i} e^{i\Delta kz} + cc \tag{9.55}$$

$$\frac{d}{dz} F_{3k}(z) = -\frac{i}{2} \kappa d_{ijk} F_{1i} F_{2j} e^{i\Delta kz} + cc \tag{9.56}$$

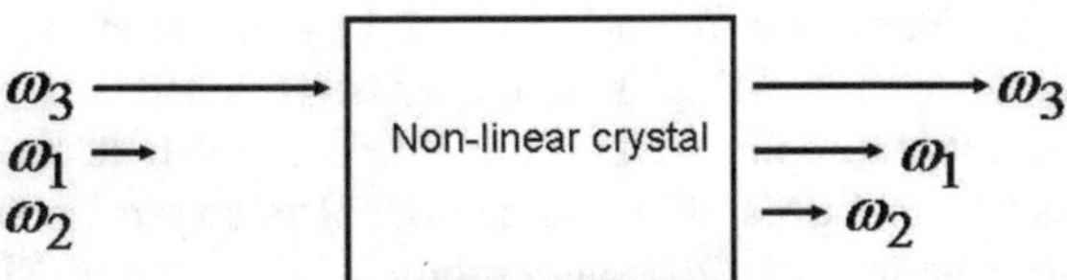

Fig. 9.6 Schematic diagram of parametric amplification involving 3 photon beams. The pump beam with frequency ω_3 and a less intense signal beam at ω_1 enter the non-linear crystal. The signal beam is amplified by energy transfer from the pump to the signal at the signal frequency. This requires the creation of a third beam at frequency ω_2 so that total photon energy is conserved. The ray at ω_2 is called the idler. The non-linear optical tensor implements the mixing among the 3 rays

This change of variable allows us to use a single coupling parameter κ:

$$\kappa = \sqrt{\frac{\mu}{\varepsilon_0} \frac{\omega_1 \omega_2 \omega_3}{n_1 n_2 n_3}}$$

and

$$\Delta k = k_3 - (k_1 + k_2) \tag{9.57}$$

In the following analysis, we impose conservation of energy:

$$\omega_3 = \omega_1 + \omega_2$$

and conservation of momentum:

$$\Delta k = 0$$

We will assume type-I collinear phase matching, so that all rays propagate in the z-direction. In this way, we can replace the tensor product by a simple algebraic product and $d_{ijk} \to d'$ at the appropriate angle relative to the optic axis for phase-matching.

$$\frac{d}{dz} F_{1i}(z) = -\frac{i}{2} \kappa d' F_{3k} F_{2j}^* \tag{9.58}$$

$$\frac{d}{dz} F_{2j}(z) = \frac{i}{2} \kappa d' F_{3k} F_{1i} \tag{9.59}$$

$$\frac{d}{dz} F_{3k}(z) = -\frac{i}{2} \kappa d' F_{1i} F_{2j} \tag{9.60}$$

Next, we assume non-depletion of the pump beam so that $F_{3k}(z) = \sqrt{\frac{n_3}{\omega_3}} E_{3k}(0)$ = constant. This leaves 2 equations only

$$\frac{d}{dz} F_{1i}(z) = -\frac{i}{2} \kappa \sqrt{\frac{n_3}{\omega_3}} E_{3k}(0) d' F_{2j}^* \tag{9.61}$$

$$\frac{d}{dz} F_{2j}(z) = \frac{i}{2} \kappa \sqrt{\frac{n_3}{\omega_3}} E_{3k}(0) d' F_{1i} \tag{9.62}$$

Simplify by defining a new parameter γ

$$\gamma \equiv \kappa F_{3k}(0) = \sqrt{\frac{\mu_0}{\varepsilon_0} \frac{\omega_1 \omega_2}{n_1 n_2}} E_{3k}(0) \tag{9.63}$$

$$\frac{d}{dz} F_{1i}(z) = -\frac{i\gamma}{2} d' F_{2j}^*(z) \tag{9.64}$$

$$\frac{d}{dz} F_{2j}^*(z) = \frac{i\gamma}{2} d' F_{1i}(z) \tag{9.65}$$

Solving for $F_{1i}(z)$:

$$F_{1i}(z) = -\frac{2i}{\gamma d'}\frac{d}{dz}F^*_{2j}(z) \tag{9.66}$$

$$\frac{d^2}{dz^2}F^*_{2j}(z) = \frac{\gamma^2(d')^2}{4}F^*_{2j}(z). \tag{9.67}$$

We choose as a trial solution: $F^*_{2j}(z) = C\cosh\frac{\gamma d'}{2}z + D\sinh\frac{\gamma d'}{2}z$.

For initial conditions we choose ray 1 to be the "signal" wave and ray 2 to be the "idler" wave. At z = 0 the idler wave intensity is zero. The nonlinear coupling builds up the idler wave intensity by generating an electromagnetic wave with frequency $\omega_2 = \omega_3 - \omega_1$. These initial conditions mean that $C = 0$ and

$$F^*_{2j}(z) = D\sinh\frac{\gamma d'}{2}z \tag{9.68}$$

Substituting this result in (9.67)

$$F_{1i}(z) = -iD\cosh\frac{\gamma d'}{2}z = F_{1i}(0)\cosh\frac{\gamma d'}{2}z \tag{9.69}$$

$$F^*_{2j}(z) = -iF_{1i}(0)\sinh\frac{\gamma d'}{2}z \tag{9.70}$$

Finally we restore the original parameters

$$E_{1i}(z) = E_{1i}(0)\cosh\left(z\frac{d'}{2}\sqrt{\frac{\mu_0}{\varepsilon_0}\frac{\omega_1\omega_2}{n_1n_2}}E_{3k}(0)\right) \tag{9.71}$$

$$E^*_{2j}(z) = -iE_{1i}(0)\sinh\left(z\frac{d'}{2}\sqrt{\frac{\mu_0}{\varepsilon_0}\frac{\omega_1\omega_2}{n_1n_2}}E_{3k}(0)\right) \tag{9.72}$$

These results show that parametric amplification occurs for both the signal and idler rays fuelled by the electric field of the pump. A more precise analysis, allowing for pump depletion would show how this occurs as the three rays propagate through the non-linear crystal. The dependence of the mixing on the non-linear optic coefficient, confirms the non-linear optical origin of this effect. The ray at ω_2 grows from 0 intensity at $z = 0$.

It can be seen from (9.71) and (9.72), that the mixing effect is strictly zero if the amplitude of both input fields is zero. To the contrary, experiment shows the presence of spontaneous down conversion fluorescence under these conditions (Ling 2008). Thus, the classical electromagnetic analysis is not capable of describing observable physics. In Chap. 5 we discussed degenerate spontaneous down conversion where $\omega_1 = \omega_2$ and its practical importance as a procedure for generation of entangled photon pairs. Similar to the case of spontaneous emission, photons from the vacuum with states that permit energy and momentum conservation stimulate parametric down conversion. Straightforward analysis (see Sect. 5.5) shows that two photons each with frequency $\omega_1 = \omega_2$ can be created from the vacuum state, while simultaneously a single photon of frequency $\omega_3 = 2\omega_1$ is annihilated. The converted intensity increases linearly with that of the pump beam, and the conversion rate is very small, on the order of one event per 10^{12} pump photons.

9.4 Summary

The non-linear optical response of materials is a field of study that is rich with phenomena and applications. Using Maxwell's equations we have focussed on the principles of two important phenomena that are used in studies of quantum photonic behavior: second-harmonic generation and parametric amplification.

Second-harmonic generation occurs as a result of the polarization of a crystal by the electric field of a traversing wave. If the material lacks a center of inversion symmetry in its structure, second harmonic distortion of the electromagnetic wave will occur. The incident pump wave and the second harmonic signal co-propagate in the non-linear material. In the case of birefringent non-linear optical crystals, there will be a specific direction where the propagation velocities of the pump and the second harmonic are the same. Propagation along this direction causes the pump and the second harmonic to remain in phase, and corresponds to the phase-matching condition. At lower conversion rates, when the intensity of the pump beam remains largely undepleted, the intensity of the second harmonic component is proportional to the square of the pump beam. This shows that two pump photons are required to produce one second harmonic photon. Under optimized conditions, nearly all of the pump light power can be converted to the second harmonic generation.

Parametric amplification is a more general technique for mixing photon energies. In its simplest form, a pump beam and a signal beam enter the NLO material. At the exit, the signal beam has been amplified, and a new optical wave appears with frequency equal to the difference between that of the pump and the signal, conserving photon energy. In contrast to the case of second harmonic generation, the amplification of the signal photon is linearly proportional to that of the pump beam. A particular case of parametric amplification is that where the amplitude of the signal beam is zero. Maxwell's equations show in this case that there is no amplified beam. Thus there should be no effect. This contradicts experiment which shows clearly that photons are emitted satisfying energy conservation. This effect is called spontaneous parametric down-conversion, with reference to spontaneous emission that occurs in light-emitting devices like lasers and LEDs. A photon created from vacuum fluctuations at the input port of the non-linear material can be amplified by parametric amplification. This results in the appearance of both signal and idler photons at the output. The analysis of this process must be carried out using quantized field theory, as we have done in Chap. 5. The result proves that the intensity of the spontaneously generated photons is linearly proportional to the pump power, in agreement with experiment, which is echoed by the classical analysis using Maxwell's equations in the regime where these can be applied.

Second-harmonic generation converts two photons with frequency $= \omega$ to one photon having frequency $= 2\omega$. Spontaneous parametric down conversion can be used to create 2 photons having frequency $= \omega$ from one photon having frequency $= 2\omega$. Although both processes are non-linear optical effects, it should be quite clear that one is not at all the inverse of the other, having their origins in completely different physical effects.

9.5 Exercises

9.1 In the case where propagation in the optical medium is lossy, show that the linear optical susceptibility is given by:

$$\chi(\omega) = \frac{Nq^2}{\varepsilon_0 m}\left(\frac{1}{\omega_0^2 - \omega^2 + i\gamma\omega}\right)$$

9.2 Using published data for KDP, calculate the phase matching condition for second-harmonic generation of a pump beam having a vacuum wavelength of 800 nm.

9.3 Refer to (9.42)–(9.44) (which explain how the non-linear optical tensor causes the electro-magnetic field of one ray to couple to fields of the two other rays). Examine the use of a nonlinear optical constant that varies periodically:

$$d_{ijk} \rightarrow \frac{d_0}{2}\left(e^{ikz} + e^{-ikz}\right)$$

(a) Show that there is a specific periodicity: $\left(= \frac{2\pi}{k_{qpm}}\right)$ that causes the exponent in (9.32)–(9.44) to separate into phase-matched and phase-mismatched terms.
(b) Show that averaging your result over a length of several periods will cause the phase-mismatched terms to average to zero, while the phase-matched terms accumulate.

This method of partial-phase matching, using an externally-imposed periodic variation in the non-linear coefficient, is called *quasi phase-matching*. It is often employed in conjunction with non-linear interactions in optical waveguides, where the interaction distance can be quite long compared to that of the periodic variation.

References

N. Bloembergen, *Non-linear Optics*, 4th edn. (World Scientific Publishing, Singapore, 1996). ISBN 981-02-2599-7

R.W. Boyd, *Non-linear Optics*, 3rd edn. (Academic Press, Burlington, 2008). ISBN 978-0-12-369470-6

Coherent, Properties of BBO (2016). https://www.coherent.com/downloads/BBO_DS.pdf

Gamdan Optics, Properties of KDP (2016). http://gamdan.com/KDP

T.F. Heinz, Second-order non-linear optical effects at surfaces and interfaces (Chap. 5), in *Non-linear Surface Electromagnetic Phenomena*, ed. by H.-E. Ponath, G.I. Stegeman (Elsevier Press, Amsterdam, 1991). http://heinz.phys.columbia.edu/publications/Pub50.pdf

A. Ling, *Entangled state preparation for optical quantum communication: creating and characterizing photon pairs from spontaneous parametric down conversion inside bulk uniaxial crystals* (Department of Physics, National University of Singapore, Thesis, 2008). https://www.quantumlah.org/media/thesis/thesis-alex.pdf

Y.R. Shen, *Principles of Non-linear Optics* (Wiley, Hoboken, 2003). ISBN 0-0471-43080-3

T. Skauli, K.L. Vodopyanov, T.J. Pinguet, A. Schober, O. Levi, L.A. Eyres, M.M. Fejer, J.S. Harris, B. Gerard, L. Becouarn, E. Lallier, G. Arisholm, Measurement of the nonlinear coefficient of orientation-patterned GaAs and demonstration of highly efficient second-harmonic generation. Opt. Lett. **27**, 628–630 (2002). http://nlo.stanford.edu/system/files/skauli_ol2002.pdf

K.K. Wong, *The Properties of Lithium Niobate, EMIS Datareviews Series No. 5* (The Institute of Electrical Engineers, London, 2002). ISBN 13: 978-0852967997

A. Yariv, *Quantum Electronics*, 2nd edn. (Wiley, New York, 1975). ISBN 0-471-97176-6

Chapter 10
Coherent States—From Single Photons to Beams of Light

Abstract A beam of light is composed of single photon states. The corresponding electric field can be expressed as a combination of coherent states that are designed to be the eigenstates of the electric field. We show that the coherent states can be expressed in terms of the number states, and develop a displacement operator that will create a coherent state from the vacuum. This development shows that the coherent states are a combination of number states distributed according to a Poisson distribution. An example of a beam of coherent light is a laser operating well above threshold. Photon emission from a laser is well described by Poisson statistics, and the second-order correlation coefficient is 1 for all times. That is, photons emitted from a laser are uncorrelated. Photons emitted from a thermal source obey Bose-Einstein statistics. Thus, they are not coherent. Correlations between photons emitted from a thermal source were first measured by Hanbury Brown and Twiss. Their experiment raised the curtain on the field of quantum optics. We show that the second-order correlation coefficient $g^2(\tau = 0)$ for photons obeying Bose-Einstein statistics is 2. The Hanbury Brown and Twiss experiment confirms this result. If one photon is emitted by a thermal source, another photon will be close at hand. Photons are bunched together. Single photons represent a third type of photon state. A single photon that encounters a beam-splitter will be either reflected or transmitted. Simultaneous detection of transmission and detection never occurs, and the photons are anti-correlated. A photon beam having this characteristic is called non-classical light.

10.1 Introduction

In the preceding chapters, we have demonstrated that the description of photon behavior using number states is both simple and powerful. Quantum mechanics shows the importance of the vacuum state, and enables the treatment of spontaneous emission and non linear optical effects such as spontaneous parametric down conversion that cannot be described using Maxwell's equations and classical electromagnetic fields.

On the other hand the purely quantum description using number states also has its limits. Important observable phenomena such as phase, and wave propagation

© Springer Nature Switzerland AG 2020

T. P. Pearsall, *Quantum Photonics*, Graduate Texts in Physics,
https://doi.org/10.1007/978-3-030-47325-9_10

in space and time are beyond the reach of the number state approach. Yet, photons are responsible for both observable quantum behavior and classical wave propagation. A logical approach would be to develop a set of new states $|\alpha\rangle$ based on various combinations of number states, since these states are complete. The electromagnetic fields can then be expressed in terms of these states. The problem can then be summarized as finding the correct combinations of number states that can be used to express the modes of the electric field. Such an approach was developed by Glauber (1963a), for which he was awarded the Nobel Prize in Physics in 2005.

We recall from Chap. 5 that the number states: $|n\rangle$ are the eigenstates of the harmonic oscillator Hamiltonian.

$$\mathcal{H} = \hbar\omega\left(a^+ a + \frac{1}{2}\right) \tag{10.1}$$

where a^+ and a are the creation and annihilation operators

$$a^+|n\rangle = \sqrt{n+1}|n+1\rangle$$

$$a|n\rangle = \sqrt{n}|n-1\rangle \text{ and } \langle n|a = \sqrt{n+1}|n+1\rangle \tag{10.3}$$

By successive application of the creation operator, we can express any of the eigenstates in terms of the vacuum state $|0\rangle$:

The number state $|n\rangle$ can be expressed in terms of the vacuum state:

$$|n\rangle = \frac{(a^+)^n}{\sqrt{n!}}|0\rangle, \text{ and its adjoint: } \langle n| = \langle 0|\frac{a^n}{\sqrt{n!}}, \tag{10.4}$$

The $|n\rangle$ states form a complete orthonormal set that describes the harmonic oscillator. It is convenient to think of the multiple operation of a^+ as a single displacement operator that translates the vacuum state to the state $|n\rangle$.

$$\mathcal{D}_{|n\rangle}(n) = \frac{(a^+)^n}{\sqrt{n!}}, \text{ and } \mathcal{D}_{|n\rangle}(n)|0\rangle = |n\rangle \tag{10.5}$$

10.2 Expansion of the Electric Field in Terms of Number States

We divide in the usual way the electric field into two complex conjugate terms:

$$\mathbf{E}(\mathbf{r}, t) = \mathbf{E}^{(+)} + \mathbf{E}^{(-)}, \tag{10.6}$$

$$\mathbf{E}^{(-)} = \left(\mathbf{E}^{(+)}\right)^*$$

where $\mathbf{E}^+(\mathbf{r}, t)$ is associated with the positive frequency terms: $e^{-i\omega t}$ and $\mathbf{E}^-(\mathbf{r}, t)$ with negative frequency terms. Next, we can separate the expression for the field into independent spatial and frequency components. We can express the spatial dependence as a family of plane wave modes:

$$\mathbf{u}_k(\mathbf{r}) = Ce^{i\mathbf{k}\bullet\mathbf{r}} \tag{10.7}$$

where C is a constant to be determined.

Summing over all modes, the electric field operator is written as:

$$\mathbf{E}(\mathbf{r}, t) = \sum_k \left(C_1 a_k \mathbf{u}_k(\mathbf{r})e^{-i\omega_k t} + C_2 a_k^+ \mathbf{u}_k^*(\mathbf{r})e^{i\omega_k t}\right) \tag{10.8}$$

where a_k and a_k^+ are the annihilation and creation operators for photons in the kth mode. Our objective is to express the eigenstates of the electric field operator as combinations of the number states.

We can limit the following developments to only the positive frequency terms. This choice leads to considerable simplification of the mathematical manipulations without sacrificing any of the physics. Keeping only the positive frequency terms:

$$\mathbf{E}(\mathbf{r}, t) = C_1 \sum_k a_k \mathbf{u}_k(\mathbf{r})e^{-i\omega_k t} \tag{10.9}$$

where the constant C_1 equals $i\left(\frac{\hbar\omega}{2}\right)^{\frac{1}{2}}$, assuring that the electromagnetic Hamiltonian has the desired units of energy. In this form the electric field is expressed as the result of the annihilation operator acting on all the modal components of the wave. We want to develop a set of states $|\alpha\rangle$ having the important property that the eigenstates of the electric field operator can be expressed in terms of these states $|\alpha\rangle$. Obviously, we want to define the $|\alpha\rangle$ states to be eigenstates of the annihilation operator. Thus,

$$a_k|\alpha_k\rangle = \alpha_k|\alpha_k\rangle \tag{10.10}$$

where α_k is a number and the eigenvalue belonging to the eigenstate of the annihilation operator. Note that the α_k are in general complex numbers, because the annihilation operator (also the creation operator) is not Hermitian.

First consider the states for a single mode k. We can expand the $|\alpha\rangle$ states in terms of the number states in the usual way,

$$|\alpha\rangle = \sum_n \langle n|\alpha\rangle |n\rangle$$

Using (10.4),

$$\langle n|\alpha\rangle = \langle 0|\frac{a^n}{\sqrt{n!}}|\alpha\rangle. \tag{10.11}$$

Substituting:

$$|\alpha\rangle = \langle 0|\alpha\rangle \sum_n \frac{\alpha^n}{\sqrt{n!}}|n\rangle. \tag{10.12}$$

Normalization gives:

$$|\langle\alpha|\alpha\rangle|^2 = 1 = |\langle 0|\alpha\rangle|^2 \sum_n \frac{|\alpha|^{2n}}{n!} \tag{10.13}$$

Note the identity:

$$\sum_n \frac{|\alpha|^{2n}}{n!} = e^{|\alpha|^2} \tag{10.14}$$

So that:

$$\langle 0|\alpha\rangle = e^{-\frac{1}{2}|\alpha|^2} \tag{10.15}$$

and

$$|\alpha\rangle = e^{-\frac{1}{2}|\alpha|^2} \sum_n \frac{\alpha^n}{\sqrt{n!}}|n\rangle, \tag{10.16}$$

which is the result we are seeking.

Next use (10.4) to substitute for $|n\rangle$:

$$|\alpha\rangle = \langle 0|\alpha\rangle \sum_n \frac{\alpha^n}{\sqrt{n!}} \frac{\left(a^+\right)^n |0\rangle}{\sqrt{n!}} = \langle 0|\alpha\rangle e^{\alpha a^+}|0\rangle = e^{-\frac{1}{2}|\alpha|^2} e^{\alpha a^+}|0\rangle \tag{10.17}$$

Following (10.5), we can now define an additional displacement operator for the states $|\alpha\rangle$:

$$\mathcal{D}(\alpha) = e^{-\frac{1}{2}|\alpha|^2} e^{\alpha a^+} \tag{10.18}$$

10.3 Coherent States of the Electric Field

10.3.1 Coherent States and the Poisson Distribution

The probability of finding a state $|n\rangle$ is:

$$|\langle n|\alpha\rangle|^2 = e^{-|\alpha|^2}\frac{|\alpha|^{2n}}{n!} \tag{10.19}$$

which we recognize as a Poisson distribution, with mean value = variance = $|\alpha|^2$

The expected number of photons in the state $|\alpha\rangle$ can be calculated directly from:

$$\langle n\rangle \equiv \bar{n} = \sum_n n e^{-|\alpha|^2}\frac{|\alpha|^{2n}}{n!} \tag{10.20}$$

But, it is much simpler to use creation and annihilation operators:

$$\langle n\rangle \equiv \bar{n} = \langle\alpha|a^+a|\alpha\rangle = \alpha^*\alpha\langle\alpha|\alpha\rangle = |\alpha|^2 \tag{10.21}$$

The variance can be calculated in the same way:

$$\langle n^2\rangle = \langle\alpha|\left(a^+a\right)^2|\alpha\rangle = \langle\alpha|a^+aa^+a|\alpha\rangle = \langle\alpha|a^+\left(a^+a+1\right)a|\alpha\rangle \tag{10.22}$$

$$=|\alpha|^4 + |\alpha|^2 = \langle n\rangle^2 + \langle n\rangle$$

$$\text{The variance } = \langle n^2\rangle - \langle n\rangle^2 = \langle n\rangle = |\alpha|^2 = \bar{n} \tag{10.23}$$

as expected for a Poisson distribution.

10.3.2 Properties of the Coherent States

The states $|\alpha\rangle$ are called the coherent states of the mode k of the electric field. They are based on a combination of number states, distributed according to a Poisson distribution with average value $|\alpha|$, where α is an eigenvalue of the annihilation operator operating on the state $|\alpha\rangle$.

The $|\alpha\rangle$ states are not orthonormal. If we consider 2 such states $|\alpha\rangle$ and $|\beta\rangle$,

$$\langle\alpha|\beta\rangle = e^{-\frac{1}{2}\left(|\alpha|^2+|\beta|^2\right)}\sum_n\sum_m \frac{(\alpha^*)^n\beta^m}{\sqrt{n!}\sqrt{m!}}\langle n|m\rangle \tag{10.24}$$

Using the orthonormal property of $|n\rangle$

$$\langle\alpha|\beta\rangle = e^{-\frac{1}{2}(|\alpha|^2+|\beta|^2)} \sum_n \frac{(\alpha^*)^n \beta^n}{n!} \langle n|n\rangle = e^{-\frac{1}{2}(|\alpha|^2-2\alpha^*\beta+|\beta|^2)} \quad (10.25)$$

Since $|\alpha|$ and $|\beta|$ are real positive numbers, the absolute value of (10.25) is not zero. Thus, the eigenstates $|\alpha|$ are therefore not mutually orthogonal.

$$|\langle\alpha|\beta\rangle| = e^{-\frac{1}{2}(|\alpha-\beta|^2)} \quad (10.26)$$

On the other hand, since we took the sum over all number states, they are complete (Glauber 1963a).

To consider completeness, we examine the expression: $\int |\alpha\rangle\langle\alpha| d^2\alpha$

$$\int |\alpha\rangle\langle\alpha| d^2\alpha = \int d^2\alpha e^{-\frac{1}{2}|\alpha|^2} e^{-\frac{1}{2}|\alpha|^2} \sum_{n,m} \frac{(\alpha)^m |m\rangle}{\sqrt{m!}} \sum_n \frac{\langle n|(\alpha^*)^n}{\sqrt{n!}} \quad (10.27)$$

Since the $|\alpha\rangle$ states are complex functions, we can write in polar coordinates (Ou 2017; Walls and Millburn 2008): $|\alpha\rangle = |\alpha|e^{i\varphi} = re^{i\varphi}$, and $d^2\alpha = rdrd\phi$. Substituting for α:

$$\int |\alpha\rangle\langle\alpha| d^2\alpha = \int_0^\infty rdr \int_0^{2\pi} d\varphi e^{-r^2} \sum_{n,m} \frac{|m\rangle\langle n|}{\sqrt{m!}\sqrt{n!}} r^m r^n e^{i(m-n)\varphi} \quad (10.28)$$

The integral over φ gives a delta function $2\pi\delta_{m-n}$, so:

$$\int |\alpha\rangle\langle\alpha| d^2\alpha = 2\pi \sum_n \frac{|n\rangle\langle n|}{n!} \left(\int_0^\infty re^{-r^2} r^{2n} dr \right) \quad (10.29)$$

Changing variables once more, let $\rho = r^2$, and $d\rho = 2rdr$

$$\int |\alpha\rangle\langle\alpha| d^2\alpha = \pi \sum_n \frac{|n\rangle\langle n|}{n!} \int_0^\infty e^{-\rho} \rho^n d\rho \quad (10.30)$$

Note that:

$$\Gamma(n+1) \equiv \int_0^\infty e^{-r} r^{n+1-1} dr \equiv n! \quad (10.31)$$

Thus,

$$\frac{1}{\pi} \int |\alpha\rangle\langle\alpha| d^2\alpha = \sum_n |n\rangle\langle n| = \mathbf{I}, \quad (10.32)$$

since the number states $|n\rangle$ form a complete, orthonormal set.

So far, we have shown that there is a set of states α each of which is composed of a sum of a group of number states $|n\rangle$ representing individual photons distributed according to a Poisson distribution. The $|\alpha\rangle$ states are not mutually orthogonal, and they are (over) complete. By design, any electric field operator can be described using the coherent states as a basis, although this description is not unique.

10.3.3 The Displacement Operator

By our intentional analogy to the analysis of number states, it is possible to create any coherent state from the vacuum through the action of a displacement operator, which we derived in (10.18):

$$|\alpha\rangle = e^{-\frac{1}{2}|\alpha|^2} \sum_n \frac{\alpha^n}{\sqrt{n!}} |n\rangle = e^{-\frac{1}{2}|\alpha|^2} \sum_n \frac{(\alpha a^+)^n}{n!} |0\rangle = e^{-\frac{1}{2}|\alpha|^2} e^{\alpha a^+} |0\rangle, \tag{10.33}$$

where: $\mathfrak{D}(\alpha) = e^{-\frac{1}{2}|\alpha|^2} e^{\alpha a^+}$.

The displacement operator (10.18) can be used to analyze the coherent states. In this section we develop this operator in a more general form to increase its range of useful applications. In the first step, we will include the complex conjugate in the exponential argument.

Note that $e^{-\alpha^* a}|0\rangle = 1$. Inserting this result,

$$\mathfrak{D}(\alpha) = e^{-\frac{1}{2}|\alpha|^2} e^{\alpha a^+} e^{-\alpha^* a}$$

and,

$$|\alpha\rangle = e^{-\frac{1}{2}|\alpha|^2} e^{\alpha a^+} e^{-\alpha^* a} |0\rangle \tag{10.34}$$

This manipulation puts the expression for $|\alpha\rangle$ in the form where we can apply the Baker-Campbell-Hausdorff theorem for operators: (for more background on the Baker-Campbell-Hausdorff Theorem, http://webhome.phy.duke.edu/~mehen/760/ProblemSets/BCH.pdf).

Let A and B be any two operators satisfying the relation: $[A, [A, B]] = 0$. Then,

$$e^{A+B} = e^{-\frac{1}{2}[A,B]} e^A \cdot e^B \tag{10.35}$$

Identifying: $A = \alpha a^+$ and $B = -\alpha^* a$, it follows that $[A, B] = |\alpha|^2$.

Thus,

$$|\alpha\rangle = e^{\alpha a^+ - \alpha^* a} |0\rangle \tag{10.36}$$

The displacement operator can now be expressed as

$$\mathcal{D}(\alpha) = e^{\alpha a^{+} - \alpha^{*} a} \tag{10.37}$$

which is the form given by Glauber in his original discussions.

$$\mathcal{D}(\alpha)|0\rangle = |\alpha\rangle \tag{10.38}$$

and its inverse

$$\mathcal{D}^{-1}(\alpha)|\alpha\rangle = |0\rangle \tag{10.39}$$

This displacement operator generates all the coherent states $|\alpha\rangle$ from the vacuum state.

There are other important properties of this displacement operator:

$$\mathcal{D}^{+}(\alpha) = \mathcal{D}^{-1}(\alpha) = \mathcal{D}(-\alpha) \tag{10.40}$$

$$\mathcal{D}^{+}(\alpha) a \mathcal{D}(\alpha) = a + \alpha \tag{10.41}$$

$$\mathcal{D}^{+}(\alpha) a^{+} \mathcal{D}(\alpha) = a^{+} + \alpha^{*} \tag{10.42}$$

10.3.4 Coherent States: Minimum Uncertainty States of the Electric Field

An important property of the $|\alpha\rangle$ states is that they represent the minimum uncertainty states for multiphoton states, that is: for beams of light. To prove that this is so, we start from (5.33) and (5.34) which are expressions for the momentum and coordinate operators for the simple harmonic oscillator:

$$\hat{p} = i\left(\frac{h\omega}{2}\right)^{\frac{1}{2}}\left(a^{+} - a\right) \tag{10.43}$$

and

$$\hat{q} = \left(\frac{\hbar}{2\omega}\right)^{\frac{1}{2}}\left(a^{+} + a\right) \tag{10.44}$$

where $\hat{p}$ and $\hat{q}$ obey the commutation relationship $[\hat{q}, \hat{p}] = i\hbar$

$$\text{The uncertainty in } \hat{q} \text{ is } \Delta q \equiv \sqrt{\langle q^2 \rangle - \langle q \rangle^2} \tag{10.45}$$

$$\begin{aligned}\langle q \rangle = \langle \alpha | \hat{q} | \alpha \rangle &= \left(\frac{\hbar}{2\omega}\right)^{\frac{1}{2}} \langle \alpha | (a^+ + a) | \alpha \rangle \\ &= \left(\frac{\hbar}{2\omega}\right)^{\frac{1}{2}} (\alpha^* + \alpha) \langle \alpha | \alpha \rangle = \left(\frac{\hbar}{2\omega}\right)^{\frac{1}{2}} (\alpha^* + \alpha)\end{aligned} \tag{10.46}$$

$$\langle q \rangle^2 = \left(\frac{h}{2\omega}\right) (\alpha^* + \alpha)^2 \tag{10.47}$$

The expectation value $\langle q^2 \rangle$ is calculated in the same way:

$$\begin{aligned}\langle q^2 \rangle = \langle \alpha | \hat{q}^2 | \alpha \rangle &= \left(\frac{\hbar}{2\omega}\right) \langle \alpha | (a^+ + a)^2 | \alpha \rangle \\ &= \left(\frac{\hbar}{2\omega}\right) \langle \alpha | (a^+ a^+ + a^+ a + a a^+ + aa) | \alpha \rangle\end{aligned} \tag{10.48}$$

$$\langle q^2 \rangle = \left(\frac{\hbar}{2\omega}\right) \langle \alpha | (a^+ a^+ + a^+ a + (a^+ a + 1) + aa) | \alpha \rangle \tag{10.49}$$

$$\langle q^2 \rangle = \left(\frac{\hbar}{2\omega}\right) (\alpha^* \alpha^* + 2\alpha^* \alpha + \alpha\alpha + 1) \langle \alpha | \alpha \rangle \tag{10.50}$$

$$\langle q^2 \rangle = \left(\frac{h}{2\omega}\right) \left((\alpha^* + \alpha)^2 + 1\right) \tag{10.51}$$

$$(\Delta q)^2 = \langle q^2 \rangle - \langle q \rangle^2 = \frac{\hbar}{2\omega} \tag{10.52}$$

In a similar fashion,

$$(\Delta p)^2 = \langle p^2 \rangle - \langle p \rangle^2 = \frac{\hbar\omega}{2} \tag{10.53}$$

The uncertainty relationship for the states $|\alpha\rangle$ is therefore:

$$\Delta q \Delta p = \frac{\hbar}{2} \tag{10.54}$$

Thus, the coherent states of the electric field are a complete set of minimum uncertainty states. This result is identical to that obtained in (5.10). This result is reassuring, but should not be surprising, since the coherent states are linear combinations of the number states, each having the same frequency, which are the minimum uncertainty eigenstates of the harmonic oscillator.

When we detect a coherent state using an efficient photodiode, the resulting signal is proportional to the intensity which is in turn determined by the squared amplitude of the electric field. The arrival rate of photons at the detector is given directly by Fermi's golden rule

$$\Gamma = \frac{2\pi}{\hbar}\langle\alpha|\mathbf{E}^{(-)}(\mathbf{r},t)\mathbf{E}^{(+)}(\mathbf{r},t)|\alpha\rangle\rho \tag{10.55}$$

The electric field operator depends on the creation and annihilation operators, and the result of (10.55) is that the count *rate* is proportional to the number of photons. Each photon arrives at the speed of light, in an orderly fashion, independent of all the others. This result implies that the count rate generated by a coherent state will also be characterized by a Poisson distribution. The average count rate of an efficient detector is simply the photon flux, which is independent of time for a coherent state. The variance in the flux is equal to the average count rate.

10.4 Photon Statistical Behavior: Correlation Versus Coherence

The best way to characterize the statistical behavior of an incoming photon beam (Fox 2006) is to measure its second-order correlation coefficient: $g^{(2)}(\tau)$, where τ is the time interval between two measurements of the incoming beam. (refer to Ou 2017, for an excellent treatment of this topic). We have introduced this correlation coefficient in the discussion of single-photon experiments in Chap. 2.

The second-order correlation coefficient is defined as:

$$g^{(2)}(\tau) = \frac{\langle\mathbf{E}^*(t)\mathbf{E}^*(t+\tau)\mathbf{E}(t+\tau)\mathbf{E}(t)\rangle}{\langle\mathbf{E}^*(t)\mathbf{E}(t)\rangle\langle\mathbf{E}^*(t+\tau)\mathbf{E}(t+\tau)\rangle} \tag{10.56}$$

In the usual case, we specify a steady-state flux of photons, so that we can set $t = 0$. We can rewrite the expression in terms of intensities which are measured by a photodetector;

$$g^{(2)}(\tau) = \frac{\langle I(\tau)I(0)\rangle}{\langle I(\tau)\rangle\langle I(0)\rangle} \tag{10.57}$$

As we have already seen in Chap. 2, when the interval time is very long (practically speaking, more than a second) the two intensities are independent of each other and

$$\langle I(\infty)I(0)\rangle = \langle I(\infty)\rangle\langle I(0)\rangle \tag{10.58}$$

Thus, $g^{(2)}(\infty) = 1$, independent of the underlying statistics.

When the interval time $\tau = 0$, we measure the autocorrelation function $g^{(2)}(0)$.

$$g^{(2)}(\tau) = \frac{\langle I(0)^2 \rangle}{\langle I(0) \rangle^2} \tag{10.59}$$

For example, the second order correlation functions for some principal photonic lineshape emission spectra are:

Lorentzian:

$$g^{(2)}(\tau) = 1 + e^{-\gamma\tau} \tag{10.60}$$

Gaussian:

$$g^{(2)}(\tau) = 1 + e^{-\gamma^2\tau^2} \tag{10.61}$$

where γ is the spectral linewidth.

Note that in both cases: $g^{(2)}(0) = 2$ and $g^{(2)}(\tau \to \infty) = 1$.

This particular measurement gives critical information about the statistical distribution of photons in the incoming beam. The Cauchy-Schwarz inequality can be applied to statistical distributions:

$$\langle x^2 \rangle \langle y^2 \rangle \geq \langle xy \rangle^2. \tag{10.62}$$

If we consider $x \equiv I(0)$ and $y \equiv 1$, then

$$\langle I(0)^2 \rangle \geq \langle I(0) \rangle^2, \text{ and } g^{(2)}(0) \geq 1 \tag{10.63}$$

For quantum states, we can express $g^{(2)}(0)$ in terms of creation and annihilation operators (Walls and Milburn 2008)

$$g^{(2)}(0) = \frac{\langle a^+ a^+ aa \rangle}{\langle a^+ a \rangle^2} \tag{10.64}$$

We recall that $\bar{n} = \langle a^+ a \rangle$ and $(\Delta n)^2 = V(n) = \left\langle (a^+ a)^2 \right\rangle - \langle a^+ a \rangle^2$ (10.65)

Rearranging terms, we have:

$$\begin{aligned} \langle a^+ a^+ aa \rangle &= \langle a^+ (aa^+ - 1) a \rangle = \langle a^+ (aa^+ a - a) \rangle \\ &= \langle a^+ aa^+ a \rangle - \langle a^+ a \rangle = \left\langle (a^+ a)^2 \right\rangle - \langle a^+ a \rangle \end{aligned} \tag{10.66}$$

By adding and subtracting the term $\langle a^+ a \rangle^2$, we deduce:

$$\langle a^+a^+aa\rangle = \langle a^+a\rangle^2 + V(n) - \langle a^+a\rangle \tag{10.67}$$

giving:

$$g^{(2)}(0) = 1 + \frac{V(n) - \langle a^+a\rangle}{\langle a^+a\rangle^2} = 1 + \frac{V(n) - \bar{n}}{(\bar{n})^2} \tag{10.68}$$

Equation (10.68) can be used to evaluate the theoretical correlation coefficient for photon states where the mean and the variance are known.

For example, consider the coherent states. The photons are distributed in a Poisson distribution with $V(n) = \bar{n}$.

$$g^{(2)}(0)_{\text{Poisson}} = 1 + \frac{\bar{n} - \bar{n}}{(\bar{n})^2} = 1 \tag{10.69}$$

This can be compared to the non-coherent thermal state which is described by the Bose-Einstein distribution. (See Loudon 2003 for an excellent treatment of this case.)

In Chap. 2, we have shown (2.6) that the average number of photons having energy $= \hbar\omega$ is:

$$\langle n(\hbar\omega)\rangle = \frac{1}{e^{\frac{\hbar\omega-\mu}{kT}} - 1} \tag{10.70}$$

The variance $V(n)$ is calculated as follows:

$$\langle(\Delta n)^2\rangle \equiv V(n) = kT\frac{\partial}{\partial\mu}\langle n(\hbar\omega)\rangle \tag{10.71}$$

$$V(n) = (kT)\left(\frac{1}{kT}\right) = \frac{\left(e^{\frac{\hbar\omega-\mu}{kT}} - 1\right)}{\left(e^{\frac{\hbar\omega-\mu}{kT}} - 1\right)\left(e^{\frac{\hbar\omega-\mu}{kT}} - 1\right)} = \frac{\langle n\rangle\left(e^{\frac{\hbar\omega-\mu}{kT}}\right)}{\left(e^{\frac{\hbar\omega-\mu}{kT}} - 1\right)} \tag{10.72}$$

Note that:

$$1 + \langle n\rangle = \frac{\left(e^{\frac{\hbar\omega-\mu}{kT}}\right)}{\left(e^{\frac{\hbar\omega-\mu}{kT}} - 1\right)} \tag{10.73}$$

Thus:

$$V(n) = \langle n(\hbar\omega)\rangle(1 + \langle n(\hbar\omega)\rangle) = \bar{n} + (\bar{n})^2 \tag{10.74}$$

The second-order autocorrelation coefficient is therefore:

$$g^{(2)}(0)_{\text{Bose–Einstein}} = 1 + \frac{V(n) - \bar{n}}{(\bar{n})^2} = 1 + \frac{(\bar{n})^2 + \bar{n} - \bar{n}}{(\bar{n})^2} = 2 \quad (10.75)$$

By comparing (10.69) and (10.75), we can see that the photons that make up an incoherent thermal field are clearly more correlated with each other than the photons that make up a coherent field, such as laser light. Glauber has addressed this apparent paradox in a landmark paper on the difference between correlation and coherence (Glauber 1963b). Coherent light is characterized by an equal probability in each time interval for detection of a photon. The presence of a photon in one interval has no influence on the probability of detection of another photon. Coherent light is emitted by a laser operating far above threshold so that spontaneous emission is a relatively negligible component of the photon field.

The correlation between photons in thermal light means that there is a significant probability of detecting multiple photons in each time interval. We refer to this feature as photon bunching. In comparison, coherent laser light is unbunched.

In Chap. 11, we shall use the same analysis to derive the second-order correlation coefficient for fermions: $g^{(2)}(0)_{\text{fermion}} = 0$.

10.5 Hanbury Brown and Twiss: Experimental Measurement of Photon Correlation

Nearly a decade before the work of Glauber established the properties of coherent light, two scientists working at the Jodrell Bank radio observatory in England developed a new kind of optical interferometer, simpler to build and to operate than the standard Michelson apparatus (Hanbury Brown and Twiss 1954). The Michelson interferometer can be used to measure the wavefront interference pattern of a distant object like a star with a finite diameter. When signals captured from two different angles are compared, the interferometer generates a fringe pattern from which the angular divergence of the source can be determined. The apparatus operates on the similar principle as that of Young's two-slit interferometer. The fringe pattern results from the interference of the *amplitudes* of the electromagnetic waves propagating along two different paths. The fringe pattern gives the first-order correlation function $g^{(1)}(0)$. If the distance to the source is known, its physical diameter can be calculated from the data. This technique at its current best leads to an angular resolution of about 4×10^{-3} arc-sec. (see Bachor and Ralph 2004 for an account). Although the signal may travel many light years to reach the Earth in pristine condition, the turbulence of the Earth's atmosphere in the last few miles corrupts the phase front of the incoming waves, and reduces the visibility of the instrument. In a similar way, any mechanical vibrations also impair resolution.

The breakthrough of Hanbury Brown and Twiss was to abandon the idea of trying to measure the phase difference between the arriving waves being viewed at two different carefully controlled angles, and just measure the correlations between the fluctuations in the signal intensities viewed by two identical detectors as a function of the difference in distance that separates them. Being radio astronomers, their experiment was first conceived using radio signals. To quote the authors:

> A new type of interferometer for measuring the diameter of discrete radio sources is described and its mathematical theory is given. The principle of the instrument is based upon the correlation between the rectified outputs of two independent receivers at each end of a baseline, and it is shown that the cross-correlation coefficient between these outputs is proportional to the square of the amplitude of the Fourier transform of the intensity distribution across the source. The analysis shows that it should be possible to operate the new instrument with extremely long baselines and that it should be almost unaffected by ionospheric irregularities.

10.5.1 Fluctuations: The Noise Is the Signal

The key contribution of the Hanbury Brown and Twiss experiment is the recognition of the singular importance of signal fluctuations and their statistical analysis as good way to measure and understand the underlying quantum-mechanical behavior of not only of photons, but as we discuss in the following chapters, of electrons and atoms: for example, distinguishing bosons from fermions. Fluctuations are sometimes referred to as noise. Hanbury Brown and Twiss realized that *the Noise is the Signal.*

As is often the case when a new simpler approach displaces accepted practice, their discovery was met with scepticism by the competition. Hanbury Brown and Twiss reasoned that their instrument could also be used to characterize visible light. But, before trying to measure the diameter of visible stars, they conceived and reported on experiment that characterized a commonly available chaotic light source: a mercury discharge lamp. Thus, their experiment could be easily repeated by a wide audience. The main objective of this experiment was not to measure the diameter of the emission area of the source, but rather to demonstrate that different photons emitted by this source were in fact correlated (Hanbury Brown and Twiss 1956). This correlation is the underlying physical principle that allows the characterization of stellar radiation, either radio or visible. By comparing the correlations in the statistical variation of *intensities* detected by 2 separated detectors, Hanbury Brown and Twiss were measuring the second-order correlation function $g^{(2)}(\tau)$. Because a measurement of $g^{(2)}(0)$ can distinguish between Bose-Einstein thermal radiation, Poisson radiation (laser light) and non-classical quantum emission, the Hanbury Brown Twiss experiment of 1956 is regarded as the starting point for the field of quantum optics. In his Nobel Prize lecture, Roy Glauber gives a stirring account of the influence of this important work both on the field and on his own research (Glauber 2005).

10.5.2 Hanbury Brown and Twiss: Experimental Measurement

The set-up shown in Fig. 10.1 enables two kinds of measurements. For example, by introducing a time-delay in one of the arms of the interferometer, correlations in emission time difference: $g^{(2)}(\tau)$, can be recorded. By displacing one of the photomultipliers so that it does not see the same emission pattern as the other, spatially-related correlation can be measured: $g^{(2)}(0)$. Hanbury Brown and Twiss reported on this latter measurement. In one case, they superimposed the beams emerging from the beam splitter, so that each detector was illuminated with the same emission pattern (although different photons!) coming from the light source. In a second measurement, they displaced one of the detectors so that a different pattern of photons would fall on each detector. In the first case, they could measure a correlation in the signal fluctuations ($g^{(2)}(0) = 2$). In the second case, no correlation was measured, ($g^{(2)}(0) = 1$). The experimental procedure is diagrammed in Fig. 10.2. To quote the authors:

> The results (…) confirm that correlation is observed when the cathodes are superimposed, but not when they are widely separated.

This result established the existence of correlations in between two photon beams coming from a thermal source that are sampled at the same point of time and space. Of course, each beam is composed entirely of different photons. The correlations are the result of interference between individual photons. If the samples are taken at different times or at different point in space, the interference

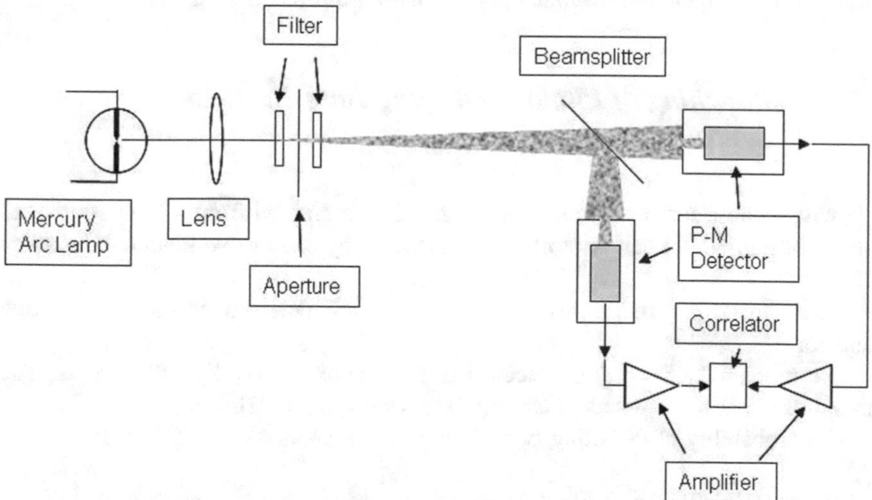

Fig. 10.1 Schematic diagram of the Hanbury Brown and Twiss experiment (Hanbury Brown and Twiss 1956). Chaotic, thermal light from a mercury discharge lamp is divided by a beam-splitter and impinges on two identical photomultiplier detectors. The signal fluctuations seen by each detector are compared in a correlator

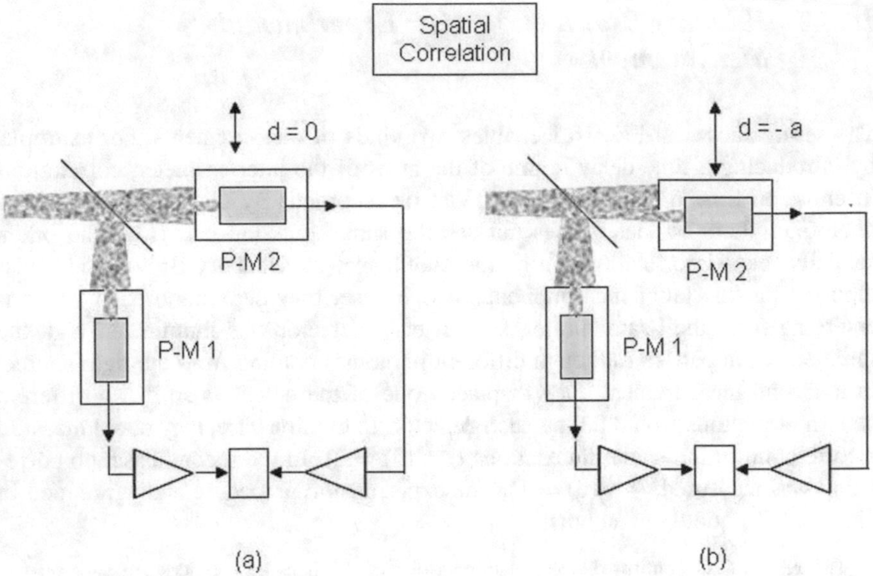

Fig. 10.2 **a** The detectors are aligned so that each is illuminated by the "same pattern" of photons. This pattern is formed by a specific physical image of the light source. **b** The detector P-M 2 is displaced so that the physical image of the light source incident on this detector is different from that incident on the detector P-M 1

disappears, and the photons are transmitted independent of each other. There is a straightforward quantum-mechanical explanation of this effect, Fano (1961).

10.5.3 Bunching of Photons Obeying Bose-Einstein Statistics

We will suppose that two photons, $|a\rangle$ and $|b\rangle$ are emitted from a light source, as shown in Fig. 10.3. Each photon can be detected by one of two identical detectors $|1\rangle$ and $|2\rangle$.

There are two possibilities for $|a\rangle$. It can be detected at one of the two detectors.

The probability that $|a\rangle$ is detected can be expressed $\langle 1|a\rangle = \langle 2|a\rangle = \alpha$. The probability that $|b\rangle$ is detected can be expressed $\langle 1|b\rangle = \langle 2|b\rangle = \beta$.

The probability of detecting two photons simultaneously is expressed:

$$P_{2\phi} = |\langle 1|a\rangle\langle 2|b\rangle + \langle 2|a\rangle\langle 1|b\rangle|^2 = \alpha^2\beta^2 + \alpha^2\beta^2 + \alpha^2\beta^2 + \alpha^2\beta^2 = 4\alpha^2\beta^2, \tag{10.76}$$

where it is seen that constructive interference occurs if the photons are in the same space at the same time. This is an example of photon bunching, i.e., $g^{(2)}(0) = 2$.

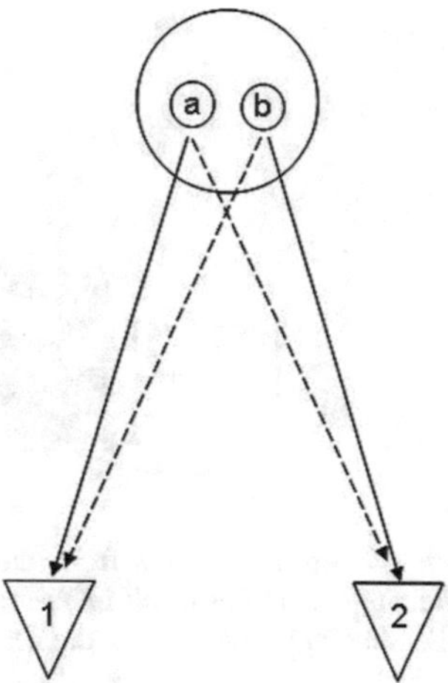

Fig. 10.3 Correlations observed in a Hanbury Brown Twiss experiment occur because each photon can be detected in one of two detectors. This uncertainty creates interference between the two possible paths for each photon. When two photons are emitted simultaneously, this leads to four terms in the expression for the detection probability. When the photons are not emitted simultaneously, there is no interference, and only two terms contribute to the probability

On the other hand, if the photons are emitted sequentially so that they do not appear in the same space at the same time, then the probabilities for detection are added independently:

$$P_{2\phi} = |\langle 1|a\rangle\langle 2|b\rangle|^2 + |\langle 2|a\rangle\langle 1|b\rangle|^2 = \alpha^2\beta^2 + \alpha^2\beta^2 = 2\alpha^2\beta^2 \qquad (10.77)$$

In this case, there are no interference terms. This situation could apply to a Poisson light source, such as a laser, i.e. $g^{(2)}(\tau) = 1$, for all values of τ.

A straightforward measurement of a photon stream will always show fluctuations of signal intensity as a function of time, for all sources. The peaks and valleys occur randomly as a function of time. We illustrate this effect in Fig. 10.4 with a sketch of signal intensity as a function of measurement time. In a real measurement, there will be a minimum detectable signal due to the noise of the apparatus. In Fig. 10.4b we show that the result of this noise floor is to put the high-intensity noise excursions in relief. If the photons are emitted by a thermal source, characterized by Bose-Einstein statistics, the high intensity excursions occur randomly as a function of time. These excursions are the signal produced by a group of photons. In between two of these peaks, there will be of course a region where fewer than average photons are emitted. This effect: grouping of photons separated be intervals of relatively sparse photons, would be an illustration of photon bunching.

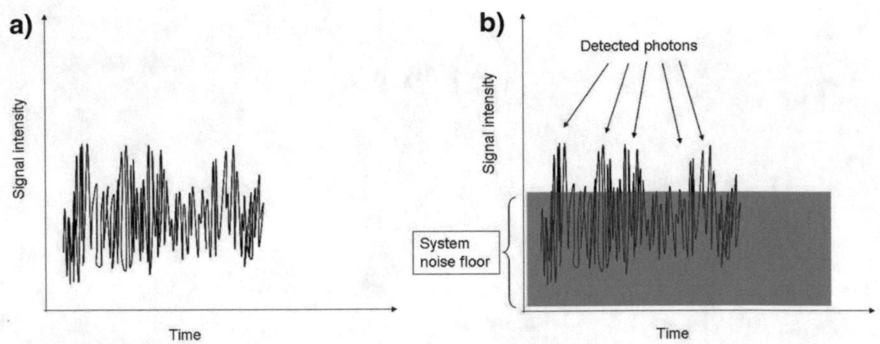

Fig. 10.4 A schematic diagram of signal intensity from an attenuated thermal source of photons. Part of the measured fluctuations is the result of the noise generated by the measurement apparatus (indicated by the shading), and part is created by fluctuations in the photon source

10.5.4 Absence of Correlations in Coherent Emission

The visible laser was invented in 1960. In 1966, following the theoretical work of Glauber on coherent states (Glauber 1963b), Arecchi, Gatti and Sona performed the Hanbury Brown Twiss experiment in the time domain in order to measure and compare photon correlations in the beam of a He–Ne laser (Poisson statistics) and the correlations using the same He–Ne laser source, but randomized to reduce coherence by diffusive scattering of a ground-glass filter, (Gaussian statistics) (Arecchi et al. 1966). Their results, in Fig. 10.5, show directly that the measured correlation coefficient of unperturbed laser light: $g^{(2)}(\tau) = 1$ for all times. This

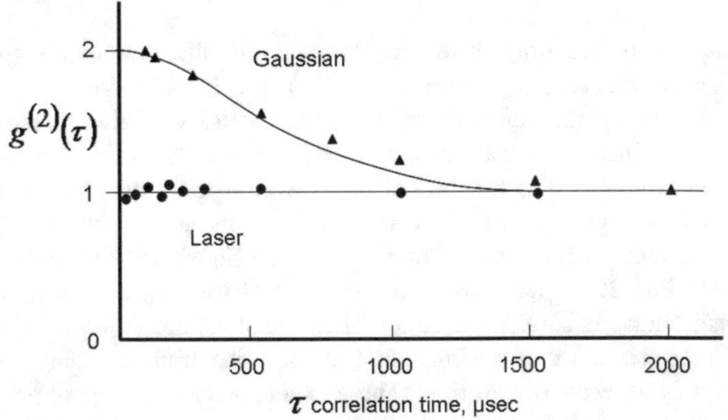

Fig. 10.5 Experimental measurements of the 2nd-order time-dependent correlation function for laser light and laser light randomized to create a Gaussian distribution of arrival times. (after Arecchi et al. 1966). The 2nd-order correlation coefficient for a Gaussian distribution is given in (10.61). The results show the absence of correlation between photons in highly coherent states corresponding to laser light. Reproduced by permission of Elsevier Publishing

shows the absence of correlation between arriving photons in highly coherent laser light and corresponds to the behavior required for Poissonian arrival statistics. On the other hand, the correlation coefficient for randomly scattered light changes as a function of increasing correlation time from $g^{(2)}(0) = 2$ to $g^{(2)}(\tau) = 1$, as expected from (10.58). This is a very elegant experiment, because it shows that correlation between photons is not a function of how the light is created. It depends only on whether the photons are in coherent states or in random states.

10.5.5 Anti-bunching in Single Photon Emission

Finally we consider the case of single photon detection. We already considered such an experiment in Chap. 2 (Fig. 10.6).

In this experiment, two identical photons are created by spontaneous parametric down conversion (SPDC). One photon serves as a herald and is detected at G, signalling the start of a detection event interval. The second photon is incident on a beam splitter and is detected either at T or at R. Detection of a photon at both T and R is not possible because a photon is an indivisible quantum, and it can be detected only once. Comparison with Fig. 10.1 shows that this experiment is a Hanbury Brown-Twiss measurement on single photons. A detection interval is defined by the time between the emission of photon pairs. This is measured by the time interval between the detection of two herald photons at detector G. The other photon of the pair passes through a beam splitter and a correlation measurement is completed by detectors T and R.

In each detection interval, the average number of photons is one: $\bar{n} = 1$. The variance is null: $V(n) = 0$. Using (10.68), the second-order correlation coefficient is:

$$g^{(2)}(0) = 1 + \frac{V(n) - \bar{n}}{(\bar{n})^2} = 0 \tag{10.78}$$

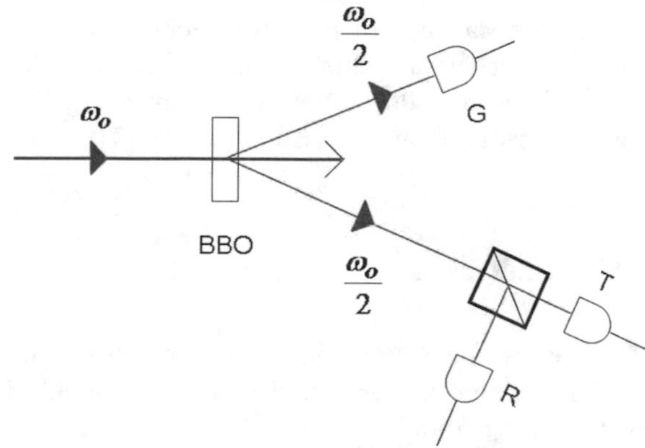

Fig. 10.6 Single photon detection experiment using emission of photon pairs from BBO. (cf. Chap. 2) The measurement interval is defined by the detection of a photon herald at G. The test photon passing through the beamsplitter is detected either at T or at R, but never at both detectors during the same measurement interval

in violation of the Cauchy-Schwarz inequality. That is, prepared single photons are anti-correlated. Grangier, and co-workers have demonstrated this non-classical state of light, (Grangier et al. 1986), in the measurement which is shown in Chap. 2, Fig. 2.16. Their result demonstrates that a photon is either entirely transmitted or reflected at the beam splitter.

This experiment has been refined and by Thorn, and coworkers, using SPDC to generate quantum photon pairs, and improvements in detection electronics (Thorn et al. 2004). Their raw experimental results give $g^{(2)}(0) < 0.02$, with additional improvement when accidental coincidences are accounted for, in violation of the classical limit of $g^{(2)}(0) = 1$ by more than 350 standard deviations. Experimental data for analysis from a similar experiment are given in Chap. 2, Exercise 2.8. The photons in these experiments are a prime example of non-classical light. These photons are anti-correlated. We refer to this as antibunching. This is a direct consequence of the quantum nature of the photon.

The above experiment is designed to assure that only one photon passes through the beam-splitter during a given time interval. This is implemented by application of SPDC, which is an infrequent event. The time interval between events, typically 1 μsec, is much longer than the detection interval, typically 1 nsec. It is this experimental situation that puts us in the quantum regime.

10.6 Summary

The coherent states are a set of functions, formed from combinations of the photon number states, with weighting given by a Poisson statistical distribution. By design, the coherent states are a complete set of eigenvectors of the electric field operator. $\mathbf{E}_k^{(+)}(\mathbf{r}, t)|\alpha\rangle = \boldsymbol{\varepsilon}_k(\mathbf{r}, t)|\alpha\rangle$, where the $\boldsymbol{\varepsilon}_k(\mathbf{r}, t)$ are the eigenvalues of the electric field operator. We have shown that the $|\alpha\rangle$ are the minimum uncertainty states of the electric field. Single-mode laser emission far above threshold is composed of a coherent state and is characterized by Poisson photon statistics.

The photons that constitute the electric field can be characterized by the correlation coefficient. The second-order auto-correlation coefficient:$g^{(2)}(0)$, is used to identify the statistics of photon detection. A value of $g^{(2)}(0) > 1$ indicates a bunched, incoherent photon stream; $g^{(2)}(0) = 1$, indicates unbunched Poissonian photons characteristic of laser emission; and $g^{(2)}(0) < 1$ signals anti-bunched, non-classical light characteristic of single photon quanta.

10.7 Exercises

10.1 Derive the expression for any number state $|n\rangle$ in terms of raising and lowering operators and the vacuum state $|0\rangle$ (10.4).

10.2 Starting from:

$$\langle \alpha | \beta \rangle = e^{-\frac{1}{2}\left(|\alpha|^2 - 2\alpha^* \beta + |\beta|^2\right)},$$

Show that:
$|\langle \alpha | \beta \rangle| = e^{-\frac{1}{2}\left(|\alpha - \beta|^2\right)}$ (10.24).

10.3 Using $A = \alpha a^+$ and $B = -\alpha^* a$, show that:
$[A, B] = |\alpha|^2$ and that $[A, [A, B]] = 0$.

10.4 Show that:

$$\mathcal{D}^+(\alpha) a |\alpha\rangle = \alpha |0\rangle$$

Use this result to demonstrate that the coherent states are eigenstates of the annihilation operator with eigenvalues $= \alpha$

$$a|\alpha\rangle = \alpha|\alpha\rangle$$

.

10.5 In Fig. 10.2 we show a diagram for a Hanbury Brown-Twiss experiment that measures spatial correlations between two streams of photons. How should this experiment be set up to measure temporal correlations in a single stream of photons?

References

F.T. Arecchi, E. Gatti, A. Sona, Time difference of photons from coherent and Gaussian sources. Phys. Lett. **20**, 27–29 (1966). http://fox.ino.it/home/arecchi/SezA/fis26.pdf

H.A. Bachor, T.C. Ralph, *A Guide to Experiments in Quantum Optics*, 2nd ed. (Wiley-VCH Verlag, Weinheim, 2004). ISBN 3-527-40393-0. https://onlinelibrary.wiley.com/doi/book/10.1002/9783527619238

Baker-Campbell-Hausdorff Theorem. http://webhome.phy.duke.edu/~mehen/760/ProblemSets/BCH.pdf

U. Fano, Quantum theory of interference effects in the mixing of light from phase-independent sources. Am. J. Phys. **29,** 539–545 (1961). https://aapt.scitation.org/doi/10.1119/1.1937827

M. Fox, *Quantum Optics, an Introduction*, (Oxford University Press, New York, 2006). ISBN 978-0-198-56673-1. https://global.oup.com/academic/product/quantum-optics-9780198566731?cc=fr&lang=en&

R.J. Glauber, Coherent and incoherent states of the radiation field. Phys. Rev. **131**, 2766–2788 (1963a). http://conf.kias.re.kr/~brane/wc2006/lec_note/Glauber-2.pdf

R.J. Glauber, Photon correlations. Phys. Rev. Lett. **10**, 84–86 (1963b). https://journals.aps.org/prl/pdf/10.1103/PhysRevLett.10.84

R.J. Glauber, One Hundred Years of Light Quanta, Nobel Prize Lecture, December 8, 2005. https://www.nobelprize.org/uploads/2018/06/glauber-lecture.pdf

P. Grangier, G. Roger, A. Aspect, Experimental evidence for a photon anticorrelation effect on a beam splitter: a new light in single photon interferences. Europhys. Lett **1**, 173–179 (1986). https://courses.physics.illinois.edu/phys513/sp2016/reading/week1/GrangierSinglePhoton1986.pdf

R. Hanbury Brown, R.Q. Twiss, A new kind of interferometer for use in radio astronomy. Phil. Mag. **45**, 663–682 (1954). https://www.tandfonline.com/doi/pdf/10.1080/14786440708520475?needAccess=true

R. Hanbury-Brown, R.Q. Twiss, Correlation between photons in two coherent beams of light. Nature **177**, 27–29 (1956). http://einstein.drexel.edu/~bob/PHYS518/brown-twiss.pdf

R. Loudon, *The Quantum Theory of Light,* 3rd ed. (Oxford University Press, New York, 2003). ISBN 0-19-850177-3. https://global.oup.com/academic/product/the-quantum-theory-of-light-9780198501770?q=Loudon%20Quantum&lang=en&cc=fr

Z.J. Ou, *Quantum Optics for Experimentalists*, (World Scientific Press, Singapore, 2017). ISBN 978–9-813-22020-1. https://www.worldscientific.com/doi/10.1142/9789813220218_0001

J.J. Thorn, M.S. Neel, V.W. Donato, G.S. Bergreen, R.E. Davies, M. Beck, Observing the quantum behavior of light in an undergraduate laboratory. Am. J. Phys. **72**, 1210–1219 (2004). https://digitalcommons.usu.edu/cgi/viewcontent.cgi?article=1797&context=psc_facpub

D.F. Walls, G.J. Milburn, *Quantum Optics,* 2nd ed. (Springer, Heidelberg, 2008). ISBN 978-3-540-28573-1. https://www.springer.com/la/book/9783540285731

Chapter 11
Quantum Fermions

Abstract Electrons and photons have common attributes and also display important differences. Populations of photons can show various degrees of correlation depending on the coherence of the source. Electrons, on the other hand show only anti-correlation, dictated by Fermi-Dirac statistics. Although electrons and photons both share wavelike behavior, they also behave like particles, and in contrast, the particle properties of electrons and photons, (for example, Fermi-Dirac statistics versus Bose-Einstein statistics) are fundamentally different. A fruitful study of single quantum electron behavior can be realized in a sample that supports 1-dimensional ballistic electron transport. While propagation of ballistic photons is the rule, creating an environment for ballistic transport of electrons requires special conditions. These conditions can be created in certain semiconductor material systems like heterostructures of AlGaAs-GaAs. A key structure for creating a one-dimensional channel is the quantum point contact. We show that the conductance of such a channel is finite and quantized in the ballistic transport regime. The quantum point contact is a fundamental building block for single electron circuits, enabling, for example, a 50% "beam-splitter" for single electrons.

11.1 Introduction

Electrons and photons have some common attributes. Each is an elementary particle. Electrons and photons display both wave and particle behavior, leading to the observation of diffraction on one hand and the transfer of momentum to other particles on the other. Photons propagate in a ballistic fashion at constant frequency until destroyed by absorption. Electrons can propagate in a ballistic manner if they are made to avoid inelastic collisions along the trajectory of interest between the source and drain.

On the other hand, there are many important and fundamental differences between electrons and photons. Electrons have mass and charge. Both mass and charge are conserved. A photon is characterized by its energy $E = \hbar\omega$ and its polarization. Photons can be created and destroyed at will, as long as energy is conserved. An electron can have a wide and continuous range of energies, and this energy can vary during the trajectory of the electron. Furthermore, observers

© Springer Nature Switzerland AG 2020

T. P. Pearsall, *Quantum Photonics*, Graduate Texts in Physics,
https://doi.org/10.1007/978-3-030-47325-9_11

in different reference frames will assign different energies, and thus different de Broglie wavelengths, to the same electron.

Neither the electron nor the photon has a well-defined size. A photon can be spatially confined to a dimension comparable to its wavelength, for example in an optical fiber. However, this confinement does not change its energy/frequency. Confinement of an electron to a space comparable to its wavelength changes its energy and introduces quantized allowed states.

The transition of an electron from one energy state to another is accompanied by the creation of a photon, conserving energy. This can occur via spontaneous or by stimulated emission. Einstein showed that this leads directly to the Bose-Einstein statistical distribution for photons, see (2.6). Because mass is conserved, there is not a comparable spontaneous emission phenomenon for electrons, although one could always inject an electron into a circuit at the source, and remove one at the drain. The fermionic statistical behavior of electrons is fundamentally different from that of photons. While correlation between photons is a reflection of Bose-Einstein statistics, Fermi-Dirac statistics require that fermions be anti-correlated because no two fermions can occupy the same state.

To further distinguish these two elementary particles, electrons interact with each other through their coulomb charge and magnetic moments. To further complicate matters, electrons can interact with their environment, gaining or losing both energy and momentum in inelastic collisions.

Despite these differences, there have been both proposals and experiments (Bocquillon et al. 2014) to study the behavior of electron quanta in solid-state circuits. The first measurements of quantum interference of ballistic electrons in vacuum were made by observation of electron diffraction (Davisson and Germer 1928). Pioneering demonstrations of ballistic quantum electron interference in solid state circuits were made at IBM labs more than 50 years later (Webb et al. 1985; Yacoby et al. 1991). A principal motivation of this work was to demonstrate for electrons an analogy to the behavior of photons in a network that creates delocalization of the quanta.

Experiments using single photons show remarkable manifestations of quantum behavior including interference and entanglement. To compare experiments using single electrons to single photon experiments, we need three basic building blocks: a ballistic transport channel, a single electron source, and a quantum beamsplitter (Parmentier 2010; Bocquillon et al. 2014). In this chapter we will analyze some basic properties of fermion behavior as it relates to electron quanta, including the basic building blocks of single-electron quantum circuits.

11.2 Operator Algebra

First, we consider the algebra for creation and annihilation operators. A comprehensive presentation has been given Landau and Lifshitz, in Chap. 9 in their classic text, Landau and Lifshitz (1958), by Schrieffer (1983), and by Baym who gives particular attention to two-body interactions, Baym (1990).

11.2.1 Boson Operator Algebra

$$\left[a, a^{+}\right] = 1, \text{ i.e.:} \tag{11.1}$$

$$\begin{aligned}\left(aa^{+} - a^{+}a\right)|n\rangle &= \left(a\sqrt{n+1}|n+1\rangle - a^{+}\sqrt{n}|n-1\rangle\right) \\ &= \left(\sqrt{n+1}\sqrt{n+1}|n\rangle - \sqrt{n}\sqrt{n}[n\rangle\right) = 1|n\rangle\end{aligned} \tag{11.2}$$

$$|n\rangle = \frac{\left(a^{+}\right)^{n}}{\sqrt{n!}}|0\rangle \tag{11.3}$$

The number of quanta in state $|n\rangle$ is given by $\left(a^{+}a\right)$

$$a^{+}a|n\rangle = n|n\rangle \tag{11.4}$$

a^{+} is the hermitian conjugate of a

$$\begin{aligned}a^{+}|n\rangle &= \sqrt{n+1}|n+1\rangle \\ \langle n+1|a^{+}|n\rangle &= \sqrt{n+1}\end{aligned} \tag{11.5}$$

So,

$$\langle n+1|a^{+} = \sqrt{n+1}\langle n|, \tag{11.6}$$

or, $\langle n|a^{+} = \sqrt{n}\langle n-1|$. That is a^{+} acts on the left like an annihilation operator.

Similarly,

$$\langle n|a = \sqrt{n+1}\langle n+1|. \tag{11.7}$$

11.2.2 Fermion Operator Algebra

A fermion state is either unoccupied, or occupied by a single quantum. Thus there are 2 states: $|0\rangle$ and $|1\rangle$.

Let us imagine 2 operators for fermions: f, an annihilation operator and f^{+}, a creation operator. In general, fermions have mass, and the fermion number is conserved, so it is forbidden to create a fermion from the vacuum. On the other hand, we could imagine a scenario where a fermion is emitted from a sea of fermions, and subsequently injected in a circuit:

$$f|1\rangle = \sqrt{n}|n-1\rangle = 1|0\rangle. \tag{11.8}$$

That is, a fermion makes a transition from an occupied state, leaving it unoccupied. An example could be an electron that makes a transition to a lower energy state, leaving the initial state empty and emitting a photon in the process.

$$f|0\rangle = 0 \tag{11.9}$$

and,

$$f^{+}|0\rangle = \sqrt{n+1}|n+1\rangle = 1|1\rangle \tag{11.10}$$

This is the reverse process where a photon is absorbed, and an electron makes a transition to an unoccupied state.

$$f^{+}|1\rangle = 0. \tag{11.11}$$

That is, you cannot have more than one particle in a state.

11.2.3 Commutators and Anti-commutators

Next, we consider commutation relationships.

For bosons:

$$\left[a, a^{+}\right] = 1 \tag{11.12}$$

Fermions, on the other hand, obey the anti-commutator relation:

$$\left\{f, f^{+}\right\} \equiv \left(ff^{+} + f^{+}f\right) = 1 \tag{11.13}$$

$$\{f, f\} = 0 \tag{11.14}$$

$$\left\{f^{+}, f^{+}\right\} = 0 \tag{11.15}$$

The anti-commutator has special importance because it contains the fermionic nature of the states.

The occupation number of $|n\rangle$ a state is given by:

$$f^{+}f|n\rangle. \tag{11.16}$$

Next, we consider a system composed of 2 fermions. There are 4 possible states:

$$|0, 0\rangle$$
$$|0, 1\rangle$$
$$|1, 0\rangle$$
$$|1, 1\rangle$$

We will label the creation and annihilation operators to specify which of the 2 states they act on. f_a acts on the state in the left-hand position and f_b acts in the right hand position.

These operators obey anti-commutator relationships:

$$\{f_a, f_a\} = \{f_a^+, f_a^+\} = 0 \tag{11.17}$$

$$\{f_a, f_b\} = \{f_a^+, f_b^+\} = 0 \tag{11.18}$$

$$\{f_a, f_b^+\} = \{f_a^+, f_b\} = 0 \tag{11.19}$$

Apply the creation operator:

$$f_a^+|0, 0\rangle = 1|1, 0\rangle \tag{11.20}$$

$$f_a^+|1, 0\rangle = 0 \tag{11.21}$$

$$f_a^+|1, 1\rangle = 0 \tag{11.22}$$

However, $f_a^+|0, 1\rangle$ is a special case, because the resulting state has 2 electrons, and exchange should be considered.

For the annihilation operator:

$$f_a|0, 0\rangle = 0 \tag{11.23}$$

$$f_a|0, 1\rangle = 0 \tag{11.24}$$

$$f_a|1, 0\rangle = 1|0, 0\rangle \tag{11.25}$$

However, $f_a|1, 1\rangle$ is a special case, because of exchange.

There are 2 electrons in the $|1, 1\rangle$ state, and it can be formed from the vacuum state in two different ways:

$$|1, 1\rangle = f_b^+ f_a^+|0, 0\rangle \tag{11.26}$$

or

$$|1, 1\rangle = f_a^+ f_b^+ |0, 0\rangle,$$

subject to the commutation relation given in (11.18):

$$f_b^+ f_a^+ = -f_a^+ f_b^+.$$

That is, the order of the operators makes a difference, and the commutation relations account for the exchange interaction between fermions.

Using, for example, the first of the two equations for the state $|1, 1\rangle = f_b^+ f_a^+ |0, 0\rangle$ given in (11.26), we consider the case of $f_a^+ |0, 1\rangle$:

$$f_a^+ |0, 1\rangle = f_a^+ f_b^+ |0, 0\rangle \tag{11.27}$$

Next, invoke exchange:

$$f_a^+ |0, 1\rangle = -f_b^+ f_a^+ |0, 0\rangle = -|1, 1\rangle \tag{11.28}$$

In a similar way,

$$f_a |1, 1\rangle = -|0, 1\rangle \tag{11.29}$$

To summarize:

$$\begin{array}{ll} f_a|0,0\rangle = 0 & f_a^+|0,0\rangle = |1,0\rangle \\ f_a|1,0\rangle = |0,0\rangle & f_a^+|1,0\rangle = 0 \\ f_a|0,1\rangle = 0 & f_a^+|0,1\rangle = -|1,1\rangle \\ f_a|1,1\rangle = -|0,1\rangle & f_a^+|1,1\rangle = 0 \end{array} \tag{11.30}$$

We will use these relationships to analyze the beamsplitter for quantum electrons in Sect. 11.8.1.

11.3 Correlation Properties of Fermions

11.3.1 Fermi Statistics and Correlation

The Hanbury Brown and Twiss experiment that we considered in Sect. 10.5 distinguishes between the arrival statistics of bosons emitted by a chaotic thermal source ($g^2(0) = 2$) and bosons emitted by a laser ($g^2(0) = 1$). Fermions, no two of which can occupy the same state, might be expected to show a different behavior.

Referring to (10.68), the second order correlation coefficient is:

$$g^{(2)}(0) = 1 + \frac{V(n) - \bar{n}}{(\bar{n})^2}, \tag{11.31}$$

where $V(n)$ is the variance of the Fermi-Dirac distribution.

The expected number of fermions was derived in (1.10) from the grand potential:

$$\langle n(E) \rangle = \bar{n} = -\frac{\partial}{\partial \mu}\Omega = \frac{1}{\left(e^{\frac{E-\mu}{kT}} + 1\right)} \tag{1.10}$$

$V(n) =$ the variance and is calculated from $\langle n(E) \rangle$ using

$$V(n) = kT\frac{\partial}{\partial \mu}\langle n(E) \rangle \tag{11.32}$$

resulting in

$$V(n) = \langle n \rangle (1 - \langle n \rangle) = \bar{n} - (\bar{n})^2. \tag{11.33}$$

The derivation of (11.33) is the subject of Exercise 11.2.

The correlation coefficient for electron quanta is zero:

$$g^2(0) = 1 + \frac{\bar{n} - (\bar{n})^2 - \bar{n}}{(\bar{n})^2} = 0 \tag{11.34}$$

As one might expect, fermions are anti-bunched. Anti-bunching of fermions is associated with destructive two-particle interference, and follows from the exclusion principle that forbids more than one identical fermion to occupy the same quantum state.

11.3.2 Experimental Measurement of the Correlation Coefficient of Fermions and Bosons

An unambiguous Hanbury Brown and Twiss experiment using fermions is challenging. Using electrons is problematic because being charged, they tend to avoid each other by electrostatic repulsion. Thus, a conclusive measurement should use neutral particles, which can nonetheless be directed toward a detector. Jeltes, and co-workers addressed this challenge by measuring and comparing correlations in the arrival times of ^{3}He (fermion) and ^{4}He (boson) atoms (Jeltes et al. 2007). In

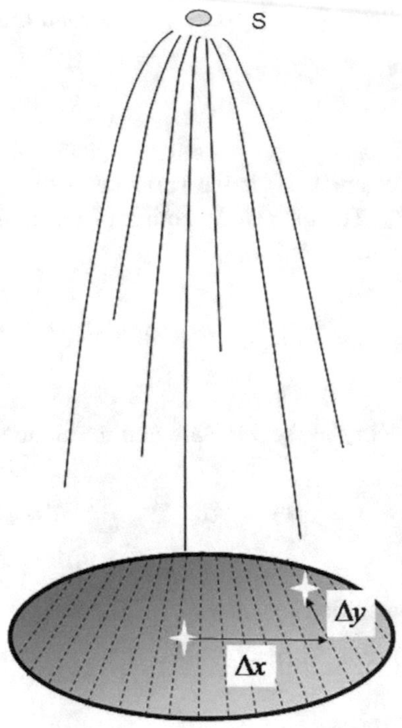

Fig. 11.1 He atoms are released at the same time from the source (S) and fall under gravity where they strike a multichannel plate (MCP) detector. The arrival of atoms is resolved in time and in spatial coordinates. The arrival statistics at a single detection element (e.g. $\Delta x = 0$ $\Delta y = 0$) are used to deduce 2nd order correlation $g^2(\tau)$ as a function of time. Adapted from Jeltes et al. 2007

each case, the particles were isolated at ultra-low temperature environment using evaporative cooling in a magnetic trap. To measure the arrival rate, the particles were released by turning off the trap and allowing the atoms to fall under the influence of gravity to a position-sensitive and time-resolved detector, as diagrammed schematically in Fig. 11.1. This is an experiment of uncommon elegance which is described briefly below. However, the reader is encouraged to take the time to study their publication given in the references to this chapter.

The falling He atoms strike a multi-channel plate (MCP) detector where detection results in an electric current. The location of the detection event in the x–y plane, and thus the separation between two detections, can be determined by the read-out delay along the two axes, implementing the original Hanbury Brown and Twiss experiment where correlation of detection events versus detector separation is measured. The resolution of separation distance is close to the correlation length (~0.8 mm) and this measurement is not conclusive. In the vertical z-direction, the situation is quite different, (refer to Schellekens et al. 2005). All He atoms reach the plate with virtually the same velocity, about 3 m s^{-1}, but at different times, depending on their original position in the trapped 3-dimensional cloud of atoms, as diagrammed in Fig. 11.2. The MCP has a time-response of 1 nsec, so the resolution in the z-direction is about 3 nm. This is 5 orders of magnitude better than the spatial resolution in the x–y plane and much less than the correlation length. Measurement of detection coincidences as a function of time interval

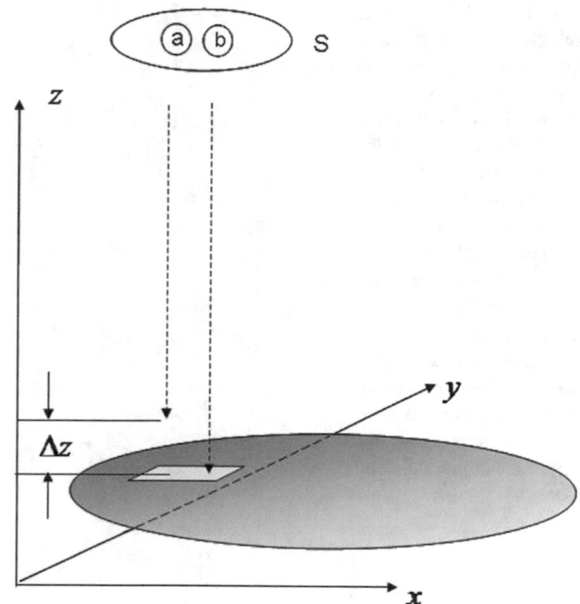

Fig. 11.2 There is a distribution in the arrival time of He atoms, which corresponds to a distribution of atom positions along the z-axis. The velocity attained by a body freely falling from the source is about 3 m s^{-1}. The time response of the detector is 1 nsec. Thus, the spatial resolution in the z-direction is on the order of 3 nm, which is much finer than the resolution in the x- or y-directions

at a single detection element yields the second-order correlation function in time: $g^2(\Delta x = 0, \Delta y = 0, \tau)$ where the time delay is proportional to the difference in position Δz of two He atoms along the z-direction.

If the two processes are indistinguishable, the wavefunction amplitudes interfere. For bosons, the interference is constructive resulting in a joint detection probability which is enhanced compared to that of two statistically-independent detection events, while for fermions the joint probability is lowered. Referring to (10.75), the probability of detecting two photons (bosons) simultaneously is expressed:

$$P_{2\text{bosons}} = |\langle 1|a\rangle\langle 2|b\rangle + \langle 2|a\rangle\langle 1|b\rangle|^2 = \alpha^2\beta^2 + \alpha^2\beta^2 + \alpha^2\beta^2 + \alpha^2\beta^2 = 4\alpha^2\beta^2, \qquad (11.35)$$

where we note the constructive interference that occurs if the bosons are in the same space at the same time. This is an example of boson bunching, i.e., $g^{(2)}(0) = 2$.

We recall from Chap. 4, (4.32), that the wavefunction of a multi-particle wavefunction for fermions must obey the Pauli exclusion principle. It is antisymmetric and must change sign under particle exchange. Thus, we must modify (11.35):

$$P_{2\text{fermions}} = |\langle 1|a\rangle\langle 2|b\rangle - \langle 2|a\rangle\langle 1|b\rangle|^2 = \alpha^2\beta^2 - \alpha^2\beta^2 - \alpha^2\beta^2 + \alpha^2\beta^2 = 0 \qquad (11.36)$$

Therefore, a Hanbury Brown and Twiss style of experiment for fermions should show anti-bunching because the fermion wavefunctions interfere destructively, and $g^{(2)}(0) = 0$.

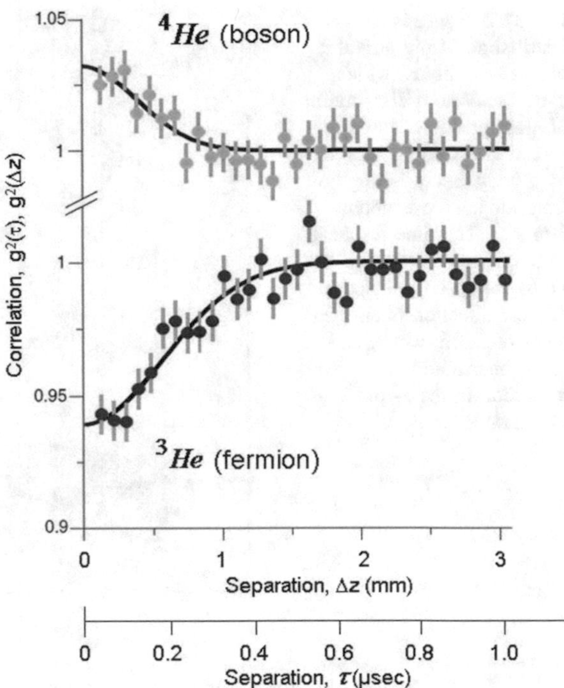

Fig. 11.3 Experimental results for second-order correlation as a function of the separation between He atoms. In order to convert the horizontal axis into time dependence, we have assumed a velocity of 3 m s^{-1} at the detector. Experimental bosonic bunching behavior for ^{4}He is seen in the upper curve, while fermionic anti-bunching behavior is seen for ^{3}He in the lower curve. Reproduced by permission of the authors and the Nature Publishing Group

He is a chemically-inert gas. Ordinary attractive or repulsive interactions between He atoms are negligible; therefore, any bunching or anti-bunching behavior can be attributed to the different quantum statistics of the two atomic species.

The experimental results, comparing the behavior of bosons and fermions, obtained by Jeltes et al. (2007) are shown in Fig. 11.3. In all measurements, the correlation coefficient $g^{(2)}(\Delta z) = 1$, for distance separations Δz longer than the coherence length. When the separation between He atoms is less than the correlation length, ^{4}He (boson) atoms show bunching, while ^{3}He (fermion) atoms show anti-bunching behavior. Although the trends toward bunching or anti-bunching show the expected behavior for bosons and fermions respectively, the measured correlation coefficients are quite far from the results expected for pure bosonic ($g^2(0) = 2$) and fermionic ($g^2(0) = 0$) behavior. C. Westbrook, the corresponding author in these studies, has provided additional information concerning the correlation measurement shown in Fig. 11.3 (Westbrook 2019). Although the resolution of the measurement in the z-direction improves the precision, the correlation coefficient is determined by an average over the entire volume of the sample, and so the resolution of the correlation coefficient is degraded by allowing many uncorrelated pairs to contribute to the measured signal. The experiment can be further improved in the case of bosonic ^{4}He by introducing a magnetic quadrupole trap in order to create an ultra-cold and spatially compact Bose-Einstein condensate. In this case, the full expected correlation coefficient for bosons, $g^2(0) = 2$ can be measured. Of course, there is not the possibility to create an analogous improvement for measurement of the anti-bunching of fermions.

11.4 Ballistic Quantum Transport

Photons propagate naturally in a ballistic manner. In a sequence of photons that propagate from a source to a detector, only the ballistic photons are detected. The exceptional photons that may be scattered unintentionally out of the direction of propagation will not contribute to the signal. When a beam splitter is inserted in the path, ballistic propagation allows us to distinguish the transmitted beam from the reflected beam. Experiments, such as those carried out by Grangier, Roger and Aspect or by Hong, Ou and Mandel, that show quantum photon behavior owe their impact and importance to the experimental ability to detect single photons (Grangier et al. 1986; Hong et al. 1987).

Measurements of quantum electron transport can be compared directly to those for quantum photons only if electrons can be made to propagate in a one-dimensional fashion without scattering, for example the transit of an electron in a vacuum environment like that of an electron microscope. These experimental conditions can also be attained by application of well-established technologies of quantum electronics, including the controlled emission of single electrons in the circuit under test (Fève et al. 2007).

Ballistic transport is achieved whenever the mean free path between scattering events is greater than the physical dimensions of the sample. One dimensional transport is achieved by confinement, initially to two-dimensions in the x–y plane using quantum well structures introduced in Chap. 3, and subsequently by imposing a strong magnetic field ($B > 1$ Tesla, typically) in order to confine the electron to a one-dimensional channel localized around the edge of the sample.

11.4.1 Conductivity in Macroscopic Materials

Most studies of ballistic electron behavior are carried out in condensed matter. Electrical conductivity in macroscopic samples at room temperature is determined by scattering properties. In this situation, we can compare characteristic scattering lengths:

$$\ell_{\text{sample}} > \ell_{\text{inelastic}} \tag{11.37}$$

Simply stated, an electron propagating between the source and the drain of a macroscopic sample (with dimensions of a cm or larger) will suffer many inelastic collisions randomizing momentum while exchanging energy with the surrounding medium. Under these conditions we can write:

$$\mathbf{j} = \sigma_{ij}\mathbf{E} \tag{11.38}$$

where $\mathbf{j}$ is the current density, $\mathbf{E}$ the electric field and σ_{ij} is the conductivity. For simplicity, we will restrict the remainder of this discussion to one dimension without loss of generality. In this macroscopic regime, the current is a stream of charged particles. The current density is the number of such particles that pass through a unit area per second. Considering for a moment a model of 1-dimensional transport, the current density is:

$$j = \sigma E = Nq^2 \frac{\tau}{m^*} E = Nq\mu E \tag{11.39}$$

where N is the concentration of "free" electrons per unit volume, μ is the electron mobility and τ is the scattering time.

$$\tau = \mu \frac{m^*}{q} \tag{11.40}$$

In solid-state materials, the mean free path (mfp) is the product of the velocity of electrons at the Fermi level and the scattering time:

$$\mathcal{L}_{\text{mfp}} = v_F \tau = \frac{m^*}{q} v_F \cdot \mu \tag{11.41}$$

Other regimes are possible. For example, in high vacuum, the scattering time for electrons may be longer than the transit time between the contacts. In this case, electron transport is ballistic. It is easier to reach the ballistic transport regime in semiconductors than in metals. For example, we can compare transport in gold to that in high-mobility GaAs where electrons are confined in a 2-dimensional electron gas (2 DEG), (see Table 11.1).

It is possible to increase the inelastic scattering time by using quantum structures to isolate carriers from the sources of scattering, or by reducing the intensity of scattering by lowering the temperature of the conductive medium. These two methods form the foundation for studies of quantum electrical transport. The

Table 11.1 Electron transport properties at low temperatures in gold and in a 2 DEG in GaAs

Material	Fermi velocity	Scattering time	$\mathcal{L}_{\text{mfp}}$	Notes
Gold	1.4×10^6 m s^{-1}	2.9×10^{-12} s	4 μm	$T = 4$ K
GaAs (2 DEG)	2.8×10^5 m s^{-1}	3.8×10^{-11} s	10 μm	$T = 200$ mK $\mu = 10^2$ m^{2}V^{-1} s^{-1}

objective is to create a ballistic transport environment for electrons in order to investigate direct analogies to the quantum photonic experiments presented in the earlier chapters of this text.

Analysis of ballistic transport in the classical regime would predict that the resistance of the conducting medium should approach zero. In the following we will see that both experiment and theory show that the conductance of a transport channel in the quantum regime is a well-defined constant, even in the case of ballistic transport.

11.5 Ballistic Conduction in Semiconductors

11.5.1 Introduction

A sample can support ballistic transport when the mean free path (11.41) is longer than the sample dimensions. The example of the previous section illustrates that ballistic transport can be achieved in semiconductors in larger samples than would be the case using metals. Even more important, the control of electron transport by application of potential barriers that can be adjusted by bias voltages cannot be applied to metal circuits, while this technology is well-developed in semiconductor circuits for many years. Although some of the first demonstrations of ballistic quantum transport were made using metal wire circuits (Webb et al. 1985), this approach has ceded to those using semiconductor materials.

Referring to (11.41), the mean free path depends on two parameters; the Fermi velocity and the carrier mobility. The Fermi velocity is a function of the concentration of electrons in the conduction band.

In three dimensions, the Fermi wave vector is the radius of a sphere in k-space that has enough states to accommodate all the free electrons:

$$2 \cdot \frac{4}{3}\pi k_F^3 = 8\pi^3 N,$$

$$k_F = \left(3\pi^2 N\right)^{\frac{1}{3}} \tag{11.42}$$

The Fermi energy is: $E_F = \frac{\hbar^2 k_F^2}{2m^*}$

and the Fermi velocity is: $v_F = \frac{\hbar k_F}{m^*}$

In a two-dimensional system, the Fermi wavector is:

$$k_F = (2\pi N)^{\frac{1}{2}} \tag{11.43}$$

The Fermi energy and Fermi velocity are of course calculated in the same way.

Thus, the Fermi velocity increases with the carrier concentration. The mean-free path increases with the mobility. In a given material, the mobility can be increased only by increasing the scattering time, which can be accomplished by lowering the carrier concentration. It might seem quite impossible to engineer a longer mean free path because of these opposing trends. The way out of this dead-end was discovered by the invention of modulation doping (Dingle et al. 1978). This breakthrough is based on structuring the semiconductor host so that the doping impurities which contribute to the free electron concentration are physically separated from the resulting free carriers which are localized in a 2-dimensional channel, created at the interface between GaAs and AlGaAs, having a reduced density of scattering centers. This structure enables at the same time high mobility and high carrier concentration leading directly to a longer mean-free path. Umansky and Heiblum have written an excellent presentation of state of the art modulation-doping physics and technology (Umansky and Heiblum 2013). Modulation doping technology is covered in more detail in Annex 11A. A major consequence is that the physics of quantum electrons in semiconductors is studied for the most part in a 2-dimensional free-electron gas (2 DEG) created by modulation doping.

11.5.2 Electronic Structure of a 2 DEG in AlGaAs/GaAs

A detailed study of the electronic properties of 2-dimensional semiconductor systems has been given in an extensive review with emphasis on Si and GaAs, but also covering other group III–V and II–VI materials, Ando et al. (1982).

The fundamental parameter for characterization of a 2 DEG is the 2-dimensional carrier concentration. In this section we will use $N_{2-d} = 2 \times 10^{11}\,\text{cm}^{-2}$, to illustrate a typical choice. Below, we determine the principal parameters for such a $Al_{0.3}Ga_{0.7}As/GaAs$ modulation-doped heterostructure.

The Fermi wavevector of an electron in this 2 DEG is:

$$k_F = \left(4\pi N_{2-d}\frac{1}{g_s}\right)^{\frac{1}{2}} = 1.12 \times 10^8\,\text{m}^{-1} \tag{11.44}$$

where g_s is the spin degeneracy = 2 in the absence of a magnetic field.

The Fermi wavelength is:

$$\lambda_F = \frac{2\pi}{k_F} = 5.60 \times 10^{-8}\,\text{m}. \tag{11.45}$$

λ_F is simply the de Broglie wavelength of an electron at the Fermi level.

For GaAs, $m^* = 0.066 m_o = 6.01 \times 10^{-32}\,\text{kg}$.

The Fermi velocity is:

$$v_F = \frac{\hbar}{m*}k_F = 1.96 \times 10^5\,\text{m}\,\text{s}^{-1} \tag{11.46}$$

$$\text{The Fermi energy} = E_F = \frac{(\hbar k_F)^2}{2m*} = 7.19 \times 10^{-3}\,\text{eV}. \tag{11.47}$$

The accumulation layer is confined in a potential well created by the heterostructure interface. The stable energy levels in the well are quantized by the confining potential. In Fig. 11.4 we show an expanded view of this well. It has a potential profile of approximate triangular shape. The energy levels for the stable confined states can be found by solving Schrödinger's equation using the methods we have presented in Chap. 3. However, numerical methods are needed to obtain an exact solution for a well having a finite potential. The solution is the subject of Exercise 11.4.

For example, Ando, Fowler and Stern have proposed a model for the potential well as an infinitely high triangle having an abrupt edge on the left and a constant electric field on the right. The solution to Schrödinger's equation for such a potential of infinite height gives the confined energies as:

$$E_n \cong \left(\frac{\hbar^2}{2m*}\right)^{\frac{1}{3}} \left(\frac{3\pi q^2 N_s}{2\varepsilon\varepsilon_o}\left(n + \frac{3}{4}\right)\right)^{\frac{2}{3}} \tag{11.48}$$

The reader is cautioned that this approximation, as attractive as it might appear, **should not be used** to determine accurate quantitative energies of confined states in the 2 DEG. Being based on an infinitely high potential well, the energy levels found from (11.48) may be as much as 2 times larger than those determined from a solution by numerical methods for a finite well, as we have already shown in Chap. 3. Accurate calculations should take into account the carrier concentration of the 2 DEG, the δ-doping level and setback, as well as the impurity concentration in the undoped GaAs layer. This calculation, which involves an iterative numerical solution of Poisson's equation, will determine the band-bending, and thus the position of the confining potential with respect to the Fermi level.

Example 11.1 In the present illustration, we can make use of (11.48) to help develop a *qualitative* description of the confined energy states. We will assume a sheet carrier doping density of $N_s = 2 \times 10^{11}$ cm^{-2}. Using (11.48), we estimate the energy of the $n = 0$ level to be 47 meV, and that of the $n = 1$ level to be 82 meV. If we further assume the carrier density of the 2 DEG to be equal to that of the doping density, we can use (11.47) to estimate the Fermi energy: $E_F = 7$ meV. Thus, the 2 DEG concentration partially fills the $n = 0$ subband from its minimum at 47–54 meV. It is clear that the $n = 1$ subband lies well above the Fermi level and remains empty. These results are illustrated in Fig. 11.4a, b. In Fig. 11.4a, we show the position of the Fermi level with respect to the energy minima of the $n = 0$ and $n = 1$ subbands. In Fig. 11.4b, we show a diagram of energy versus wavevector in the plane of the 2 DEG. The 2 DEG occupies all states up to the Fermi wavevector.

Having determined the position of the subbands with respect to the Fermi energy, the potential profile can be integrated into a diagram for the entire $Al_{0.4}Ga_{0.7}As$/GaAs structure, which is treated in the appendix to this chapter.

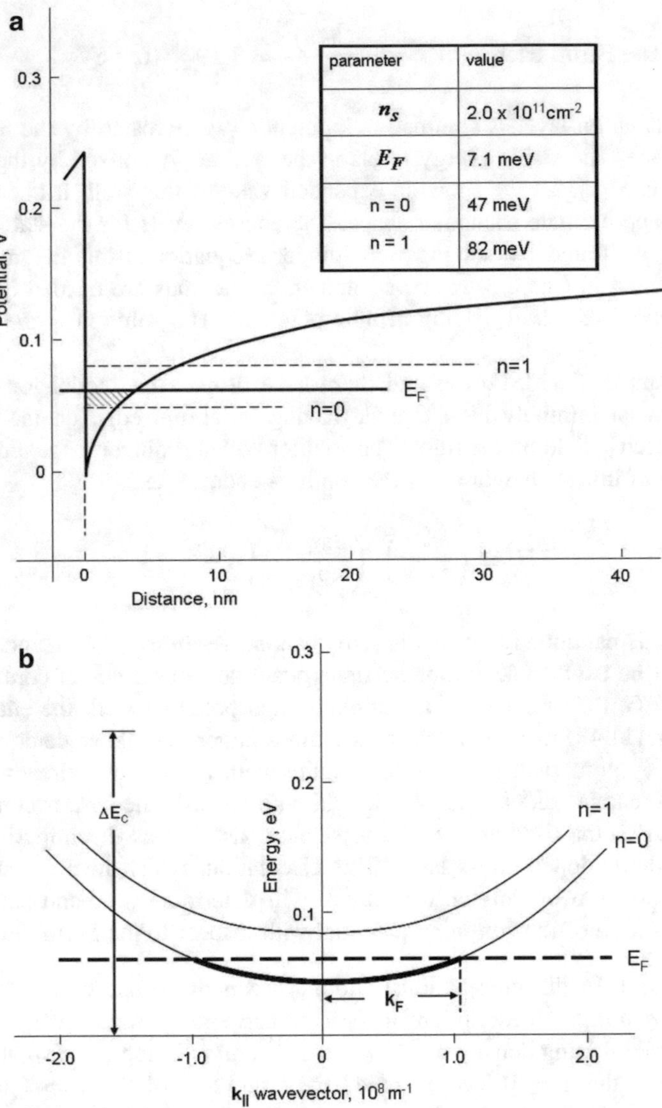

Fig. 11.4 **a**. Potential profile of the conduction band for a heterostructure composed of $Al_{0.4}Ga_{0.7}As/GaAs$. Using (11.48), we have estimated energies for the lowest energy subbands in the 2 DEG. The energy separation between these two subbands is much larger than the Fermi energy, so only the lowest energy subband is occupied by electrons. **b**. Energy-wavevector diagram in the plane of the 2 DEG for the $Al_{0.4}Ga_{0.7}As/GaAs$ structure considered in Example 11.1. Using (11.44) gives the Fermi wavevector and determines the Fermi energy. The states occupied by electrons lie below the Fermi level and are indicated by the bold line

11.6 Quantum Fermion Conductance

11.6.1 The Landauer Model

The key condition of the ballistic transport regime is that the mean free path (mfp) for inelastic scattering become greater than the distance between the source and drain contacts.

$$\mathcal{L}_{\text{mfp}} > \ell_{\text{sample}} \tag{11.49}$$

In this situation, the mobility is no longer a useful concept for describing electron behavior. On the other hand, the conductance (which is the reciprocal of the ohmic resistance) remains a key transport parameter that can be both calculated and measured directly. In the classical regime, the absence of inelastic scattering would lead to an infinitely large mobility and the conductance would increase without limit. However, when the behavior of individual electrons is analyzed, we can show that the conductance in the absence of any scattering (both inelastic and elastic) remains finite and quantized.

A major objective is to create and study the behavior of ballistic electrons in a 1-dimensional channel, this being the direct analogue of single photon propagation. The injecting contact for a single electron is simply and directly modeled as a potential barrier that separates the electron reservoir from the ballistic channel. An electron is either reflected or scattered by this barrier or is transmitted in this 1-dimensional collision process. In the case of ballistic transport, no energy is given up by the electron. The conductivity is determined by the transmission coefficient for this event. In the limit of single electron ballistic transport, the 1-dimensional channel that supports the transport behaves like a quantum state. The Pauli exclusion principle assures that only two electrons (taking into account spin degeneracy) can occupy this channel at the same time. This situation is distinct from photon behavior where there is no definite limit on the occupation number of a channel.

Landauer was the first to carry out this calculation for localized scattering by impurities in metallic conductors (Landauer 1957). Each scattering center is treated as an obstacle. When treatment of this problem is reduced to 1-dimensional transport, each obstacle can be characterized by a potential barrier with a transmission and reflection coefficient. To achieve a single-electron emitter, it suffices to assure a ballistic transport regime confined to one dimension with a single obstacle that can be treated as a potential barrier. In Chap. 3, we have already discussed single-electron propagation in the presence of a potential barrier in the context of tunneling. We can use the same approach in the present case to model the injecting contact for single electrons as the simple potential barrier shown in Fig. 11.5.

The electron reservoir can be characterized by its chemical potential: μ_R, and the electrons in the conduction channel by a chemical potential: μ_C. At equilibrium, $\mu_R = \mu_C$, and the net transport of electrons across the barrier is zero. Application of an exterior potential $= qV$ upsets the balance, and creates a net flux in the direction of lower chemical potential.

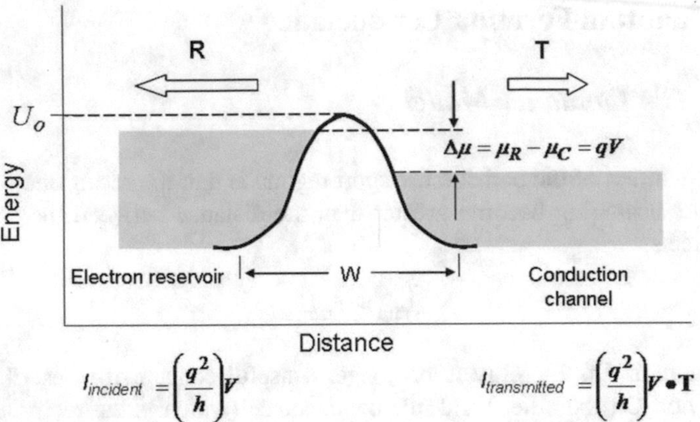

Fig. 11.5 A schematic diagram of a 1-dimensional model for a single electron that encounters an obstacle, represented by a potential U_o and a width W while propagating from an electron reservoir to a conducting channel under the influence of an applied potential V. Either transmission past the obstacle or reflection occurs

The current transported across the barrier is the product of the electronic charge multiplied by the electron velocity and the number of electrons per unit length that are able to traverse the barrier at constant energy E between an occupied state on the left-hand side of the barrier and an unoccupied state on the right hand side.

$$I_T = n(E)qv_F \cdot \mathbf{T} \tag{11.50}$$

where $\mathbf{T}$ is the transmission coefficient.

If $\rho(E)$ is the 1-d density of electrons per unit energy per unit length,

$$n(E) = \Delta E\rho(E) = (E_F + qV - E_F)\rho(E) = qV\rho(E) \tag{11.51}$$

$$\rho(E) = \frac{\partial\rho}{\partial k}\frac{\partial k}{\partial E} = \frac{g_s}{2\pi}\frac{1}{\hbar v_F} \tag{11.52}$$

The state degeneracy is given by g_s.

$$n(E) = qV\frac{g_s}{2\pi\hbar v_F} \tag{11.53}$$

The current carried by an electron is:

$$I = \frac{g_s q^2}{h}V, \tag{11.54}$$

independent of any material-specific parameters.

The current that passes the barrier includes the tunneling transmission probability:

$$I_T = \frac{g_s q^2}{h} V \cdot \mathbf{T} \tag{11.55}$$

The conductance is defined as:

$$G = \frac{d}{dV} I_T = g_s \frac{q^2}{h} \mathbf{T} \tag{11.56}$$

in units of Ω^{-1}, and is known as the Landauer formula.

The value of the fundamental quantum of conductance is

$$G_{\text{quantum}} \equiv \frac{q^2}{h} = \frac{1}{25812.81} \Omega^{-1} \tag{11.57}$$

For the case of electrons in the absence of a magnetic field each state is spin-degenerate, so that $g_s = 2$

$$G = g_s \frac{d}{dV} I_T = \frac{2q^2}{h} \mathbf{T} \tag{11.58}$$

If the channel is fully open, $\mathbf{T} = 1$, and the result shows that a perfectly ballistic channel has a finite conductance given by (11.56). This is a revolutionary concept. In the absence of collisions, the resistance is not zero, but is a precise value determined by a ratio of fundamental constants. The conductance can be increased by adding additional channels in parallel. The overall conductance for j channels is the arithmetic sum:

$$G_{\text{tot}} = j \frac{2q^2}{h} \tag{11.59}$$

11.6.2 Transmission Coefficient

The transmission coefficient should be thought of as a response of a propagating electron though a medium. This can be a more general situation than electron propagation over an abrupt barrier that is presented in Chap. 3. The transmission coefficient can be represented by a function that varies between 0 and 1. In a non-dissipative environment like ballistic transport, the sum of the transmission and reflection coefficients is equal to unity.

$$\mathbf{T} + \mathbf{R} = 1 \tag{11.60}$$

Without specifying the actual functional dependence of the barrier potential profile on distance we can expand the potential in a Taylor's series around $x = x_o$:

$$V(x) = \sum_{n=0}^{\infty} \frac{1}{n!}(x - x_o)^n \left(\frac{d^n}{dx^n}\right) V(x)|_{x=x_o} \tag{11.61}$$

Without loss of generality we can set $x_o = 0$;

$$V(x) = U_o + V'(0)x + \frac{1}{2}V''(0)x^2 + \frac{1}{6}V'''(0)x^3 + \cdots + \frac{1}{n!}\frac{d^n}{dx^n}V(x)|_{x=0}x^n \tag{11.62}$$

If the potential can be modeled as a symmetric function about $x = 0$, then it follows that the derivatives of all the odd powers in x are equal to zero. Keeping the lead remaining terms:

$$V(x) \cong U_o + \frac{1}{2}V''(0)x^2 + \text{ higher order terms} \tag{11.63}$$

There are 2 cases. When $V(x)$ is concave upward, we have a parabolic confinement potential which has the well-known solutions of the harmonic oscillator. When $V(x)$ is concave downward, we have a parabolic barrier. Although Schrodinger's equation for this barrier is simple to write down, the solution for the transmission coefficient as a function of incident energy in closed form is not possible. This impasse is unfortunate because (11.63) can be used to model the potential for many types of barriers.

Notwithstanding the complex nature of exact determination of the transmission coefficient, we know already how the transmission coefficient should appear. At incident energies less than U_o, the transmission coefficient should approach 0. For energies greater than U_o, the transmission coefficient should approach unity.

The sigmoid function is a well-known mathematical expression that has this characteristic.

$$S(x) = \frac{1}{1 + e^{-x}} \tag{11.64}$$

This function appears frequently in analysis of statistics, and the reader will recognize that the Fermi-Dirac distribution is a version of this function.

In the place of an exact expression for the transmission coefficient, we can construct a useful approximation using the sigmoid expression:

$$\mathbf{T}(E) \cong \frac{1}{1 + e^{-2\pi\left(\frac{U_o - E}{E_o}\right)}} \tag{11.65}$$

Here, U_o is the barrier height and E_o is an adjustable parameter that determines the rate of change between $\mathbf{T} = 0$ and $\mathbf{T} = 1$. It is straightforward to relate E_o to the thickness of the barrier. Thus, only two parameters, the barrier height and its 1-dimensional thickness are needed to evaluate (11.65). In Fig. 11.6, we show an example of this approximation.

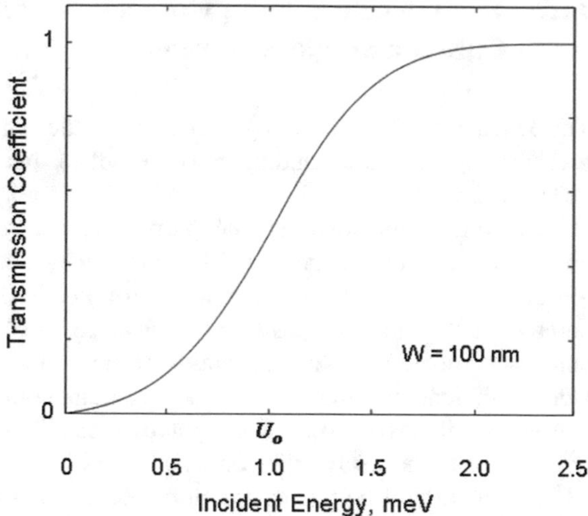

Fig. 11.6 Transmission coefficient for single-electron transmission over a 1-dimensional barrier having a potential of 1 meV, and a width at the base of 100 nm, calculated using the sigmoid approximation (11.65). Transmission probability does not become substantial until the barrier energy is exceeded

The sigmoid function is in general a satisfactory substitute for a numerical solution of transmission coefficient for this problem. However, it fails in the limit of thin barriers when $E_o \gg (U_o - E)$.

We can compare the approximation using the sigmoid function to the numerical solution for the transmission coefficient. Since an expression in closed functional form is not possible, we use the transfer matrix method developed in Chap. 3 and we divide the same parabolic potential into a series of piece-wise elements. An example of such a comparison is shown in Fig. 11.7. A more detailed comparison for other conditions is the subject of Exercise 11.6.

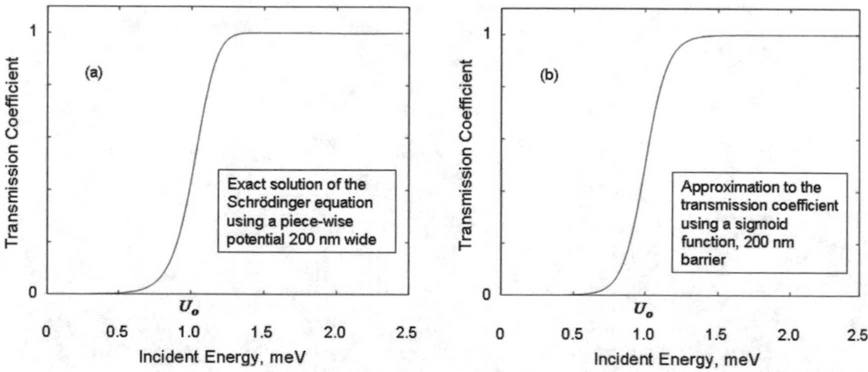

Fig. 11.7 **a** Transmission coefficient calculated using the transfer-matrix method developed in Chap. 3 to solve Schrödinger's equation. The barrier is parabolic in shape with a height of 1 meV and a width of 200 nm. **b** Approximation of the transmission coefficient using a sigmoid function with the same parameters. The approximation is excellent for the parameters of this example, as is confirmed by visual inspection. The approximation diverges from the exact solution as the barrier thickness is reduced

11.7 The Quantum Point Contact: A Voltage Controlled Gate for Single Electrons

The Landauer model for single electron scattering is an important conceptual breakthrough in understanding how to isolate and characterize single electron behavior. The analysis is based on conduction in only one dimension. If we could realize a one-dimensional conduction path that has a single barrier, we would be able to verify some aspects of the Landauer model. If in addition we could control the barrier height, that is, its transmission coefficient, we would have an ideal component for studying quantum conductance. Such a system: a one-dimensional conductance pathway having a single barrier with a voltage-controlled transmission coefficient, is embodied in the quantum point contact, or QPC. Since its conception, it has become a ubiquitous element in all experimental studies of quantum electron behavior in semiconductors.

The structure of a QPC is straightforward. Modulation doping is used to create a 2 DEG with a mean-free-path greater than several μm. If we constrict the 2 DEG in one of its free dimensions, the result is a 1-dimensional pathway in the remaining direction. If we can create a potential barrier in this constriction, we will have two important elements of the QPC. If we create this barrier electronically, by applying a potential difference relative to the 2 DEG, then the barrier height and its transmission coefficient can be controlled. These are the key elements of a QPC. A schematic diagram of such a QPC showing the electron potential energy along the vertical axis and the 2 spatial dimensions along the x- and y-directions is displayed in Fig. 11.8.

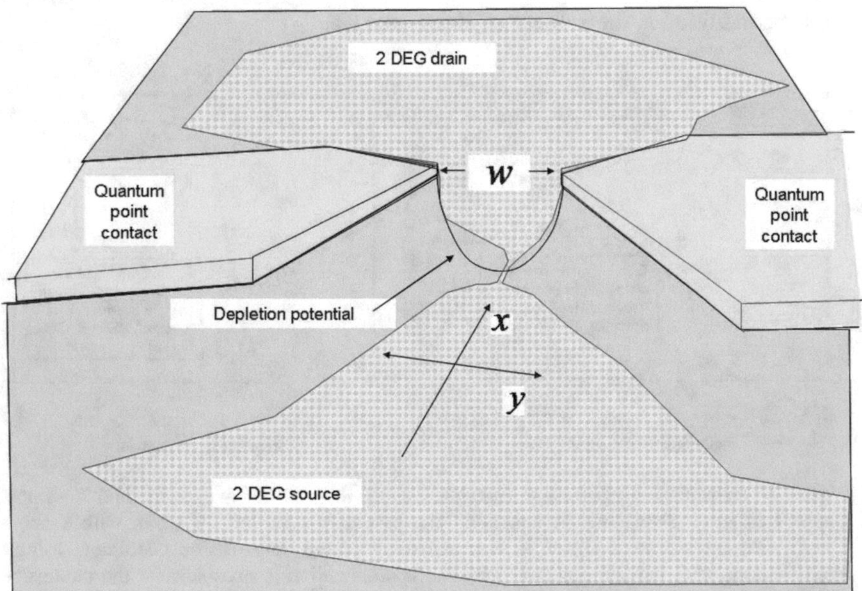

Fig. 11.8 Schematic diagram of a quantum point contact, QPC, showing potential energy in the vertical direction as a function of x- and y-coordinates. Application of a negative potential between the QPC and the 2 DEG causes local depletion of the 2 DEG. The channel width w decreases with increasing depletion

11.7.1 Quantum Conductance

A quantum point contact is a structure that enables the injection of single electrons in a well-defined quantum state. The quantum point contact can be adapted to the control of electrons in a 2-d, high-mobility electron gas (2 DEG) by using it to impose a variable constriction on the flow of electrons from one 2 DEG to another. The electrons in each 2 DEG can be characterized by a chemical potential μ. When two separate 2 DEG are in equilibrium, their chemical potentials are equal to each other, and equal to $\frac{E_F}{q}$. Under this condition, no net transport of electrons from one 2 DEG to the other will occur. If a potential difference between the respective 2 DEG can be established, then net transport can occur, provided that there exist one or more conducting channels between the two regions. This structure is one of the building blocks needed to build and study single-electron quantum devices.

The conductance of such a channel can be determined from basic transport considerations (Wharam et al. 1988). In the presence of a voltage across the channel, a difference in chemical potentials is established, the electrons gain energy qV, and thus velocity δv.

$$\Delta E = q(\mu_1 - \mu_2) = q\Delta V = \frac{1}{2}m^*(v_F + \delta v)^2 - \frac{1}{2}m^* v_F^2 = m^* v_F \delta v + \frac{1}{2}m^* \delta v^2 \tag{11.66}$$

If the change δv is small compared to the Fermi velocity

$$q\Delta V \cong m^* v_F \delta v \tag{11.67}$$

The current in the jth channel is expressed:

$$I_j = n_j q \delta v, \text{ where } n_j \text{ is the number of carriers per unit length} \tag{11.68}$$

The one-dimensional density of states is:

$$\mathcal{N}(E) = \frac{g_s}{h}\left(\frac{m^*}{2E}\right)^{\frac{1}{2}} \tag{11.69}$$

where g_s is the degeneracy each state. For the case of electrons, $g_s = 2$.

$$n_j = \int_0^{E_F} \mathcal{N}(E)dE = \frac{g_s}{h}\left(\frac{m*}{2}\right)^{\frac{1}{2}} 2E_F^{\frac{1}{2}} = \frac{g_s}{h}\left(\frac{m*}{2}\right)^{\frac{1}{2}} 2\left(\frac{m*}{2}\right)^{\frac{1}{2}} v_F = \frac{g_s}{h} m * v_F \tag{11.70}$$

and,

$$I_j = g_s \frac{q}{h} m^* v_F \delta v = g_s \frac{q^2}{h} \Delta V = G \Delta V \tag{11.71}$$

In the absence of a magnetic field $g_s = 2$, and the conductance of a channel is thus:

$$G = \frac{2q^2}{h} = \frac{1}{12.9\,\mathrm{k}\Omega}, \tag{11.72}$$

which is identical to (11.58), provided the transmission coefficient is unity. If multiple channels are present, $G = j\left(\frac{2q^2}{h}\right)$, where j is the number of channels.

11.7.2 Controlled Depletion of a 2 DEG

The application of a negative bias voltage between the 2 DEG and the $Al_{0.3}Ga_{0.7}As$ confinement layer will lift the conduction band that contains the 2 DEG upward to a higher potential. When the conduction band energy is increased above the Fermi energy, the 2-dimensional conduction channel is depleted. We can use this feature to create a 1-dimensional channel in a 2 DEG. In Fig. 11.9 we show a schematic diagram of potential as a function of distance in a structure having two electrodes separated by a gap. Application of a negative potential between the electrode and the 2 DEG will locally raise the energy of the conduction band. In Fig. 11.9a, the negative bias voltage V_{G1} is large enough to lift the conduction band potential above the Fermi level everywhere between the point contacts, pinching off the channel. In Fig. 11.9b, a less negative bias voltage, V_{G2} is still large enough to deplete the 2 DEG under the point contacts, but is not large enough to deplete the 2 DEG in the intermediate region between the contacts. This enables a 1 dimensional channel for conduction in the point contact region. Transport of an electron through this bottleneck can still take place (perpendicular to the plane of the paper). In Fig. 11.9c, the bias voltage V_{G3} is small enough that no bending of the conduction band occurs under the point contacts, and the 2 DEG exists uniformly beneath the contact region.

The applied voltage creates a confinement potential which is a function of the x-direction and at the same time, a potential barrier U, the height of which increases as the confinement becomes stronger. As the potential V_G increases in magnitude, the barrier height U increases, and width of the potential well decreases, causing the separation in energy between the quantized states in this well to increase. In Fig. 11.10, we look at the same situation in the x-direction of electron transport, perpendicular to the confinement y-direction. In this direction, the result of the applied voltage is to create a potential barrier in the x-direction of transport.

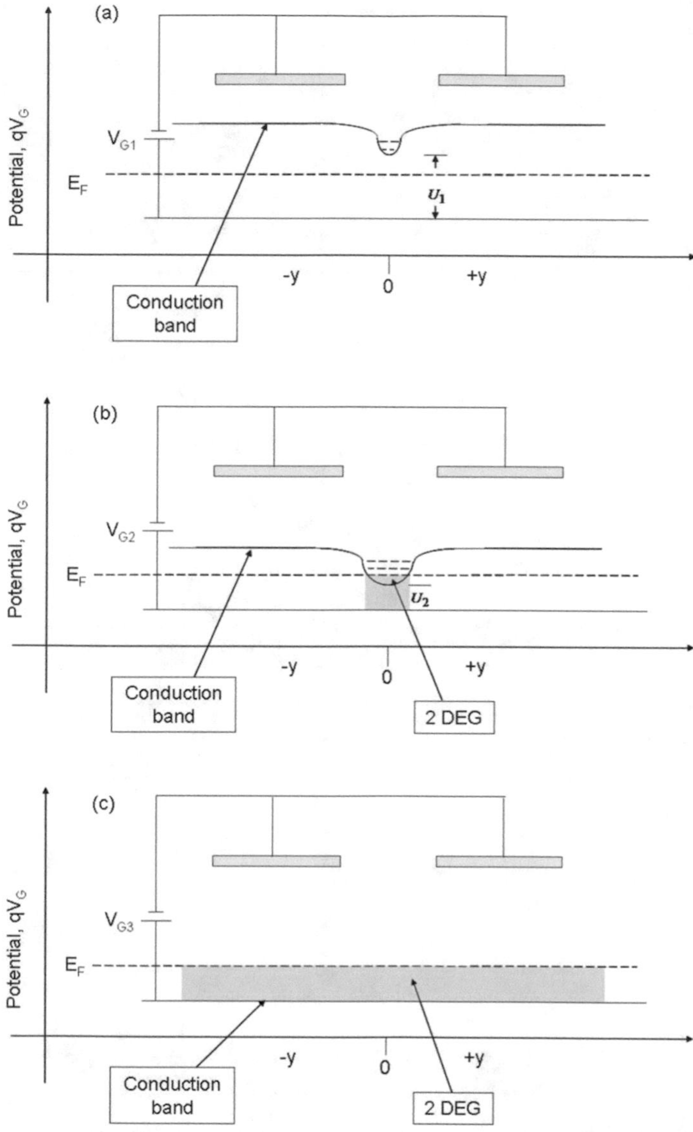

Fig. 11.9 Application of a bias voltage between the point contact and the 2 DEG causes bending of the conduction band. The conduction band potential is shown a function distance y between two quantum point contacts. The x-coordinate is $x = 0$. **a** V_{G1} is sufficiently negative to raise the conduction band energy everywhere above the Fermi level, $U_1 > E_F$, pinching off conduction between source and drain. The channel between the gate electrodes may become so narrow that quantized energy levels can be resolved. **b** V_{G2} is less negative than V_{G1}. The conduction band potential under the gate electrodes is raised, depleting electrons from the conduction band in this region. The potential of the conduction band is also raised in the channel region by an amount U_2, but not above the Fermi level, creating a channel in between the two contacts. This channel that can transport electrons is more narrow than the spacing between gate electrodes. **c** V_{G3} is less negative than V_{G2} to the point that negligible band-bending occurs and the conduction between source and drain is unrestricted

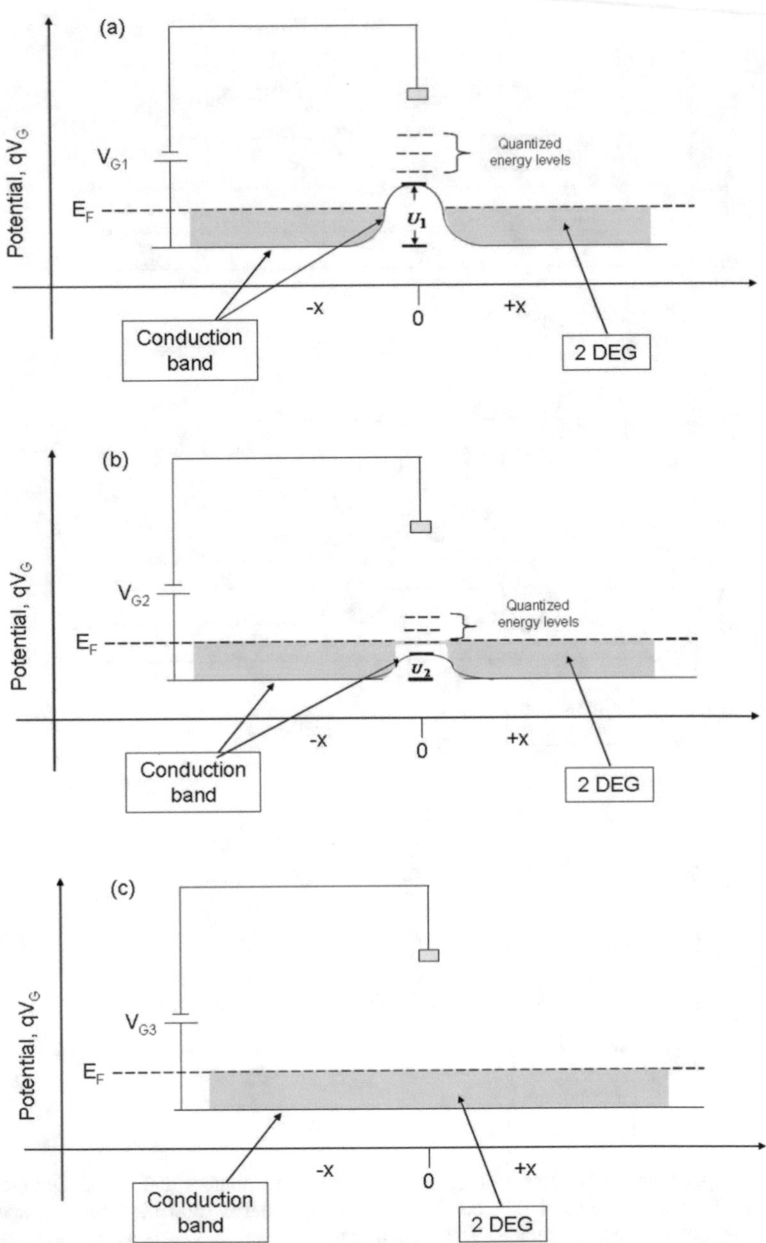

Fig. 11.10 The source lies on the left-hand side of this diagram, and the drain on the right. Imposition of a gate voltage, as illustrated in Fig. 11.9, creates a potential barrier between the source and drain. A crucial case is illustrated in Fig. 11.10b. **a** V_{G1} is sufficiently negative to raise the peak height of the conduction band above the Fermi level, stopping conduction and pinching off the channel. **b** Although the barrier height, U_2, lies below the Fermi level, conduction in the *x*-direction across the barrier can occur only if one of the quantized energy levels also lies at or below the Fermi level. **c** At sufficiently-low gate voltage, there is no impediment to conduction between source and drain

11.7.3 Analysis of the Quantum Point Contact Structure

A quantum point contact (QPC) can be realized experimentally by placing a split electrode gate over the 2 DEG. This electrode pair is used to apply a gate voltage between the electrode and the 2 DEG, and it depletes the 2 DEG of electrons, thus creating a quasi 1-dimensional bottleneck, separating the source and drain. The width w of this bottleneck is determined by the gate voltage. A schematic diagram of a QPC has been shown in Fig. 11.8.

The effect of carrier depletion is to create a potential with a minimum in the y-direction in the gap and simultaneously a maximum in the x-direction between the contacts. This situation is called a saddle point, and the 2 dimensional potential is called a saddle-point potential, as shown in Fig. 11.11

For convenience, we will set the coordinate of the saddle point at $(x, y) = (0, 0)$. The potential well is symmetric about the origin along the both the x- and y-axes. Because of this symmetry, we can use (11.62) to write the saddle potential as a polynomial expression to the 3rd order as for the QPC structure as:

$$P(x, y) = U(x) + V(y) \cong \left(U_o - \frac{1}{2}Kx^2\right) + \frac{1}{2}m^*\omega^2 y^2, \tag{11.73}$$

The solution to Schrödinger's equation using this potential gives the allowed states for an electron in the QPC.

$$-\frac{\hbar^2}{2m^*}\frac{d^2}{dr^2}\Psi(x, y) = (E - P(x, y))\Psi(x, y) = (E - V(y) - U(x))\Psi(x, y) \tag{11.74}$$

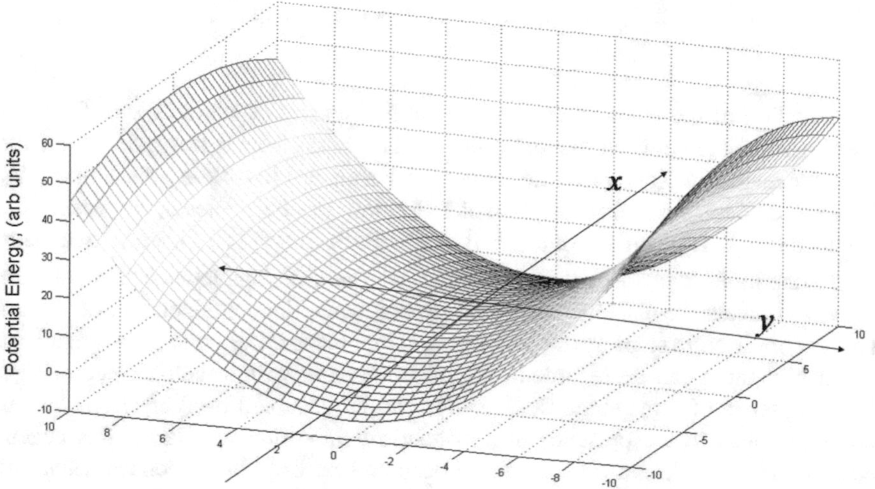

Fig. 11.11 The 2-dimensional symmetric potential created by a quantum point contact forms a saddle-point structure. The curvature is negative in the x-direction, forming a barrier of height U, while the potential in the y-direction creates a ladder of discrete stationary energy states, see (11.78) and (11.79)

The potential is separable in x and y, and so the wavefunction can be expressed as a simple product.

$$\Psi(x, y) = \psi(x)\phi(y) \tag{11.75}$$

Considering first the solution in the x-direction:

$$-\frac{\hbar^2}{2m^*}\frac{d^2}{dx^2}\psi(x) - \frac{1}{2}Kx^2\psi(x) = (E - U_o+)\psi(x) \tag{11.76}$$

This is the problem of the transmission of a particle over a potential barrier that we have already considered in Sect. 11.6.2. The transmission coefficient can be approximated by a sigmoid function and is given in (11.65). For the purposes of our discussion here, it is important to remember that the transmission coefficient is close to unity for electron energy greater than U_o.

Next, we set $x = 0$, and consider the situation along the y-axis in the gap between the point contacts. The potential well is given by:

$$V(y) = \frac{1}{2}m^*\omega_o^2 y^2 + U_o \tag{11.77}$$

$$-\frac{\hbar^2}{2m^*}\frac{d^2}{dy^2}\varphi(y) + \frac{1}{2}m^{*2}\omega_o^2 y^2\varphi(y) = \left(E_y - U_o\right)\varphi(y) \tag{11.78}$$

This is the equation for a quantum harmonic oscillator. The solution gives a series of states with quantized energies:

$$\frac{\hbar\omega_o}{2}, \frac{3\hbar\omega_o}{2}, \frac{5\hbar\omega_o}{2} \cdots E_n = \hbar\omega_o\left(n + \frac{1}{2}\right) \tag{11.79}$$

Quantization of the stable states means that the density of states is zero for energies intermediate between the quantized levels. Thus, electron transport is enabled only for quantized states that lie below the Fermi level. As the gate voltage V_G is reduced, more quantized levels will fall below the Fermi level, increasing the conductance between the source and drain. Following (11.72), each quantized level constitutes a channel that contributes $2\frac{q^2}{\hbar}$ to the conductance. Because the quantized levels are discrete, we would expect to see the conductance of the QPC increase in discrete steps as the gate voltage is increased, (i.e., made less negative).

Each channel can be regarded as a quantum state, with a well-defined energy. In the absence of a magnetic field, a state can be occupied by 2 electrons having opposite spins. In the presence of a magnetic field sufficiently large, this degeneracy can be lifted, and a state can be occupied by only one electron. Thus, the QPC is a structure that is ideally suited to control the transport of single electrons in suitable semiconductor materials such a modulation-doped AlGaAs/GaAs heterostructure.

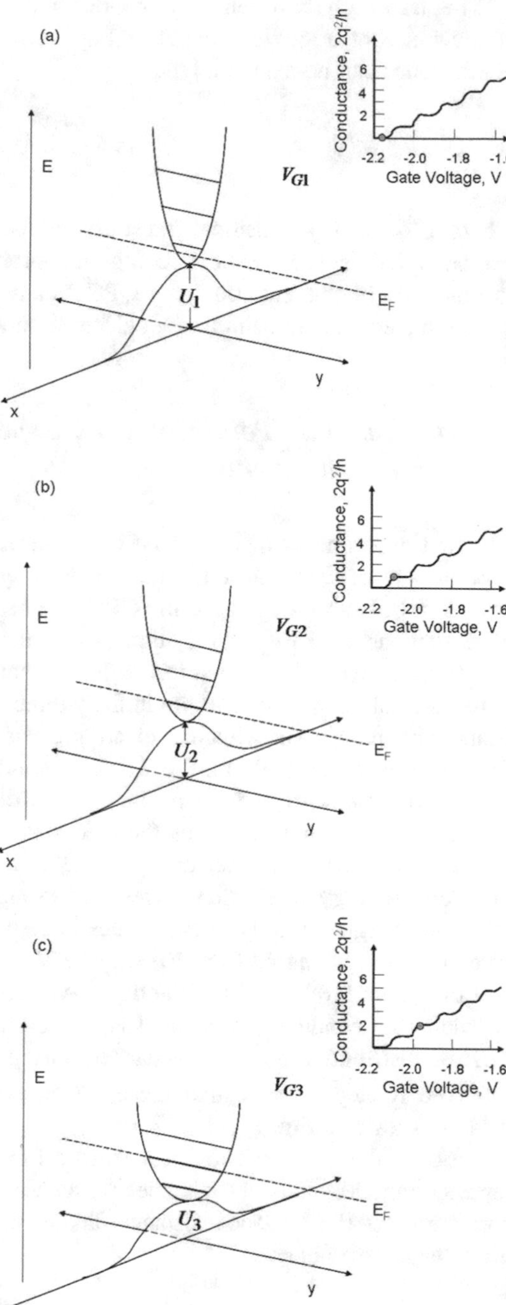

Fig. 11.12 Conductance in the channel between source and drain requires a finite density of states at the Fermi level. Here we illustrate how the gate voltage applied at the point contact can be used to change the width of the potential well of the saddle potential and align the allowed states of the well with the Fermi level. **a** All the allowed states lie above the Fermi level. Thus the density states in the region of the point contact is null and the conductance is zero, as shown by the orange dot in the inset on the upper right. **b** The gate voltage is increased in the positive direction to about $V_G = -2.1$ V. The lowest confined state of the potential well is aligned with the Fermi level and a single channel is opened with conductance $G = \frac{2q^2}{h}$. **c** The gate voltage is increased further to $V_G = -1.95$ V. The second confined state of the potential well is aligned with the Fermi level, and now two channels are open with a total conductance of $G = \frac{4q^2}{h}$

The transition between zero conductance, shown in Fig. 11.12a, and the conductance of a single channel in Fig. 11.12b is a most interesting case. The conductance can be expressed as:

$$G = \frac{2q^2}{h}\mathbf{T} \tag{11.80}$$

where **T** is the 1-dimensional transmission coefficient of the Landauer model. **T** can be varied between 0 and 1. Using an appropriate gate voltage, **T** can be tuned to equal ½. In this situation, the QPC can be designed into a beamsplitter for quantum electrons, transmitting an electron on average 50% of the time.

11.7.4 Quantum Point Contacts: Experimental Measurements

The first experiments that resolved the quantized behavior of conductance using a point contact were carried out independently (Wharam et al. 1988; van Wees et al. 1988). Both groups used the QPC as a beam-splitting component. van Wees et al. demonstrated the quantization of conductance in a $Al_{0.3}Ga_{0.7}As$ and GaAs quantum-point contact structure similar to that under discussion. Point contacts with a spatial opening of 250 nm in the *y*-direction were fabricated using electron-beam lithography. The effective electronic width of the confining channel when fully-open is larger: 360 nm. Electrons are confined in the *y*-direction by applying a negative bias voltage with respect to the drain. As shown above, the corresponding electric field confines the electrons to a narrow channel by creating a saddle potential well in the region of the points. The lateral confinement results in discrete energy levels that correspond to eigenstates of the potential well. At all other energies, the density of states is zero. Measuring at $T = 400$ mK, the gate voltage is increased from $V_G = -2.2$ V, causing the quantized states to intersect the Fermi level, one after the other. As each level falls below the Fermi level, an additional conductance channel is opened. Note that the Fermi energy of the electron distribution remains unchanged during the measurement. van Wees, et al. observed 16 steps in the conductance, quantized in steps of $\Delta G = \frac{2q^2}{h} = \frac{q^2}{\pi\hbar}$; their data are shown in Fig. 11.13.

Reading the measured resistance From Fig. 11.13a, we find $R_1 = 13.5$ kΩ, representing the quantized channel resistance plus the contact resistance. After correction for the contact resistance, taking the reciprocal of channel resistance gives the conductance:

$$G_1 = \frac{1}{12.9\,\text{k}\Omega} \cong \frac{2q^2}{h} \tag{11.81}$$

The clear resolution of the steps in channel conductance indicates that the transmission over the barrier along the *x*-direction in the channel is close to the

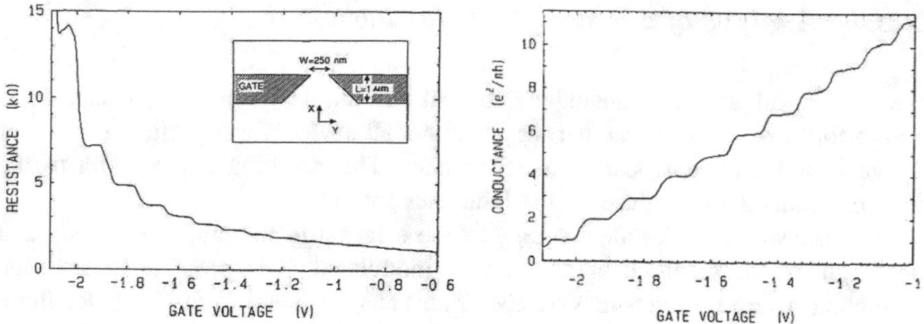

Fig. 11.13 Measurements of quantized conductance using a quantum point contact applied to a 2-dimensional electron gas in a modulation-doped structure of $Al_{0.3}Ga_{0.7}As$ and GaAs (van Wees et al. 1988). The measurements shown here were taken at a temperature of $T = 400$ mK and in the absence of an applied magnetic field. **a** Measurement of resistance for current in the channel formed by the point contact on the surface of the sample. Note that the electronic width of the channel is 360 nm. **b** The same data are plotted as the conductance which is the arithmetic reciprocal of the resistance (Reproduced from van Wees et al. (1988) by permission of the American Physical Society)

classic regime (i.e., there are no transmission resonances). Thus, the length of the transmission barrier U along this direction is considerably greater than the point contact separation and much greater than the Fermi wavelength, $\lambda_F = 42$ nm. This experimental measurement of quantized conductance is an unambiguous confirmation of the Landauer formula (11.80) for single quantum transmission over a 1-dimensional barrier. A complete, a rigorous, and very readable treatment of quantum conductance and quantum point contacts has been given by Beenakker and van Houten (1991). This review is an essential resource for the study of quantum electron transport. These authors have subsequently written an excellent introductory article dealing only with the QPC (van Houten and Beenakker 1996).

11.8 A Single Electron Beamsplitter: Fermion Interference at a Potential Barrier

A one-dimensional potential barrier is an important element in the study of quantum fermion behavior. In addition it is a fundamental building block for design and fabrication of single electron circuits. From this standpoint, the single most important part of relationship between conductance and gate voltage is the transition from $G = 0$ to $G = \frac{2q^2}{h}$. This transition is controlled by setting and maintaining the gate voltage to obtain the desired transmission, coefficient, for example, $\mathbf{T} = ½$. From an experimental standpoint, such a barrier, having a well-defined transmission coefficient is straightforward to fabricate using widely-available epitaxial film technology.

11.8.1 Analysis of a Beamsplitter for Fermions

The theoretical analysis, analogous to that for single photons in Chap. 5, is straightforward. A potential barrier is an ideal device for injecting quanta of charge into a 1-dimensional ballistic channel. The resulting contribution to the channel conductance is given by the Landauer formula.

In a case where the incident energy of the electron is maintained constant, and the height of the potential barrier can be modulated, it is possible to envisage a situation where the transmission coefficient can be tuned to 50%. Under these circumstances, a transmission barrier can also be used in an analogous way to an optical beam splitter.

We will consider the behavior of fermions incident on opposite sides of a quantum "beam splitter" with a transmission coefficient of 50%, which could be realized by using a potential barrier engineered to give the desired transmission coefficient. Because of the linearity of the governing Schrödinger's equation, transmission properties are the same in each direction. We recall from Sect. 5.6 that the interference of two identical photons incident simultaneously on opposite sides of a beam splitter results in interference of the wavefunctions so that photons exit in pairs on one side or the other of the beam splitter. In the case of fermions, interference also occurs, but the results are quite distinct from those measured for photons.

Referring to Fig. 11.14, we represent the single electron beam splitter by a box with inputs at a and b and exits at c and d. We represent the state of the fermions at the entry by $|1, 1\rangle_{ab}$,

indicating that an occupied fermion state is incident on each side of the beam splitter. According to (5.5), this incident state is transformed by the beam splitter as follows:

$$|1, 1\rangle_{ab} \to \frac{1}{2}\left\{\left(c^+c^+ - d^+d^+\right)|0, 0\rangle_{cd} + \left(c^+d^+ - d^+c^+\right)|0, 0\rangle_{cd}\right\} \tag{11.82}$$

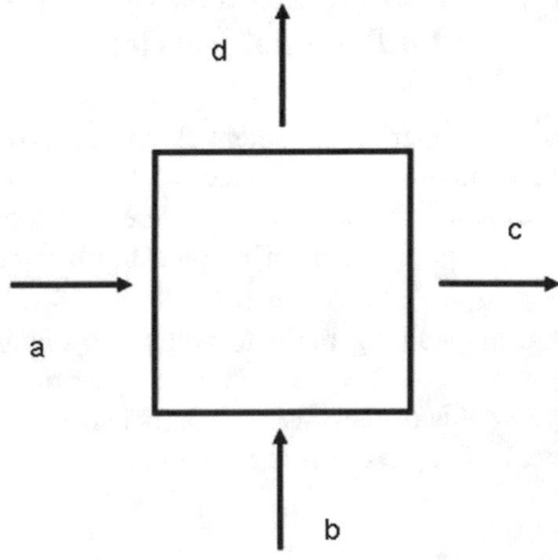

Fig. 11.14 A schematic representation of the passage of two fermions through a beam splitter with a transmission coefficient of 50%

Making use of fermion algebra:

$$c^+c^+|0,0\rangle = c^+|1,0\rangle = 0 = d^+d^+|0,0\rangle \tag{11.83}$$

$$c^+d^+|0,0\rangle = c^+|0,1\rangle = -1|1,1\rangle. \tag{11.84}$$

and

$$d^+c^+|0,0\rangle = 1|1,1\rangle. \tag{11.85}$$

Thus:

$$|1,1\rangle_{ab} \to -\frac{1}{2}(2|1,1\rangle_{cd}) = -|1,1\rangle_{cd} \tag{11.86}$$

That is, single fermions such as electrons incident simultaneously on opposite sides of a transmission barrier, exit separately from opposing sides of the barrier, rather than showing the pairing behavior observed for single photons (which are bosons). This is a result of the anti-correlation behavior of fermions.

11.8.2 Creating an Extended One-Dimensional Channel: Role of the Magnetic Field

In this section we will describe the design of an electron beam-splitter using a QPC, because this particular application is the basic element in nearly all single quantum electron experiments carried out in GaAs 2 DEG samples.

A beam-splitter should produce both a transmitted and a reflected stream of quanta, obeying the basic relationship: $\mathbf{T} + \mathbf{R} = 1$. Just as in the case of quantum photons, both the transmitted and reflected streams of electrons are needed to perform correlation measurements. Conceptually, it is easy to imagine detecting the transmitted stream of electrons that pass through a QPC. But it is not so straightforward to separate the reflected stream from the incident stream on the other side. An attractive solution would be to restrict the quantum electrons to a 1-dimensional channel, not just at the QPC, but throughout the channel. The application of a magnetic field perpendicular to the plane of the 2 DEG achieves precisely this result.

The effect of the magnetic field is to confine the electrons in the 2 DEG to a set of energy levels:

$$E_n = n\hbar\omega_c = \hbar\frac{qB}{m^*}, \text{ where } \omega_c \text{ is the cyclotron frequency.} \tag{11.87}$$

When the magnetic field is sufficiently large the magnetic energy levels or Landau levels dominate the energy structure for electrons. This regime is reached when:

$$\mu B > 1 \tag{11.88}$$

For the case of a 2 DEG in GaAs, with a mobility of $\mu = 10^6$ cm^2-V^{-1}s^{-1}, the high magnetic field regime is reached when the magnetic field exceeds $B = 0.1$ Tesla, which is quite a modest magnetic field. If the magnetic field is further increased to the level of a few Tesla, the quantum Hall regime obtains. In this environment, the interior region of the sample behaves like an insulator, and electrons are free to move only around the edge of the sample. Since the electrons are already confined to the 2 DEG, this additional restriction creates a 1-dimensional linear channel around the circumference of the sample. An additional feature of this magnetic field regime is that all backward scattering in the 1-dimensional electron channel is suppressed (Büttiker 1988). Electrons can propagate only in single file around the edge of the sample, referred to as chiral conduction.

The physics of this transformation is the basis of the quantum Hall effect (QHE). It has been the subject of three Nobel Prize awards. While a worthy treatment of QHE is outside the focus of this text, an excellent introductory description has been given by Halperin (1986), and an equally good in-depth treatment by Tong (2016).

11.9 Single Fermion Circuits: Some Examples

A basic quantum-electron beamsplitter is shown in Fig. 11.15. The applied magnetic field assures that both the transmitted beam and the reflected beam are separated from the incident beam and conserved by confining electron flow to edge states, creating two distinct arms for experimental measurement.

11.9.1 Hanbury-Brown Experiments for Electrons

Elegant and convincing experiments that measure noise and correlations in a stream of electron quanta have been carried out by several groups, Henny et al. (1999), Oliver et al. (1999), using sample designs that mimic for electrons the Hanbury Brown Twiss setup for photons. The basic idea of the measurement is to use a beam-splitter to separate a 1-dimensional electron beam into two components which are compared using an electronic correlator. A diagram of the device used by Henny, et al. is shown in Fig. 11.16 (Oberholzer et al. 2000).

In Chap. 2, (Fig. 2.14) we have considered a similar experiment for single, countable photons. In the case where only one particle passes the beam splitter during a measurement interval, be it a photon or an electron, it is either transmitted or reflected. Thus the transmitted and reflected particle streams are fully

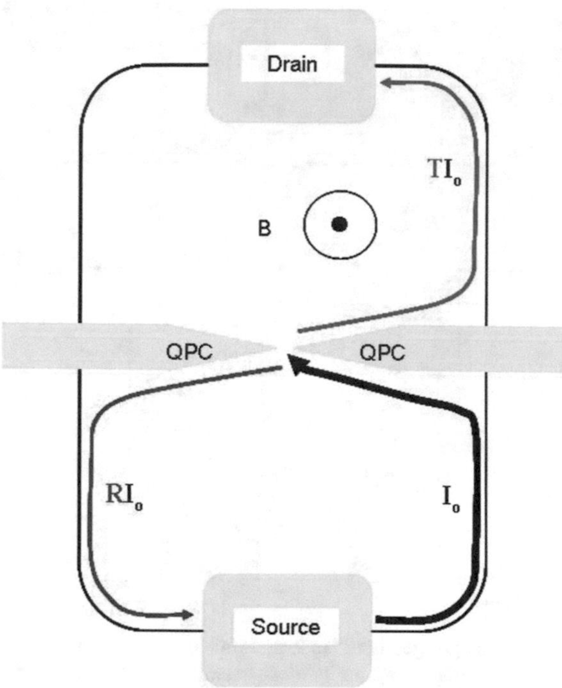

Fig. 11.15 Schematic diagram of a quantum point contact (QPC) functioning as an electron beam splitter. The sample physical boundary is shown in black. The applied magnetic field is directed out the plane of the paper. The field is large enough to assure chiral and 1-dimensional conduction in Landau level edge states around the periphery of the sample. An appropriate bias voltage with respect to the drain is applied to the QPC contacts. This redefines the periphery electrostatically, creating a conducting pathway between the points of the QPC contacts. The incident electron stream $\mathbf{I}_o$ is shown as a heavy black line, The transmitted beam T $\mathbf{I}_o$ is shown in blue and the reflected beam R in red: $T\,\mathbf{I}_o + R\,\mathbf{I}_o = \mathbf{I}_o$

anti-correlated. In the present case, technology to detect single electrons was not available, and this required statistical analysis of the electron stream, that is, measurement of fluctuations in the current, to determine the level of correlation or anti-correlation between the transmitted and reflected streams.

The transmitted and reflected currents are measured separately, amplified and compared in a correlator. The autocorrelation function of the transmitted beam measures the fluctuations (that is, the noise) in the number of electron particles in the stream. These fluctuations increase proportionally to the incident current, I_o. The cross-correlation between the transmitted and reflected beam behaves differently. If electrons in the transmitted and reflected beams are anti-correlated, then the cross-correlation must decrease as the current is increased (Oberholzer et al. 2000).

With reference to this measurement, the auto correlation function is expressed:

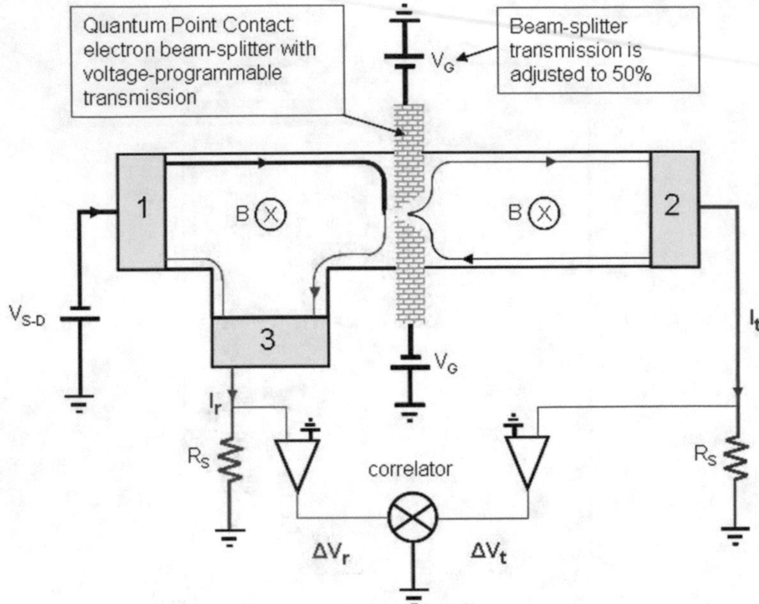

Fig. 11.16 Sketch of a sample geometry to enable measurements of correlation between two electron streams created by impinging an incident beam on an electron beam-splitter. Electrons are introduced through ohmic contact (1) into a degenerate 2-dimensional electron gas. The sample temperature is 4 K or less. The beam-splitter is achieved using a quantum point contact, set to transmit 50% of the electrons toward contact (2). A magnetic field in the quantum Hall regime is necessary to ensure orderly chiral transport along the edges of the sample, as shown. The sample size is on the order of a few μm assuring that transport occurs in the ballistic regime (From Oberholzer et al. (2000), reproduced by permission)

$$\left\langle (\Delta n_t)^2 \right\rangle = \mathbf{T}(1-\mathbf{T})\left[\frac{\mathbf{T}}{1-\mathbf{T}}\left\langle (\Delta n)^2 \right\rangle + \langle n \rangle\right] \tag{11.89}$$

Since $\left\langle (\Delta n)^2 \right\rangle$ and $\langle n \rangle$ are both positive numbers, the auto correlation is ≥ 0.

The cross-correlation is

$$\langle \Delta n_r \Delta n_i \rangle = \mathbf{T}(1-\mathbf{T})\left[\left\langle (\Delta n)^2 \right\rangle \langle n \rangle\right] \tag{11.90}$$

Comparing (11.90) with (11.33),

$$\langle \Delta n_r \Delta n_i \rangle = \mathbf{T}(1-\mathbf{T})[V(n) - \langle n \rangle] = \mathbf{T}(1-\mathbf{T})\left[-\langle n \rangle^2\right] \tag{11.91}$$

where: $V(n) = \langle n \rangle - \langle n \rangle^2 =$ the variance of the Fermi-Dirac distribution.

Since $\langle n \rangle^2$ is always positive, it is immediately clear that the cross-correlation between two fermion streams is negative.

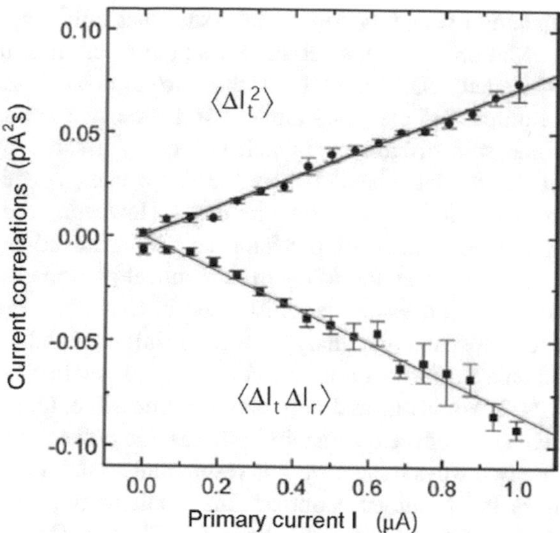

Fig. 11.17 Auto correlation of the transmitted electron stream $\langle \Delta I_t^2 \rangle$ (highlighted in red) and the cross-correlation of the transmitted and reflected electron streams $\langle \Delta I_t \Delta I_r \rangle$ (highlighted in blue) passing through a 50–50% beam splitter, implemented by a quantum point contact (QPC). The auto-correlation increases with current, while the cross-correlation decreases with increasing current, reflecting quantum fermion behavior (Reproduced from Oberholzer, Capri School 2006, with permission of the author)

Henny et al. (1999) and Oberholzer et al. (2000) demonstrated that the edge channel which transports electrons from the contact (1) to the OPC is fully occupied. Thus there is no uncertainty in the occupation number and the shot noise is null. We can use this information to simplify (11.88) and (11.90) as follows:

$$\begin{aligned} \left\langle (\Delta n_t)^2 \right\rangle &= \mathbf{T}(1-\mathbf{T})\langle n \rangle, \\ \text{and } \langle \Delta n_r \Delta n_i \rangle &= -\mathbf{T}(1-\mathbf{T})\langle n \rangle \end{aligned} \tag{11.92}$$

Since the total current in the circuit is proportional to the average number of carriers, the autocorrelation and cross-correlation are equal, but opposite in sign and directly proportional to the current. Experimental results of Henny and Oberholzer, shown in Fig. 11.17, display precisely this behavior, Oberholzer (2006).

Parmentier has given in-depth treatment of the Hanbury Brown and Twiss measurement for electrons, including sample structure, experimental conditions, and analysis of fluctuations (Parmentier 2010).

11.9.2 Simultaneous Incidence on a Beam-Splitter: Quantum Interference Between Electrons

Next, we illustrate the behavior of two electron quanta incident on opposite sides of a 50–50% beam splitter. A sufficient requirement for the observation of the interference of quantum wavefunctions is to assure that only single electrons are incident on the beam splitter. This requires a single-electron emitter. Two such components are needed, one for each of the two sources. Interference at the quantum level requires that electrons emitted from the two sources be indistinguishable: the same spin orientation (which is happily assured by the magnetic field used to create the 1-d channel) and the same energy which is achieved to a good approximation by process technology. However, in this experiment we are not in the same domain of precision as that achieved by using spontaneous parametric down-conversion to create 2 identical photons in the case of the Hong-Ou-Mandel (HOM) measurement discussed in Chap. 5.

We have already presented the analysis of the result of simultaneous incidence of 2 indistinguishable fermions on a particle beam splitter in Sect. 11.8.1, using quantum operators. There is one and only one result possible. One fermion exits in each of the exit channels of the beamsplitter; that is, the cross-correlation is 100%. Note that there are two ways to achieve this result, and so the interference of wave functions entangles the fermions. Contrast this result to that of the case of the quantum interference of bosons that is described in Chap. 5. Quantum interference of bosons on a beamsplitter produces two distinct results, each with equal probability: 2 bosons in one exit channel or 2 bosons in the other exit channel. However, like the case for fermions, the bosons are entangled also. These simple observations illustrate that quantum fermion interference on a particle beam-splitter is not a simple analogue to the HOM measurement for photons. There are similarities and differences, and the differences are important.

11.9.2.1 A Single Electron Emitter

A key component is the single-electron emitter. Fève et al. have developed and demonstrated an electron emitter capable of releasing a single electron on electrical command, which we refer to as a quantum electron pump (Fève et al. 2007). Two such pumps can be controlled independently, so that there can be a fixed time delay between the electron emission events. In the experiment which we describe below, this electrically-programmable time delay permits the synchronisation of the time of arrival of two electrons on either side of a beam-splitter.

The quantum electron pump consists of a quantum dot (Q-D), which is a potential well with quantum confinement in all three dimensions. A simple square-wave potential is used to raise or lower the potential of the Q-D with respect to the transmission potential of an adjacent QPC. When the highest occupied level rises above the QPC transmission energy, an electron is emitted across the barrier into a channel. The Q-D thus acquires a net positive charge, and its potential immediately decreases, cutting off further electron emission until the next cycle of the square wave excitation. The energy of the emitted electron is essentially determined by the potential of the QPC which remains constant during the experiment.

We can use a semi-classical estimate to get an idea of how this electron pump functions. First, assume a 2 DEG with concentration $n_s = 2 \times 10^{15}\ \mathrm{m}^{-2}$. Thus, the Fermi velocity $v_F = 1.9 \times 10^5\ \mathrm{ms}^{-1}$. Next, assume a square-wave excitation with frequency $f = 2\,\mathrm{GHz}$. The corresponding period is $\frac{1}{f} = 500\,\mathrm{ps}$. If the channel region of the sample were sufficiently long, the spacing between electron wavepackets would be $\Delta d \approx 5 \times 10^{-10} \cdot 1.9 \times 10^5 \approx 90$–$100\ \mu\mathrm{m}$. Further, let us suppose that the half-width of the electron wavepacket were much less than the period, for example: 50 ps. This would correspond to a spatial width of $\Delta w = 9.5\,\mu\mathrm{m}$. This simple estimate shows that the resulting electron train would consist of 1-electron wavepackets well-separated one from another. It also shows that when an electron wavepacket enters a circuit of dimensions of a few microns, it will occupy the entire circuit volume. And finally, there is only one wavepacket on each side of the beamsplitter at any one time, followed by a considerably longer time period where no electrons are present. These would be the right conditions to measure quantum interference of single electron wavepackets incident on opposite sides of an electron beam splitter.

11.9.2.2 The Noise is the Signal

The Hong-Ou-Mandel (HOM) experiment (see Chap. 5, Sect. 5.6) for photons measures the cross-correlation between the arrival of two photons at a beamsplitter. When the wavefunctions of two bosons interfere, the output consists of two photons that exit together in one or the other of the two exit directions. As a result, the cross-correlation $\langle I_1 I_2 \rangle$ between the signals received at the detectors 1 and 2 decreases to zero at full interference (see Fig. 5.10). In the present case involving fermions, the situation is different. The measurement of the cross-correlation of the single electrons incident on opposite sides of a beamsplitter reaches a maximum at full interference, as we have already demonstrated in Sect. 11.8.1. Such an experiment would not represent the straightforward analogue of the HOM measurement for photons. The elegance of the HOM experiment is that a null signal is measured when boson interference occurs, and zero is a specific number. Thus, a HOM experiment for fermions should be redesigned to identify and measure the quantity that does go to zero when 2 electrons interfere at a beamsplitter. This quantity is the noise of the transmitted electron signal.

The noise in a one-dimensional stream of electrons occurs only because of uncertainty in the arrival time of quanta. Where there is no uncertainty, the noise is zero. Henny, et al. used this principle for the auto-correlation and cross-correlation measurements in the Hanbury Brown and Twiss described in Sect. 11.9.1 (Henny et al. 1999). In the present case of two-electron interference there is one case where there is no uncertainty, and therefore total noise suppression: the simultaneous incidence of identical fermions. This situation is schematically indicated in Fig. 11.18a. All other outcomes (see Fig. 11.18b) resulting from non-simultaneous incidence are uncertain to some degree and generate noise.

We can represent the normalized wavefunctions of the two incident electrons as φ_1 and φ_2. If the two electrons are indistinguishable, they have the same quantum numbers, and can differ at most by a phase factor. That is:

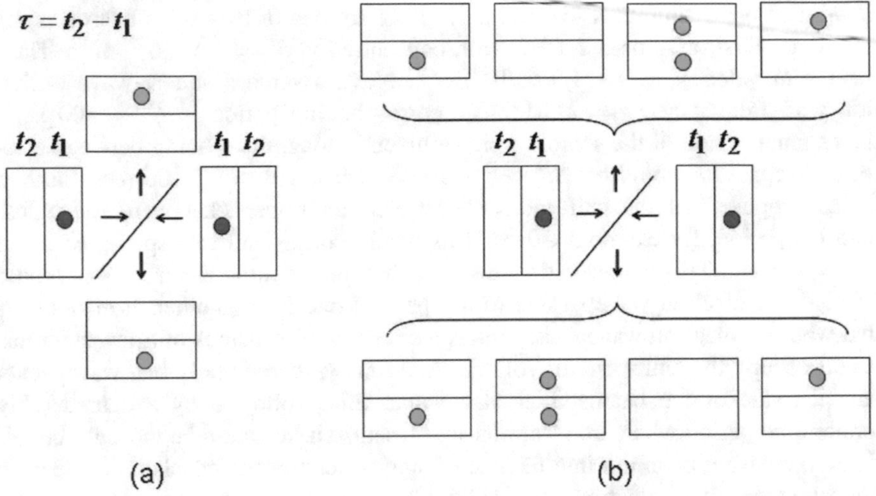

Fig. 11.18 Schematic diagram of the outcomes of two indistinguishable fermions incident on opposite sides of a quantum point contact beam splitter. The blue-colored icons represent incident electrons, while the green-colored icons represent transmitted or reflected electrons. The rectangular boxes represent intervals of time, with the duration of 1 interval in this being equal to the time delay between the arrival of the 2 interfering electrons. **a** Simultaneous incidence. Only one outcome is possible. There is no uncertainty and therefore no associated noise. **b**. Non simultaneous incidence. One entry configuration can result in four distinguishable outcomes with equal probability. This uncertainty creates noise in the measured output signal

$$\phi_2(k) = \phi_1(k)e^{ikv_F\tau} \tag{11.93}$$

where v_F is the Fermi velocity, and τ is the time delay between the arrival time of the two quanta.

This experiment, which defines the fundamental nature of quantum electrons on the basis of the noise of the current, has been discussed and analyzed (Jonckheere et al. 2012). The noise in the current as a function of the delay in the arrival times of the two electrons can be compared to the noise in the current when only one electron is incident on the beam splitter. This is, of course, the situation of the Hanbury Brown and Twiss (HBT) measurement. Jonckheere, et al. have shown that the noise in the one arm of the interferometer: $S_3(\tau)$, relative to the noise of a single fermion in the HBT measurement can be expressed as:

$$\frac{S_{33}}{2S_{\text{HBT}}} = 1 - \left| \frac{\int_0^\infty dk\varphi_1(k)(1-f_k)\varphi_2^*(k)(1-f_k)}{\int_0^\infty dk|\varphi_1(k)|^2(1-f_k)^2} \right|^2, \tag{11.94}$$

where f_k is the Fermi distribution function. Substituting for $\varphi_2(k)$,

$$\frac{S_{33}}{2S_{\text{HBT}}} = 1 - \left| \frac{\int_0^\infty dk|\varphi_1(k)|^2(1-f_k)^2 e^{-ikv_F\tau}}{\int_0^\infty dk|\varphi_1(k)|^2(1-f_k)^2} \right|^2 \tag{11.95}$$

It is seen straight away that in the case of simultaneous arrival of 2 fermions when $\tau = 0$, the noise $S_{33} = 0$.

At the other extreme, when $\tau \gg t\left(\frac{1}{2}\right)$ the half-width in time of the wave packet, the second term in (11.94) averages to 0:

$$\int_0^{\infty} dk|\varphi_1(k)|^2(1 - f_k)^2 e^{-ikv_F\tau} \to 0 \tag{11.96}$$

The electrons are independent of each other. Each of the 2 electrons contributes the same level of noise as it would in a HBT measurement so that,

$$S_{33} = 2S_{\text{HBT}} \tag{11.97}$$

11.9.2.3 Experimental Measurement of Quantum Interference

Bocquillon and co-workers demonstrated the interference of quantum wavefunctions of two electrons in an elegant experiment that illustrates the behavior analysed by Jonckheere et al. (Bocquillon et al. 2013). The structure of the sample used to measure two-electron interference is shown in Fig. 11.19. It can be seen that the layout is quite similar to that used for measuring correlations in the experiment of Henny et al. (1999) shown previously in Fig. 11.16. The key distinguishing difference is the replacement of each of the two ohmic contact sources by a single electron pump, synchronized so that the contact controlled by QPC-2 injects a single fermion quantum at a controlled delay τ relative to the emission controlled by QPC-1. The two single electron sources are placed 3 μm distant from the beamsplitter.

The electron concentration of the 2 DEG used in their experiment is $n_s = 2.1 \times 10^{15}\,\text{m}^{-2}$. Measurements are taken at low temperature: $T = 100\,\text{mK}$ in a high magnetic field: $B = 2.68T$.

The current I_3 is detected and amplified as shown Fig. 11.19. Signal averaging takes place over a period much longer than the repetition rate of electron emission. The noise in the measured current constitutes the signal of interest. The data obtained by Bocquillon, et al. are shown in Fig. 11.20. There is a clear dip in the noise level near, but not exactly at the time delay $\tau = 0$. We also can see that the noise minimum does not reach zero. These two results indicate that the 2 electron quanta are not totally indistinguishable. Since the spins are aligned by the magnetic field, it can only be the case that the energies of the electrons are not the same. This difference is likely caused by a difference in the transmission potential of the controlling gates QPC1 and QPC2. Since the emission time from the single-electron pump depends on the transmission potential of the QPC, a difference in the transmission potential of the QPC gates creates both a difference in the energies of the emitted electrons and affects the time delay.

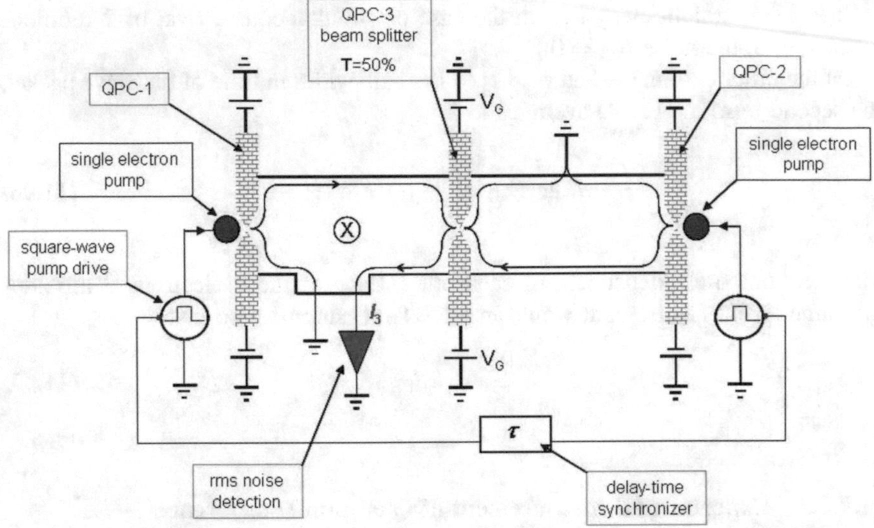

Fig. 11.19 Schematic diagram of the quantum electron circuit used to measure correlations between single electrons incident on opposite sides of a beam-splitter implemented by a quantum-point contact. The circuit layout is similar to that used to measure electron correlations in the fermion analogue to the Hanbury Brown and Twiss experiment discussed in Sect. 11.9.1. A contact to ground between QPC-3 and QPC-2 eliminates back reflections from reaching QPC-2. The key difference is the use of single-electron pumps to enable synchronized single-electron incidence on the beamsplitter. Note that 3 QPC devices are used to implement this device, demonstrating importance of the QPC as a fundamental "building block" of quantum electron circuits

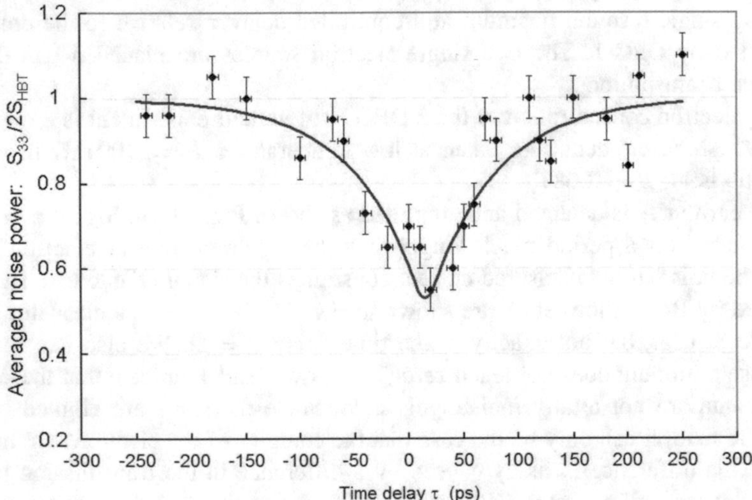

Fig. 11.20 The measured noise resulting from the incidence of two electrons on a beam-splitter depends on the difference in the time of arrival τ. The minimum in the noise indicates simultaneous incidence. The solid line has been calculated using an exponential lineshape function (11.98). The data have been taken by Bocquillon et al. (2013) and are reproduced here by permission of the authors

We have fit a simple linewidth function to the data using the expression proposed by Bocquillon based on (11.94):

$$\frac{S_{33}}{2S_{\mathrm{HBT}}} = 1 - \gamma e^{-\frac{|\tau-\tau_0|}{\tau_{\frac{1}{2}}}} \tag{11.98}$$

The solid line in Fig. 11.20 was calculated using $\gamma = 0.5$, $\tau_o = 15\,\mathrm{ps}$, and $\tau_{\frac{1}{2}} = 62\,\mathrm{ps}$. This function reproduces the minimum in the noise at simultaneous incidence to the two electrons in a convincing manner. Bocquillon et al. (2013) call this minimum the *Pauli Dip*, in reference to the requirements of the Pauli exclusion principle.

11.9.2.4 Mach-Zehnder Interferometer: A Single-Particle Quantum Interference Device

In thin-film optics, the Mach-Zehnder interferometer (MZI) is a classical device in which a monochromatic single-mode photon beam passes through a beam splitter and propagates in two arms. In one arm there is a component that retards the phase of one photon beam relative to the other, effectively changing the propagation length of one arm relative to the other. A schematic of this device for photons is diagrammed in Fig. 11.21. Perturbations in the optical length of the test arm relative to the reference arm will result in a modulation of the output intensity.

Interference occurs when the beams are recombined at a second splitter. If they are in phase, 100% of the signal is recovered. If they are 180° out of phase, 0% of the beam is transmitted. For intermediate values of the phase difference, partial transmission is observed. This configuration is often used to modulate a beam of light. The visibility of the interferometer is defined as:

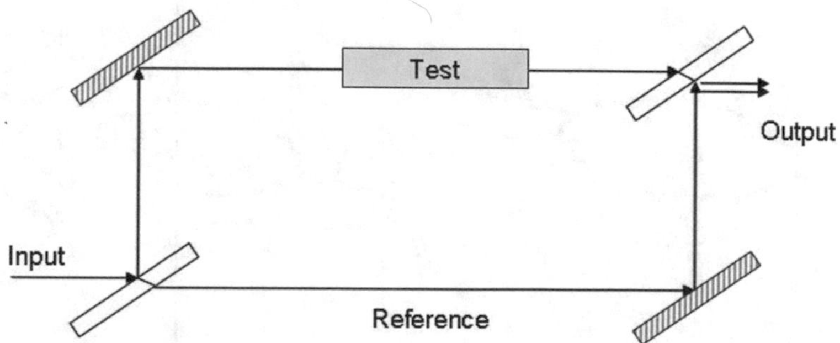

Fig. 11.21 The Mach-Zehnder interferometer for photons. The input beam is split into two equal components that propagate through separate arms of the interferometer prior to recombination at the output

$$v \equiv \frac{I_{max} - I_{min}}{I_{max} + I_{min}} \tag{11.99}$$

MZI designs for photons are commercially available having visibilities $v > 0.99$ for applications such as sensors, optical modulators and switches.

The MZI can also be used to study single photons. In this case, the length of one arm can be changed using a "trombone arm delay line" (see Fig. 5.11). The photon interferes with itself, demonstrating the non-local character of the single photon wavefunction.

Electron interference can be measured using a MZI. Marton, et al. presented an early example of ballistic electron propagation in a MZI device that used sequential 15 nm gold foils as beamsplitters and as mirrors to guide the electrons beams in a high-vacuum environment (Marton et al. 1954). As in the optical case, output intensity depends on the phase difference between the electron wavefunction in the two beams that are made to interfere at the output. There are choices for introducing this phase difference. In addition to the method used by Marton, another possibility is to impose a magnetic field in a direction perpendicular to the plane of the interferometer. This results in a phase difference proportional to the magnetic field. This idea was first described by Ehrenberg and Siday (1949), and rediscovered more than a decade later by Aharonov and Bohm (1959).

The Aharonov-Bohm (A-B) effect, as it has become to be known, is a thought experiment. A beam of electrons is incident on a MZI. A 1-dimensional magnetic field **B** is threaded through the interior of the interferometer. The vector potential **A** for the magnetic field lies in the plane perpendicular to the magnetic field. The electrons propagate around both sides of this solenoidal field and recombine on the other side. A diagram of this situation is shown in Fig. 11.22a

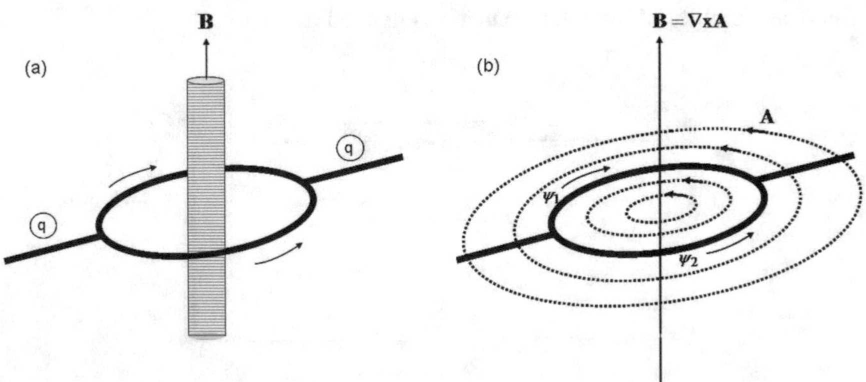

Fig. 11.22 Schematic diagram of the Ehrenberg-Siday or Aharonov-Bohm thought experiment. **a** An electron particle wave propagates on either side of an extended microscopic conducting solenoid that produces a magnetic field **B** that is directed perpendicular to the plane of propagation and non-zero only inside of the solenoid. **b** The magnetic field is accompanied by its vector potential which is non-zero and which lies in the plane of propagation

The electron wavefunction ψ is divided into two branches, representing the probability density for the two transmission paths around the solenoid region. Note that the propagation direction along the right-hand path is parallel to the vector potential **A**, whereas for the left-hand branch propagation is anti-parallel to **A**. This distinction leads to a phase difference and thus to quantum interference between the two components of the wavefunction when the two paths merge at the exit.

The Hamiltonian for the Schrödinger equation of the electron wave function in the presence of the vector potential is:

$$\mathcal{H} = \frac{1}{2m}(\mathbf{P} - q\mathbf{A})^2 \tag{11.100}$$

The solution to (11.100) has the form $\psi = \psi^o e^{-i\varphi}$, where:

$$\phi = \frac{q}{\hbar}\oint \mathbf{A} \cdot ds \tag{11.101}$$

The phase $\phi = \phi + 2n\pi$, assuring that the electron wavefunction is single-valued around the ring. The interference between the two components is given by the difference in the accumulated phase in each branch. We recall that the integral of the vector potential along a closed path is equal to the magnetic flux enclosed by the path.

$$\phi_1 - \phi_2 = \frac{q}{\hbar}\oint \mathbf{A} \cdot \mathbf{ds} = 2\pi\frac{q}{h}\Phi = 2\pi\frac{BA}{\Phi_o}, \tag{11.102}$$

where **B** is the magnetic field, A is the area enclosed by the MZI and $\Phi_o = \frac{h}{q} = 4.14 \times 10^{-15}$ webers.

The phase difference can be varied by changing the magnetic field, or by keeping the field constant and changing the area of the region where the magnetic field is non-zero. This interference is periodic resulting in A-B oscillations with period $\frac{h}{q}$ in the current exiting the structure.

Testing the physics of this effect has posed a major challenge to experimentalists because of the need to invent nanofabrication technologies with the capacity to create samples in which the effect could be measured. Some fifty years later, Pozzi et al. (2016) are perhaps the first to have demonstrated interference in a sample structure that corresponds directly to the original thought experiment. The straightforward analysis of the A-B effect depends on the assumption that the magnetic field is confined to the interior of the interference path. Experimentalists, however, typically work with uniform large area magnetic fields into which the entire sample is immersed. It is now understood that the passage of electron wavefunctions in opposing directions around an enclosed magnetic flux will result in observable quantum interference, although the analysis becomes more complicated. A-B oscillations can be observed in solid-state samples under the influence of a uniform magnetic field, starting with the experiments of Webb et al. (1985) cited at the beginning of this chapter.

The MZI for quantum electrons in a 2 DEG consists of two chiral circuits that assure two separate 1-dimensional paths from the source to the drain. A diagram of such a circuit is shown in Fig. 11.23a. The establishment of a chiral circuit requires the presence of a magnetic field perpendicular to the plane of transport. The resulting MZI for a 2 DEG is thus already in the required configuration to observe A-B oscillations.

Ji, and co-workers at the Weizmann Institute have designed, fabricated and measured the quantum transport properties of such a circuit (Ji et al. 2003).

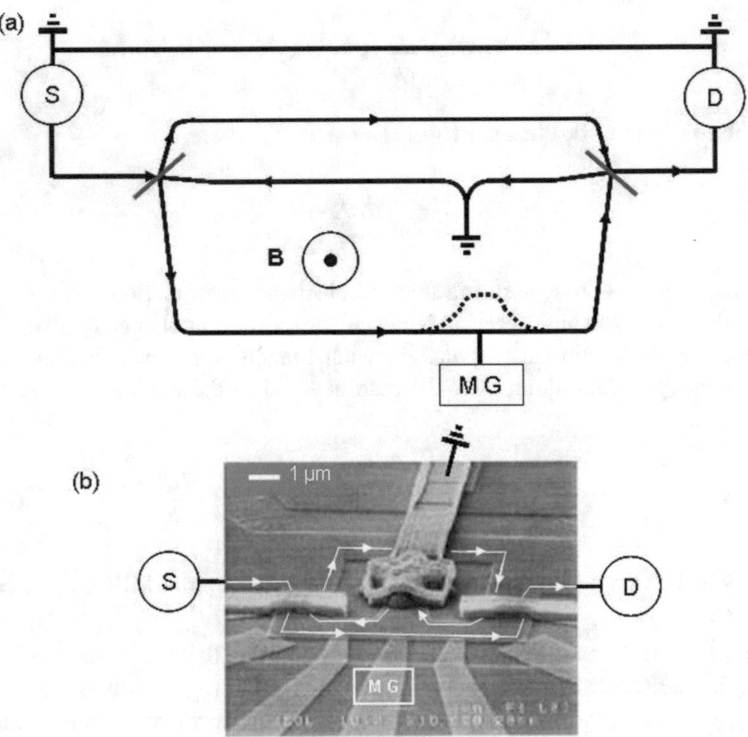

Fig. 11.23 A Mach-Zehnder interferometer for electron interference. **a** A schematic diagram of an electron MZI. This structure is more complex than that of a MZI for photons shown in Fig. 11.21. A chiral circuit is required for each branch of the interferometer. As shown previously in Fig. 11.19, a contact to ground between the drain D and the source S prevents back reflections from reaching S. The area enclosed by the MZI can be varied applying an electrostatic potential via the modulation gate (M G). A potential applied to this gate changes the perimeter of the electron path as indicated by the dotted line, thus changing the area *A* of the circuit. **b** A scanning electron micrograph of an MZI realized by Ji, et al. in a GaAs/AlGaAs 2 DEG. The annotations show the paths of chiral conduction in the 2 branches of the interferometer and the placement of the modulation gate. The contact to ground is achieved via an air bridge. The upper MZI path from S to D passes under this bridge (Adapted from Ji et al. (2003) by permission from Springer Nature)

An annotated scanning electron micrograph of their MZI sample is shown in Fig. 11.23. An important feature of their design is the inclusion of a modulation gate (M G). This is an electrostatic contact that implements local depletion of the 2 DEG, which changes the transport path, effectively modulating the area and thus the flux contained inside the perimeter of the MZI.

Example 11.2 Estimate the periodicity of A-B oscillations in a MZI fabricated in a GaAs/AlGaAs 2 DEG sample, such as that used by Ji et al. (2003) For simplicity, assume an area $A = 10\,\mu\text{m}^2$ and a magnetic field $\mathbf{B} = 1T$. The flux would be $\Phi = 10^{-11}$ webers. The variation of the magnetic field or area required to create one complete A-B oscillation is approximately 4 parts in 10^4. In the experiment of Ji, et al. a magnetic field of $\mathbf{B} = 5.5T$ was used. The area enclosed by the electron path was $A = 4.5 \times 10^{-11}\,\text{m}^2$. In this particular case the variation of the magnetic field corresponding to one A-B oscillation would be about $\Delta\mathbf{B} = 1.5 \times 10^{-5}$.

In this experiment, modulation of the area by applying an electric field is relatively straightforward. Ji et al. (2003) used this approach to capture A-B oscillations with visibility $\mathcal{V} > 0.6$. A result is shown in Fig. 11.24.

To achieve this result, data were taken at extremely low temperature, (T = 20 mK). At higher temperatures (T = 100 mK), the visibility decreased to near zero, showing that the phase coherence between the two transport paths was nearly non-existent under this latter condition. This result in itself is an extremely powerful comment on ballistic transport of electrons. The modulation-doped 2 DEG structure is effective at elimination of ionized-impurity scattering.

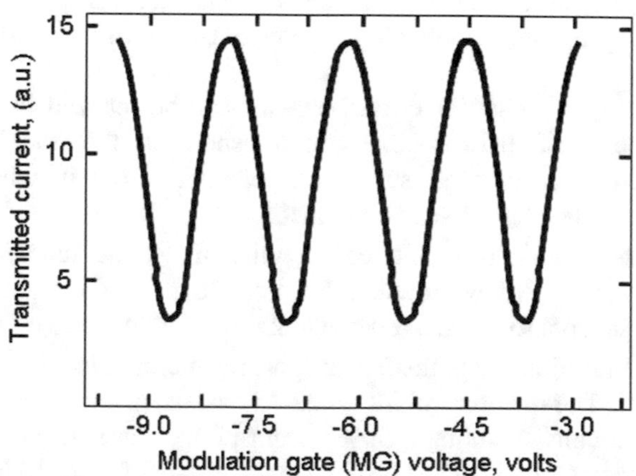

Fig. 11.24 Experimental recording of A-B oscillations of electrons traversing a MZI in a magnetic field of 5.5 T at T = 20 mK. The flux enclosed by the circuit is varied by changing the area of the circuit using an electric field applied to the modulation gate shown in Fig. 11.23. Each period corresponds to a change in the flux by one flux quantum $\frac{h}{q}$ (Adapted from Ji et al. (2003) by permission from Springer Nature)

This opens the curtain to quantum electron transport in samples having a dimensions less than 10 μm. However, modulation-doping does not eliminate all scattering. In particular, electron-electron interactions are not affected by modulation-doping. In fact, electron-electron scattering is most likely responsible for the degradation of visibility with rising temperature (Marguerite et al. 2016). Study of electron-electron interactions using the MZI is an area of current experimental and theoretical interest (Neder et al. 2006; Tewari et al. 2016; Duprez et al. 2019).

11.10 Summary

It has taken decades of painstaking research and engineering to develop solid-state samples in which electrons can be made to propagate in a ballistic fashion, analogous to photon propagation. There have been important unexpected discoveries along the way, such as the quantum Hall effect. Electrons and photons have common attributes and also display important differences. Populations of photons can show various degrees of correlation depending on the coherence of the source. Electrons, on the other hand show only anti-correlation, dictated by Fermi-Dirac statistics.

We have presented some key experiments on the ballistic propagation of boson and fermion isotopes of He that show these properties. Experiments involving ballistic electrons require careful sample preparation that confines the electron to a 2 DEG with very high mobility at very low sample temperatures. While propagation of ballistic photons is the rule, creating an environment for ballistic transport of electrons requires special conditions. Forcing the electron to pass through a narrow gap restricts the electron to one-dimensional transport in this region, and is a first step toward establishing extended 1-dimensional ballistic transport of electrons.

A fruitful study of single electron behavior can be realized in a sample that supports 1-dimensional ballistic electron transport. In this situation, an electron that encounters an obstacle such as a potential barrier is either transmitted or reflected. Landauer has first shown theoretically that the conductance of an electron in this situation is quantized in units of $\frac{q^2}{h}$, the fundamental unit of quantum conductance. A key structure for creating a one-dimensional channel is the quantum point contact. We have shown that the conductance of such a channel is finite and quantized in the ballistic transport regime in units of $\frac{q^2}{h}$, verifying the Landauer model. The quantum point contact is a fundamental building block for single electron circuits, enabling a 50% “beamsplitter” for single electrons.

While the geometry of Fig. 11.16 resembles that of Fig. 10.6, it is important to remember that electrons and photons are very different fundamental particles. For example, additional interactions in quantum fermion experiments are the coulomb repulsion and the exchange interactions between electrons. In order to get meaningful results from the sample geometries shown in Figs. 11.16 and 11.19 or Fig. 11.23 a number of special conditions must be satisfied:

- Ballistic, linear one-dimensional electron transport
- A degenerate, 2-dimensional electron gas with a mean-free-path for collisions that is greater than the sample dimensions
- Measurements at low temperature: $T < 100$ mK or even very low temperature ($T < 10$ mK)
- Nano-fabrication of an operational electron beam splitter: the quantum point contact (QPC)
- Programmable single fermion emission
- Measurements at high magnetic field in the regime of the quantum Hall effect (QHE).

These conditions can be obtained readily in a state-of-the-art laboratory equipped for magneto-transport studies, but they also raise a point of reflection regarding the potential application of results in practical electronic devices.

The Mach-Zehnder interferometer (MZI) is a device that is widely used in classical and quantum optics measurements. Commercially-available devices can be used in a wide variety of operating conditions. Integrated photonics is used to fabricate monolithic MZI with dimensions of a few cm, which are used to modulate optical fiber telecommunications signals at multiple GHz rates. Quantum electron MZI devices are much smaller with dimensions of a few microns. To be operational, however, they must be housed in a dilution refrigerator in the presence of a superconducting magnet capable of imposing magnetic fields of several Tesla. Thus, it is not yet possible to infer the nature of technology applications of quantum fermion devices based on a simple analogy to today's guided wave photonic devices, even though some of the underlying physics is based on shared wavelike behavior. Electrons and photons are also particles, and in contrast, the particle properties of electrons and photons, (for example, Fermi-Dirac statistics versus Bose-Einstein statistics) are fundamentally different.

11.11 Exercises

11.1 Develop the four complementary relationships analogous to (11.28) for the fermion annihilation operator.

11.2 Prove that the variance in state occupation for quanta distributed according to the Fermi-Dirac law is given by $V(n) = \langle n \rangle (1 - \langle n \rangle) = \bar{n} - (\bar{n})^2$.

11.3 Calculate the Fermi energy and the Fermi wavelength for a 3-d electron concentration of $n = 1 \times 10^{18}\,\text{cm}^{-3}$. Compare your values to the Fermi energy and Fermi wavelength for an equivalent 2 DEG with a sheet electron concentration of $n_s = 1 \times 10^{12}\,\text{cm}^{-2}$.

11.4 Develop a numerical solution to Schrödinger's equation for a free-electron confined to a triangular potential well of finite depth. Model the potential profile numerically as you wish, and treat the problem like a case of resonant tunnelling using the transfer matrix method to determine the energies of the confined states in the well region (see Fig. 11.25). Compare your results to those obtained by using Ando's formula for a triangular well (11.47) having the same parameters.

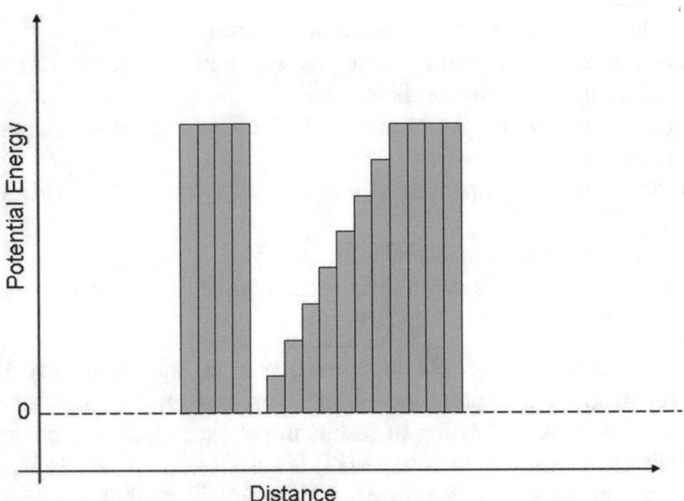

Fig. 11.25 A piecewise model of a triangular potential well. Make use of the observation in Chap. 3 that the resonant energy levels of the potential well are close to those obtained by solving for the resonant tunnelling energies through the same structure

11.5 Derive the expression for the dependence of E_o on barrier width for the simple harmonic oscillator.

11.6 Use the transfer matrix method to calculate the transmission coefficient of a 1-dimensional free electron over a barrier. A barrier with a parabolic potential profile can be assumed, as shown in Fig. 11.26. Compare your

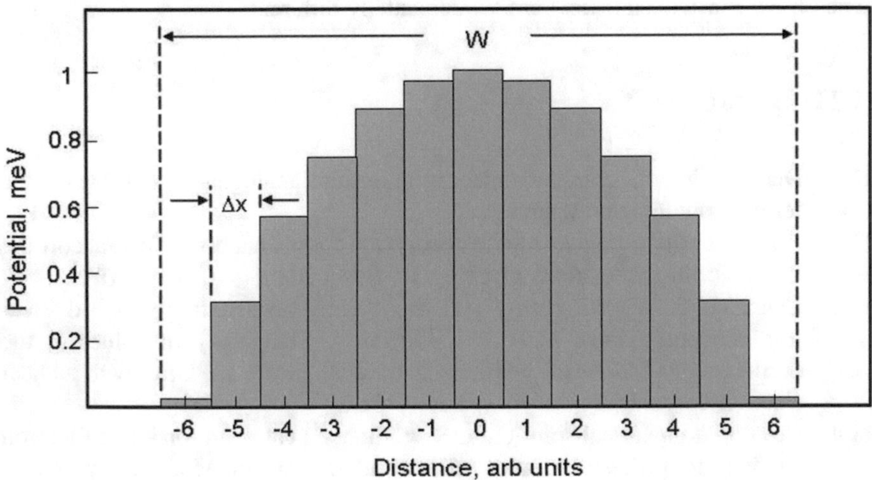

Fig. 11.26 An example of a piecewise potential barrier with a parabolic dependence of energy on distance

result to that using a sigmoid function having the same parameters of peak energy and width.

11.7 Compare the result for the analytical solution to Schrödinger's equation for transmission coefficient across a single barrier in the thin-barrier limit to the results that which one gets by evaluation of the sigmoid function: $\mathbf{T}(E) \cong \frac{1}{1+e^{-2\pi\left(\frac{U_o-E}{E_o}\right)}}$, using the same parameters for the barrier height E_o and barrier width U_o.

11.8 Assuming a 2 DEG carrier concentration $n_s = 5 \times 10^{15}\,\mathrm{m}^{-3}$, calculate the energy levels in a parabolic well with $W = \lambda_F$. What constraints does the energy spacing place on the measurement temperature?

11.9 Current is carried only by electrons at the Fermi level. van Wees et al. (1988) showed that all quantized levels lying at or below the Fermi level contribute to the measured conductivity of the QPC. Show how it is possible that the energy levels that lie below the Fermi level in the center of the QPC can support electrons carrying current at the Fermi level.

11.10 Calculate the size of the electron wavepacket by calculating the Fermi velocity and using the time half-width and other parameters from the measurement of Bocquillon et al. (2013). Compare to the Fermi wavelength.

Annexe 11.A Electron Transport and Modulation-Doping in Semiconductors

The environment for electron transport in semiconductors is different from that of electrons in metals. The electrons that carry current are "donated" to the conduction band (or holes to the valence band) by ionized impurities, be they intentionally introduced or residual. The maximum impurity concentration that can be sustained in GaAs without disrupting crystalline structure is about one in 10,000, or a carrier concentration in the range of $N_D \approx 10^{18}\,\mathrm{cm}^{-3}$. This situation results in a Fermi velocity: $v_F \approx 5 \times 10^7\,\mathrm{cm\,s}^{-1}$. The mean-free path can be estimated using the Fermi velocity and the mobility:

$$\mathcal{L}_{\mathrm{mfp}} = \frac{m^*}{q} v_F \cdot \mu \tag{11.103}$$

The mobility of electrons in GaAs is a function of temperature. At room temperature, scattering by lattice vibrations, (polar optic phonons) limits the mobility. These vibrations help to screen out the scattering by the ionized impurities that contribute carriers to conduction. The effect of lattice vibrations can be reduced to a negligible level by lowering the sample temperature. At the same time, the ionized impurities are no longer screened and ionized impurity scattering rate increases, eventually becoming the dominant scattering mechanism for

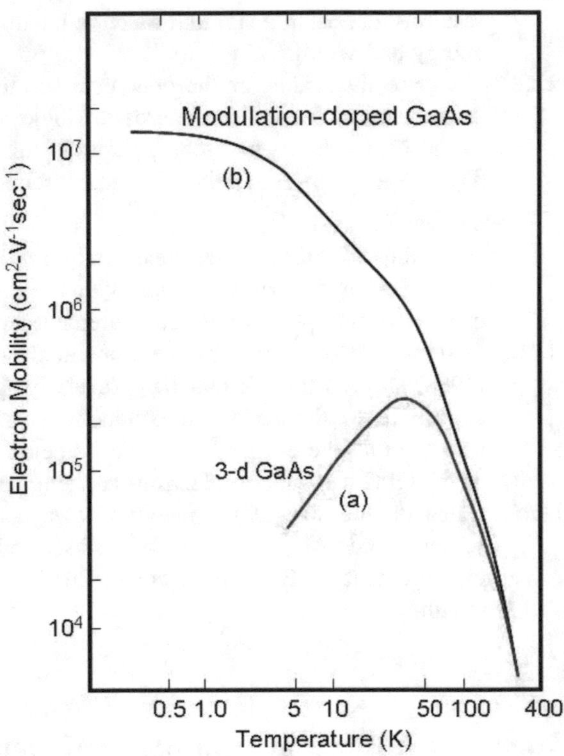

Fig. 11.27 Comparison of mobility of electrons in very high purity GaAs, having a total impurity concentration less than $N_{imp} = 10^{14}$ cm^{-3}. **a** 3-d high-purity GaAs, Stillman and Wolfe (1976). **b** Modulation-doped GaAs (Pfeiffer et al. 1989)

temperatures $T < 77$ K. The resulting electron mobility for GaAs having less than 10^{14} cm^{-3} total impurities can be seen in Fig. 11.27 (Stillman and Wolfe 1976).

Annexe 11.A.1 Modulation Doping and Electron Mobility

Further reduction in the impurity concentration from that achieved by Stillman and Wolfe is difficult to imagine because of thermodynamic requirements at the temperature of crystal growth. The breakthrough was discovered by Ray Dingle who proposed and demonstrated an astonishingly simple and effective solution: physically separate the electron charge carriers from the ionized impurity donors (Dingle et al. 1978). This procedure is called modulation doping, and it consists of restricting the ionized impurities to a wider band gap material such as $Al_{0.3}Ga_{0.7}As$, directly adjacent to a lower band gap material (GaAs). Intentionally introducing a high-level of impurity doping, localized in the wider bandgap material lowers its conduction band to near the Fermi level and creates an accumulation channel for electrons at the interface between $Al_{0.3}Ga_{0.7}As$ and GaAs.

This procedure reduces the scattering potential of ionized impurities on the electrons accumulated in this conduction channel. The effect on carrier mobility at low temperatures is dramatic, raising the mobility as shown in Fig. 11.27b, by nearly two orders of magnitude above the maximum value for the mobility in high-purity GaAs (Pfeiffer et al. 1989). Significant increases in electron mobility above this figure have been demonstrated by creating symmetric modulation–doped channel structures (Umansky et al. 2009).

The mean free path depends on both the mobility and the Fermi velocity. In bulk GaAs samples, high mobility is achieved by reducing the impurity level, but the Fermi velocity is diminished at the same time. Modulation doping separates the electron charge carriers from the ionized impurities, reducing scattering and maintaining a high carrier concentration, and at the same time a high Fermi velocity. This leads directly to a longer mean-free path (Weisbuch and Vinter 1991).

High mobility semiconductor samples suitable for ballistic quantum electron experiments are grown by molecular-beam epitaxy, (Umansky and Heiblum 2013). This short sentence encompasses decades of careful study and experiments that identify sources of electron scattering and development and fabrication of specific structures that minimize these events.

Figure 11.28 is a series of cartoon sketches that shows how the various components of the $Al_{0.3}Ga_{0.7}As/GaAs$ modulation-doped structure are combined. Note the critical role played by the interfaces between the different materials. These structures are developed using the principles laid out in Umansky and Heiblum (2013).

The carrier concentration of the 2 DEG depends primarily on the level of modulation doping, but it can be further increased or decreased by applying an external potential between the gate electrode and GaAs. A positive bias will

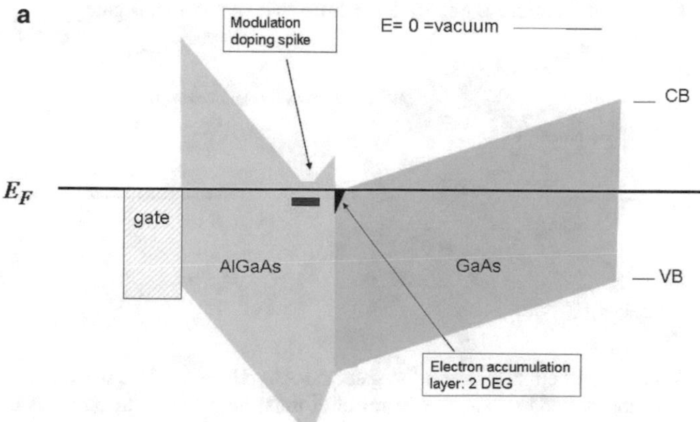

Fig. 11.28a A complete modulation-doped $Al_{0.3}Ga_{0.7}As/GaAs$ structure at equilibrium

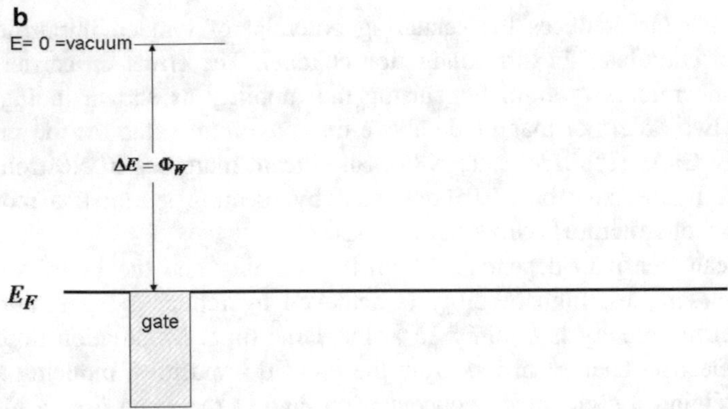

Fig. 11.28b The gate electrode is a highly-conducting metal such as gold. The Fermi level lies in the conduction band at an energy below the vacuum level given by the work function Φ_W of the metal

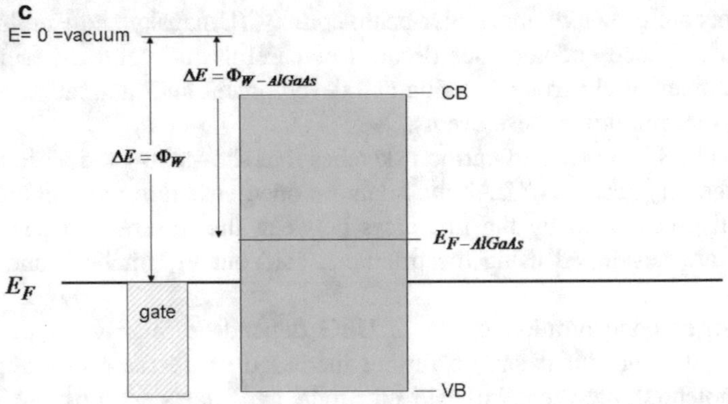

Fig. 11.28c Introduce the $Al_{0.3}Ga_{0.7}As$ layer. It has a different work function and its Fermi level will be situated at a different energy level from that of the metal gate

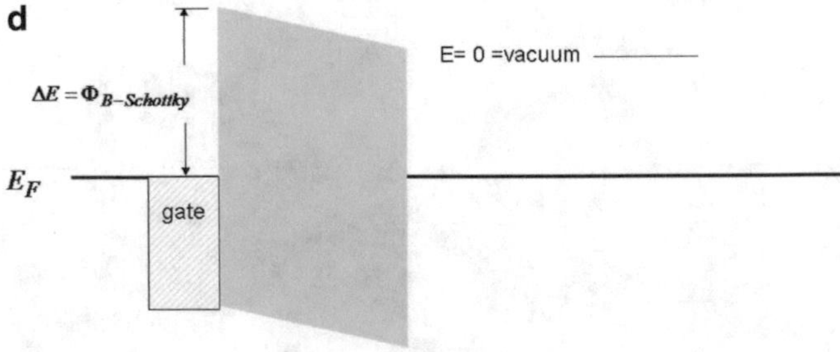

Fig. 11.28d Make the junction between the gate and $Al_{0.3}Ga_{0.7}As$. The work function must be continuous across the boundary, and the Fermi level must be constant throughout the structure. This creates an electric field that causes band bending to occur and local depletion of the electron concentration at the metal-semiconductor junction. The difference in energy between the two conduction bands is the Schottky barrier. The impurity concentration of $Al_{0.3}Ga_{0.7}As$ is so low that it behaves like an insulator, and the band bending has close to a linear dependence on distance

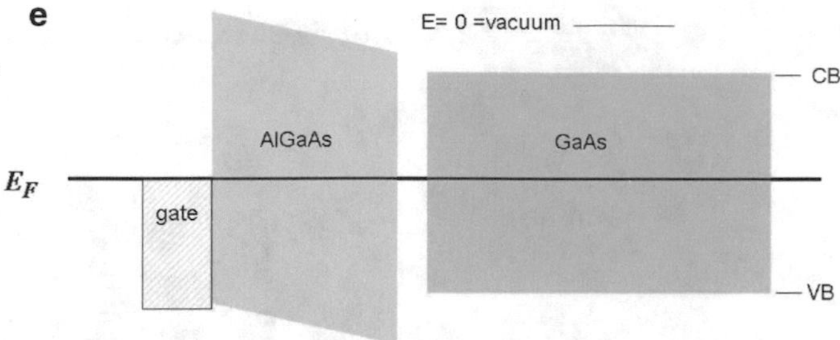

Fig. 11.28e Introduce the GaAs layer. Align its Fermi level to that of $Al_{0.3}Ga_{0.7}As$

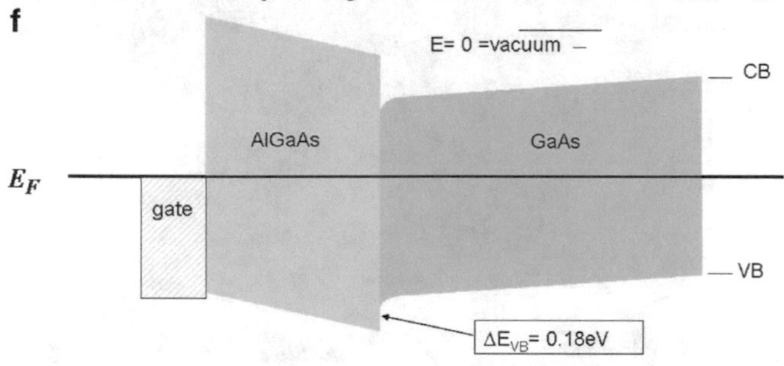

Fig. 11.28f Make the junction between $Al_{0.3}Ga_{0.7}As$ and GaAs. The valence bands of the two materials are offset by an energy $\Delta E_{VB} = 0.18$ eV related to the difference in the work functions of the respective materials. This creates an electric field at this interface which causes band bending at the interface between these two materials, lowering the conduction band of GaAs relative to that of $Al_{0.3}Ga_{0.7}As$ by about 0.25 eV

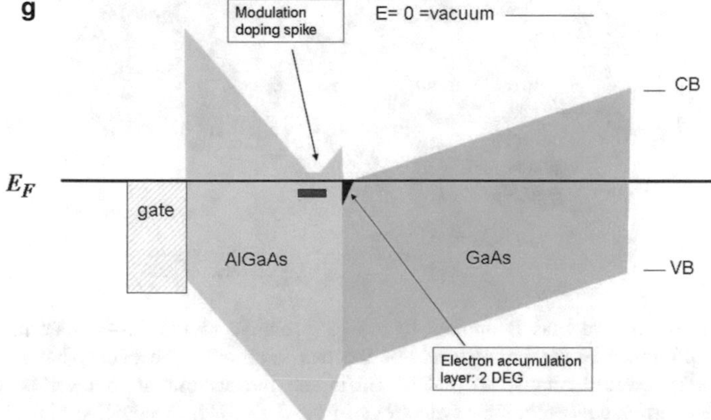

Fig. 11.28g Introduce modulation doping in the $Al_{0.3}Ga_{0.7}As$ layer. At a sufficient level, (e.g. $N_{\text{mod-doping}} = 5 \times 10^{11}\ \text{cm}^{-2}$) or greater. The conduction band in the $Al_{0.3}Ga_{0.7}As$ layer is lowered to nearly the Fermi level. This has the effect of lowering the energy bands at the interface which is typically less than 100 nm distance from the modulation-doping region. The conduction band of GaAs is bent downwards to the point that it lies beneath the Fermi level at the interface between the two materials. The result is the creation of a thin accumulation layer of nearly free electrons, forming a 2-dimensional electron gas, or 2 DEG. The 2 DEG forms a high mobility channel at the interface between $Al_{0.3}Ga_{0.7}As$ and GaAs. Ando, Fowler and Stern have given a detailed treatment of the accumulation region (Ando et al. 1982)

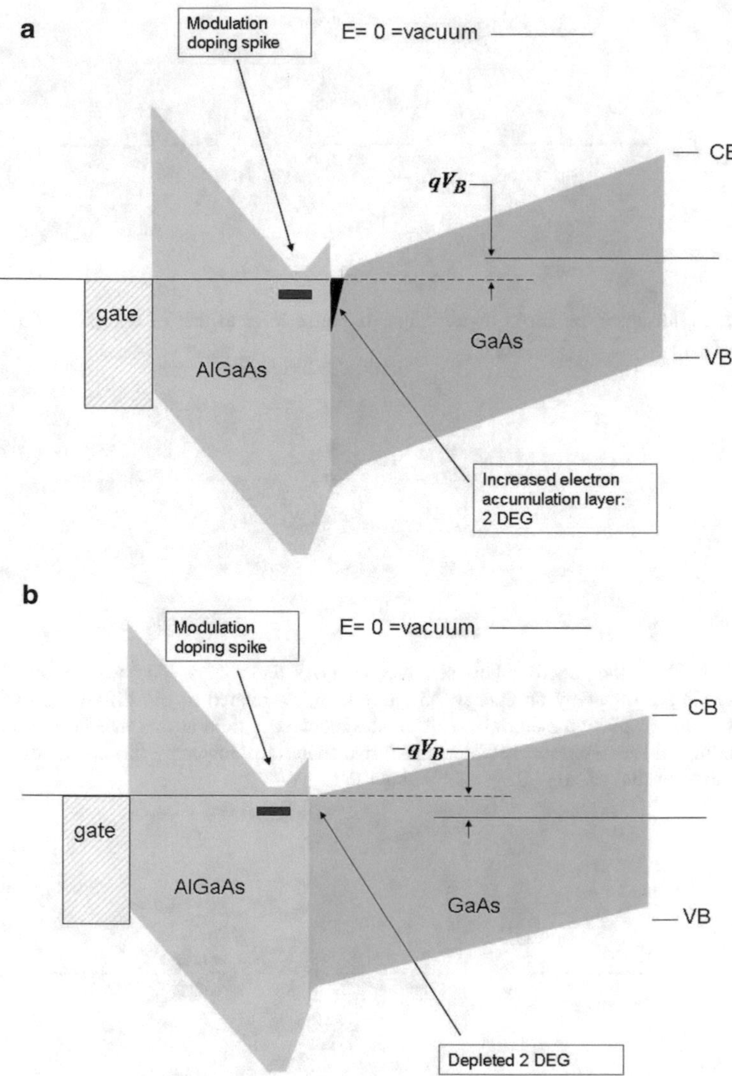

Fig. 11.29 **a** A positive bias is applied to the gate relative to the GaAs material. This bias maintains a difference in the Fermi levels of the two materials. The energy-band structure of $Al_{0.3}Ga_{0.7}As$ is lowered relative to that of GaAs and the accumulation layer is augmented, enhancing the carrier concentration of electrons in the 2 DEG. **b** A negative bias is applied to the gate relative to the GaAs material. This bias maintains a difference in the Fermi levels of the two materials. The energy-band structure of $Al_{0.3}Ga_{0.7}As$ is raised relative to that of GaAs and the accumulation layer is depleted, lowering the carrier concentration of electrons in the 2 DEG. A sufficiently large negative bias will deplete all of the carriers, resulting in pinch-off of the 2 DEG

increase the concentration and a negative bias can be used to raise the conduction band of GaAs above the Fermi level, resulting in pinch-off of the conduction channel created by the 2 DEG. Each case is illustrated in Fig. 11.29a, b.

Two critical parameters of this design are the impurity doping concentration and the set-back distance of the doping layer from the interface. The conduction electrons are confined to a channel where they constitute a 2-dimensional electron gas (2 DEG).

The strong n-type δ-doping of the $Al_{0.3}Ga_{0.7}As$ contributes electrons to the conduction band. Some of the charge diffuses to the interface and is confined to 2 dimensions by the band bending of the two materials. The electrons in the 2-d electron gas (2 DEG) experience reduced ionized impurity scattering, because the ionized impurities are isolated from the 2 DEG by a 70 nm spacer layer. The scattering potential of the residual impurities in the high-purity GaAs channel is screened out by the extremely high 2 DEG electron concentration. There is a range of choices for doping and setback. Umansky and Heiblum (2013) have demonstrated a range of such 2 DEG structures where the choice of design parameters optimizes transport properties for particular experiments.

Annexe 11.A.2 Electron Mean-Free Path in Modulation-Doped Semiconductors

We can compare the mean-free path for electrons in GaAs for 3-d materials and modulation-doped materials. The mean-free path of an electron in the 2-dimensional channel of a modulation-doped $Al_{0.3}Ga_{0.7}As/GaAs$ structure at $T = 4\,K$ can be estimated using (11.103): $\mathcal{L}_{mfp} \approx 150\ \mu m$, taking the mobility from Fig. 11.27, and assuming a 2-dimensional sheet carrier density of $N_{2-d} = 10^{12}\,cm^{-2}$. This length is at least an order of magnitude larger than typical dimensions of quantum fermion samples. One could expect to measure ballistic electron effects under these conditions in samples having an overall dimensions of ~10 μm or smaller.

This result can be compared to that for high purity GaAs at $T = 50\,K$ where the mobility is at a maximum: $\mathcal{L}_{mfp} \approx 0.07\ \mu m$, again taking the mobility from Fig. 11.5, and assuming a 3-dimensional carrier density: $N_{3-d} = 10^{13}\,cm^{-3}$. One would not expect to measure ballistic electron effects in high mobility GaAs samples having dimensions of ~1 μm. Detection of ballistic effects might be possible in samples using this material having dimensions less than 0.01 μm.

Modulation doping is a key enabling technology for establishing a channel for ballistic electron transport.

References

Y. Aharonov, D. Bohm, Significance of electromagnetic potentials in the quantum theory. Phys. Rev. **115**, 485–491 (1959). http://www.cchem.berkeley.edu/chem122/docs/aharonov%20and%20bohm%201959.pdf

T. Ando, A.B. Fowler, F. Stern, Electronic properties of two-dimensional systems. Rev. Mod. Phys. **54**, 437–672 (1982). https://s3.amazonaws.com/academia.edu.documents/35253173/Ando_Electronic_properties_of_2D_systems_1982.pdf?AWSAccessKeyId=AKIAIWOWYYGZ2Y53UL3A&Expires=1556198210&Signature=mvr%2B34lzk6zlvwQXzs0tO99cBV4%3D&response-content-disposition=inline%3B%20filename%3DAndo_et_al_Electronic_properties_of_2D_s.pdf

G. Baym, *Lectures on Quantum Mechanics* (Perseus Books, New York, 1990). ISBN: 0-8053-0667-6. https://idoc.pub/documents/lectures-on-quantum-mechanics-gordon-baym-6nq8go2051nw

C.W.J. Beenakker, H. van Houten, Quantum transport in semiconductor nanostructures, in Chapter 1 in *Solid State Physics*, ed. by H. Ehrenreich, D. Turnbull, , vol. 44 (Academic Press, New York, 1991), pp. 1–228. https://arxiv.org/pdf/cond-mat/0412664.pdf

E. Bocquillon, V. Freulon, J.-M. Berroir, P. Degiovanni, B. Plaçais, A. Cavanna, Y. Jin, G. Fève, Coherence and indistinguishability of single electrons emitted by independent sources. Science **339**, 1054–1057 (2013). https://hal.archives-ouvertes.fr/hal-00954138/document

E. Bocquillon, V. Freulon, F.D. Parmentier, J.-M. Berroir, B. Plaçais, C. Wahl, J. Rech, T. Jonckheere, T. Martin, C. Grenier, D. Ferraro, P. Degiovanni, G. Fève, Electron quantum optics in ballistic chiral conductors. Ann. Physik **526**, 1–30 (2014). https://onlinelibrary.wiley.com/doi/pdf/10.1002/andp.201300181

M. Büttiker, Absence of backscattering in the quantum Hall effect in multiprobe conductors. Phys. Rev. B **38**, 9375–9389 (1988). https://sites.fas.harvard.edu/~phys191r/References/e3/buttiker1988.pdf

C.J. Davisson, L.H. Germer, Reflection of electrons by a crystal of nickel. Proc. Natl. Acad. Sci. U.S.A. **14**, 317–322 (1928). https://www.ncbi.nlm.nih.gov/pmc/articles/PMC1085484/

R. Dingle, H.L. Störmer, A.C. Gossard, W. Wiegmann, Electron mobilities in modulation-doped semiconductor heterojunction superlattices. Appl. Phys. Lett. **33**, 665–667 (1978). https://aip.scitation.org/doi/abs/10.1063/1.90457

H. Duprez, E. Sivre, A. Anthore, A. Aassime, A. Cavanna, A. Ouerghi, U. Gennser, F. Pierre, Macroscopic electron quantum coherence in a solid-state circuit. Phys. Rev. X **9**, 021030 (2019). https://journals.aps.org/prx/pdf/10.1103/PhysRevX.9.021030

W. Ehrenberg, R.E. Siday, The refractive index in electron optics and the principles of dynamics. Proc. Roy. Soc. Sec. B **62**, 8–21 (1949). https://iopscience.iop.org/article/10.1088/0370-1301/62/1/303/meta

G. Fève, A. Mahé, J.-M. Berroir, T. Kontos, B. Plaçais, D.C. Glattli, A. Cavanna, B. Etienne, Y. Jin, An on-demand coherent single-electron source. Science **316**, 1169–1172 (2007). http://www.phys.ens.fr/~placais/publication/2007_Science_Injecteur.pdf

P. Grangier, G. Roger, A. Aspect, Experimental evidence for a photon anti-correlation effect on a beam splitter: a new light in single photon interferences. Europhys. Lett. **1**, 173–179 (1986). https://courses.physics.illinois.edu/phys513/sp2016/reading/week1/GrangierSinglePhoton1986.pdf

B.I. Halperin, The quantized Hall effect. Sci. Am. **254**, 52–60 (1986). https://pdfs.semanticscholar.org/eba6/824465a30420f5ef4d2b30456d55cbe0b05c.pdf

M. Henny, S. Oberholzer, C. Strunk, T. Heinzel, K. Ensslin, M. Holland, C. Schönenberger, The fermionic Hanbury Brown and Twiss experiment. Science **284**, 296–298 (1999). https://epub.uni-regensburg.de/3370/1/ScienceHBT.pdf

C.K. Hong, Z.Y. Ou, L. Mandel, Measurement of subpicosecond time intervals between two photons by interference. Phys. Rev. Lett. **59**, 2044–2046 (1987). http://quantumagic.narod.ru/Articles/Mandel_1987.pdf

T. Jeltes, J.M. McNamara, W. Hogervorst, W. Vasse, V. Krachmalnicoff, M. Schellekens, A. Perrin, H. Change, D. Boiron, A. Aspect, C.I. Westbrook, Comparison of the Hanbury Brown-Twiss effect for fermions and bosons. Nature **445**,. 402–405 (2007). https://www.google.com/url?sa=t&rct=j&q=&esrc=s&source=web&cd=20&ved=2ahUKEwjptN_ohf_eAhUKRBoKHVCgCp04ChAWMAl6BAgDEAI&url=https%3A%2F%2Fwww.lcf.institutoptique.fr%2Fcontent%2Fdownload%2F7855%2F46200%2Ffile%2F2007%2520nature05513%2520bosons%2520fermions.pdf&usg=AOvVaw2UBVoY4pQ8E6l_Q31Ova2z

Y. Ji, Y. Chung, D. Sprinzak, M. Heiblum, D. Mahalu, H. Shtrikman, An electronic Mach-Zehnder interferometer. Nature **422**, 415–418 (2003). https://heiblum.weizmann.ac.il/wp-content/uploads/2016/03/nature01503.pdf

T. Jonckheere, J. Rech, C. Wahl, T. Martin, Electron and hole Hong-Ou-Mandel interferometry. Phys. Rev. B **86**, 125425 (2012). http://www.cpt.univ-mrs.fr/~jonckheere/papers/jonckheere_rech_wahl_martin_PRB2012.pdf

R. Landauer, Spatial variation of currents and fields due to localized scatterers metallic conduction. IBM J. Res. Dev. **1**, 223–231 (1957). http://citeseerx.ist.psu.edu/viewdoc/download?doi=10.1.1.91.9544&rep=rep1&type=pdf

L.D. Landau, E.M. Lifshitz, *Quantum Mechanics* (Addison-Wesley, Reading, 1958). ISBN: 0-08-020940-8

A. Marguerite, C. Cabart, C. Wahl, B. Roussel, V. Freulon, D. Ferraro, C. Grenier, J.-M. Berroir, B. Plaçais, T. Jonckheere, J. Rech, T. Martin, P. Degiovanni, A. Cavanna, Y. Jin, G. Fève, Decoherence and relaxation of a single electron in a one-dimensional conductor. Phys. Rev. B **94**, 115311 (2016). http://www.phys.ens.fr/~placais/publication/2016_PRB_Marguerite_decoherence-relaxation.pdf

L. Marton, J. Arol Simpson, J.A. Suddeth, An electron interferometer. Rev. Sci. Inst. **25**, 1099–1104 (1954). https://doi.org/10.1063/1.177094525

I. Neder, M. Heiblum, Y. Levinson, D. Mahalu, V. Umansky, Unexpected behavior in a two-path electron interferometer. Phys. Rev Lett. **96**, 016804 (2006) https://www.researchgate.net/profile/Izhar_Neder/publication/7289645_Unexpected_Behavior_in_a_Two-Path_Electron_Interferometer/links/542aac130cf277d58e875bd1/Unexpected-Behavior-in-a-Two-Path-Electron-Interferometer.pdf

S. Oberholzer, Hanbury Brown and Twiss-type experiments in electronic conductors. Capri Spring Sch. Trans. Nanostruct., 2–6 April 2006. http://www.capri-school.eu/capri06/lectures/Oberholzer_2.pdf

S. Oberholzer, M. Henny, C. Strunk, C. Schönenberger, T. Heinzel, K. Ensslin, M. Holland, The Hanbury Brown and Twiss experiment with fermions. Phys. E Low-dimens. Syst. Nanostruct. **6**, 314–317 (2000). http://citeseerx.ist.psu.edu/viewdoc/download?doi=10.1.1.310.6656&rep=rep1&type=pdf

W.D. Oliver, J. Kim, R.C. Liu, Y. Yamamoto, Hanbury Brown and Twiss–type experiment with electrons. Science **284**, 299–301 (1999). http://www2.optics.rochester.edu/~stroud/cqi/stanford/5_hanbury_brown.pdf

F. Parmentier, Short-time Hanbury-Brown and Twiss correlation of a single-electron beam in ballistic conductors, Physics [physics]. Université Pierre et Marie Curie - Paris VI, 2010. https://tel.archives-ouvertes.fr/tel-00556458v2/document

L. Pfeiffer, K.W. West, H.L. Störmer, K.W. Baldwin, Electron mobilities exceeding 10^7 cm^2/Vs in modulation-doped GaAs. Appl. Phys. Lett. **55**, 1888–1890 (1989). https://www.sp.phy.cam.ac.uk/drp2/2DEGmobility.pdf

G. Pozzi, C.B. Boothroyd, A.H. Tavabi, E. Yücelen, R.E. Dunin-Borkowski, S. Frabboni, G.C. Gazzadi, Experimental realization of the Ehrenberg-Siday thought experiment. Appl. Phys. Lett. **108**, 083108 (2016). http://rafaldb.com/papers/J-2016-Appl-Phys-Lett-Ehrenberg-Siday.pdf

M. Schellekens, R. Hoppeler, A. Perrin, V. Gomes, D. Boiron, A. Aspect, C.I. Westbrook, Hanbury Brown-Twiss effect for ultracold quantum gases. Science **310**, 648–651 (2005). https://arxiv.org/pdf/cond-mat/0508466.pdf

J.R. Schrieffer, *Theory of Superconductivity*, vol. 3, Revised edn. (Benjamin/Cummings, Reading, 1983). IBIN: 0-8053-8502-9

G.E. Stillman, C.M. Wolfe, Electrical characterization of epitaxial layers. Thin Solid Films **31**, 69–88 (1976). https://www.sciencedirect.com/science/article/pii/0040609076903552

S. Tewari, P. Roulleau, C. Grenier, F. Portier, A. Cavanna, U. Gennser, D. Mailly, P. Roche, Robust quantum coherence above the Fermi sea. Phys. Rev. B **93**, 035420 (2016). https://hal-cea.archives-ouvertes.fr/cea-01366658/document

D. Tong, *Quantum Hall Effect*. TIFR-Infosys Lecture (Tata Institute of Fundamental Research, Mumbai, India, 2016). http://www.damtp.cam.ac.uk/user/tong/qhe.html

V. Umansky, M. Heiblum, MBE growth of high-mobility 2 DEG, in Chapter 6 in *Molecular Beam Epitaxy: From Research to Mass Production*, ed. by M. Henini (Waltham, Elsevier Inc., 2013), pp. 121–137. ISBN-13: 978-0123878397. https://heiblum.weizmann.ac.il/wp-content/uploads/2016/03/MBE_Chapter.pdf

V. Umansky, M. Heiblum, Y. Levinson, J. Smet, J. Nübler, M. Dolev, MBE growth of ultra-low disorder 2 DEG with mobility exceeding 35×10^6 cm^2 V^{-1} s^{-1}. J. Cryst. Growth **311**, 1658–1661 (2009). https://www.sciencedirect.com/science/article/pii/S0022024808009901

H. van Houten, C.W.J. Beenakker, Quantum point contacts. Phys. Today **49**, 22–27 (1996). https://arxiv.org/pdf/cond-mat/0512609.pdf

B.J. van Wees, H. van Houten, C.W.J. Beenakker, J.G. Williamson, L.P. Kouwenhoven, D. van der Marel, C.T. Foxon, Quantized conductance of point contacts in a two-dimensional electron gas. Phys. Rev. Lett. **60**, 848–850 (1988). https://openaccess.leidenuniv.nl/bitstream/handle/1887/3316/172_020.pdf?sequence=1

R.A. Webb, S. Washburn, C.P. Umbach, R.B. Laibowitz, Observation of h/e Aharonov-Bohm Oscillations in normal-metal rings. Phys. Rev. Lett. **54**, 2696–2699 (1985). https://scholarcommons.sc.edu/cgi/viewcontent.cgi?referer=, https://scholar.google.com/&httpsredir=1&article=1093&context=phys_facpub

C. Weisbuch, B. Vinter, *Quantum Semiconductor Structures* (Academic Press, San Diego, 1991). ISBN: 0-12-742680-9. http://bookregin.info/quantum-semiconductor-structures-fundamentals-and-applications-new-books-c-weisbuch-borge-vinter-to-read.pdf

C.I. Westbrook, Private communication (2019)

D.A. Wharam, T.J. Thornton, R. Newbury, M. Pepper, H. Ahmed, J.E.F. Frost, D.J. Hasko, D.C. Peacock, D.A. Ritchie, G.A.C. Jones, One-dimensional transport and the quantisation of the ballistic resistance. J. Phys. C: Solid State Phys. **21**, L209–L214 (1988). https://www.researchgate.net/profile/Trevor_Thornton/publication/230968130_One-dimensional_transport_and_the_quantisation_of_the_ballistic_resistance/links/5898bad8a6fdcc32dbdd0616/One-dimensional-transport-and-the-quantisation-of-the-ballistic-resistance.pdf

A. Yacoby, U. Sivan, C.P. Umbach, J.M. Hong, Interference and dephasing by electron-electron interaction on length scales shorter than the elastic mean free path. Phys. Rev. Lett. **66**, 1938–1941 (1991). https://journals.aps.org/prl/abstract/10.1103/PhysRevLett.66.1938

Index

© Springer Nature Switzerland AG 2020
T. P. Pearsall, *Quantum Photonics*, Graduate Texts in Physics,
https://doi.org/10.1007/978-3-030-47325-9

K

L

M

N

O

P

Q

Printed in the United States
by Baker & Taylor Publisher Services